157
Advances in Biochemical Engineering/Biotechnology

Aims and Scope

This book series reviews current trends in modern biotechnology and biochemical engineering. Its aim is to cover all aspects of these interdisciplinary disciplines, where knowledge, methods and expertise are required from chemistry, biochemistry, microbiology, molecular biology, chemical engineering and computer science.

Volumes are organized topically and provide a comprehensive discussion of developments in the field over the past 3–5 years. The series also discusses new discoveries and applications. Special volumes are dedicated to selected topics which focus on new biotechnological products and new processes for their synthesis and purification.

In general, volumes are edited by well-known guest editors. The series editor and publisher will, however, always be pleased to receive suggestions and supplementary information. Manuscripts are accepted in English.

In references, Advances in Biochemical Engineering/Biotechnology is abbreviated as *Adv. Biochem. Engin./Biotechnol.* and cited as a journal.

More information about this series at http://www.springer.com/series/10

Georg Reifferscheid • Sebastian Buchinger
Editors

In vitro Environmental Toxicology - Concepts, Application and Assessment

With contributions by

S. Belkin · A. Brinke · M. Brinkmann · S. Buchinger · N. Egetenmeyer · T. Elad · B.I. Escher · N. Haase · M. Heringa · H. Hollert · A. Jahnke · C. Kienle · Y. Oda · J. Oehlmann · T.G. Preuss · S. Schäfer · W. Schulz · K.E.C. Smith · P. Steinberg · G.D.A. Umbuzeiro · E.L.M. Vermeirssen · M. Wagner · S.C. Weiss · G. Witt · E. Zeiger

Springer

Editors
Georg Reifferscheid
Federal Institute of Hydrology
Koblenz, Germany

Sebastian Buchinger
Federal Institute of Hydrology
Koblenz, Germany

ISSN 0724-6145 ISSN 1616-8542 (electronic)
Advances in Biochemical Engineering/Biotechnology
ISBN 978-3-319-45906-6 ISBN 978-3-319-45908-0 (eBook)
DOI 10.1007/978-3-319-45908-0

Library of Congress Control Number: 2017938647

Printed on acid-free paper

This Springer imprint is published by Springer Nature
The registered company is Springer International Publishing AG
The registered company address is: Gewerbestrasse 11, 6330 Cham, Switzerland

Preface

In recent years, considerable technological progress has been made in the development of in vitro bioanalytical methods and instruments for the elucidation of toxic effects of compounds of both natural and anthropogenic origin.

Such methods, which are progressively applied in toxicologyand environmental science, allow the detection of cytotoxicity as well as the investigation of specific, sublethal effects of chemicals and chemical mixtures of different complexity. In toxicology, in vitro bioanalytical tools have so far mainly been used to generate scientific knowledge, to elucidate the chemical causes of effects, and to provide data in support of environmental monitoring.

It is widely accepted that in vitro methods add substantial value to the field of ecotoxicology because of their efficiency, their high throughput capacity, and their ability to obtain mechanistic information about toxicity and basic data on possible toxicological risks in different environmental compartments. However, the use and interpretation of test results in regulation is challenging and still under discussion. Although reduction, replacement, and refinement of in vivo toxicologyis always being called for by society and regulatory stakeholders, one main reason for regulatory obstruction is that options for in vitro/in vivo extrapolation of effects are still missing.

This book gives an overview of the current state of the art of in vitro bioassays in the field of (eco)toxicology with special focus on effects of very high concern and reasonable areas of application. Furthermore, selected chapters address topics related to the application of in vitro bioassays in environmental sciences, such as passive sampling/passive dosing and effect-directed analysis. A special chapter describes the possibilities of linking results of in vitro assays to in vivo effects by making use of physiologically-based pharmacokinetic modeling. According to the basic test principles, the underlying concepts of the various techniques are shown.

The book exemplifies the use of in vitro approaches in different fields of application. It discusses the potential, current limitations, research needs,and regulatory perspectives of some selected bioanalytical tools and of in vitro bioassays in general.

Georg Reifferscheid
Sebastian Buchinger

Contents

Adv Biochem Eng Biotechnol (2017) 157: 1–58
DOI: 10.1007/10_2016_2

Published online: 3 February 2017

Endocrine Disruption and In Vitro Ecotoxicology: Recent Advances and Approaches

Martin Wagner, Cornelia Kienle, Etiënne L.M. Vermeirssen, and Jörg Oehlmann

Abstract Endocrine-disrupting chemicals (EDCs) are man-made compounds interfering with hormone signaling. Omnipresent in the environment, they can cause adverse effects in a wide range of wildlife. Accordingly, Endocrine Disruption is one focal area of ecotoxicology. Because EDCs induce complex response patterns in vivo via a wide range of mechanisms of action, in vitro techniques have been developed to reduce and understand endocrine toxicity. In this review we revisit the evidence for endocrine disruption in diverse species and the underlying molecular mechanisms. Based on this, we examine the battery of in vitro bioassays currently in use in ecotoxicological research and discuss the following key questions. Why do we use in vitro techniques? What endpoints are we looking at? Which applications are we using in vitro bioassays for? How can we put in vitro data into a broader context? And finally, what is the practical relevance of in vitro data? In critically examining these questions, we review the current state-of-the-art of in vitro (eco)toxicology, highlight important limitations and challenges, and discuss emerging trends and future research needs.

Keywords Bioanalytical tools, Bioassay, Effect-directed analysis, Endocrine-disrupting chemical, Mechanism of action, Risk assessment

---

M. Wagner (✉) and J. Oehlmann
Department Aquatic Ecotoxicology, Goethe University Frankfurt, Max-von-Laue-Str. 13, Frankfurt 60438, Germany
e-mail: wagner@bio.uni-frankfurt.de

C. Kienle and E.L.M. Vermeirssen
Swiss Centre for Applied Ecotoxicology, Eawag-EPFL, Dübendorf, Switzerland

Contents

1 Introduction[1]

1.1 Endocrine Disruption: What Is It All About?

The Metazoan endocrine systems essentially and intricately regulate physiological processes in both the short and the long term, including behavior, development, metabolism, reproduction, and stress response. Endocrine systems are highly diverse, with neurotransmitters and neurohormones as ancestral mechanisms of paracrine/endocrine communication in invertebrates [1]. As more complex organizations evolved, endocrine systems were complemented with endocrine cells (annelids) and glands (mollusks, arthropods) secreting 'true hormones' [1]. In spite of their diversity, some components of the invertebrate endocrine systems are remarkably conserved and resemble their vertebrate counterparts [2]. Steroid receptors, for example, have evolved from a common ancestor present before the origin of bilaterally symmetric animals. These receptors presumably already controlled reproduction [3, 4].

Ecotoxicological research has demonstrated in field and laboratory experiments that chemicals can interfere with the endocrine systems, a phenomenon known as Endocrine Disruption (ED). Effects resulting from such disruption and causative compounds, are manifold and have been extensively reviewed elsewhere [2, 5–14]. This includes a comprehensive discussion of ED in mollusks [15], fish [16],

[1]This is an extended and updated version of the introduction published in Wagner [368].

amphibians [17], reptiles [18], and birds [19]. Recent advances in ecotoxicology cover ecotoxicogenomics [20], early exposures [21], population and ecosystem sustainability [22], and the impact of additional stressors, such as climate change [23, 24]. In the light of the wealth of available studies on ED in wildlife species, only selected, well-established examples are discussed here.

1.1.1 Invertebrates

Organotin-induced imposex[2] development in mollusks is one of the best-documented incidences of ED in the field [25]. In the early 1970s, Blaber [26] and Smith [27] observed penis-like structures in female gonochoristic marine caenogastropods from the field. A decade later, tributyltin (TBT) used in naval antifouling paints was implicated in the occurrence of imposex [28, 29]. Numerous field and laboratory studies have conclusively linked organotin exposure to imposex in more than 250 mollusk species (see reviews [15, 25, 30]). Because imposex is irreversible and may result in female sterility, some of the gastropod populations have been locally eradicated (see [31]). However, legislatory action has resulted in declining organotin levels and slow recovery of some caenogastropod populations [32–36]. Despite the clear-cut evidence of TBT toxicity, scientific debate about the underlying mechanism is ongoing. Amongst several available theories (reviewed in [30, 37]), disruption of androgen signaling[3] [38–42] and the retinoid pathway[4] [43–49] appears to be the most promising. Putting aside the putative dispute, a recent mechanistic study suggests that both pathways might be involved in imposex induction [50].

ED in caenogastropod mollusks is not restricted to organotin compounds but is induced by several estrogenic, androgenic, and antiandrogenic substances (for review see [15]). One prominent example is the plastic monomer Bisphenol A (BPA)[5]: In a molluskan model, BPA exposure induced so-called superfemales, that is, females with additional reproductive organs, enlarged glands, and escalated reproduction [51]. Because effects were observed at the lowest concentration studied, a follow-up study established BPA effects at even lower, nanogram per liter levels [52]. Heavily criticized by industry-funded scientists [53], Oehlmann and colleagues [54] replicated the original findings and reported that inadequate experimental conditions (elevated temperatures) masked the BPA effects.

[2]Imposex is defined as the imposition of male reproductive characteristics, for example, penis development, on female individuals.

[3]For example, female testosterone levels might be increased by aromatase inhibition. See Fernandes et al. [369] for a comprehensive review on molluskan steroid biosynthesis.

[4]Organotin compounds have been shown to be potent agonists of the retinoid receptors RXR and RAR.

[5]BPA found to be estrogenic in the 1930s [370] but was abandoned as synthetic estrogen in favor of the more potent diethylstilbestrol (DES). Today, it is mainly used as building block of polycarbonate plastics to produce food and beverage containers [130].

Unsurprisingly, subsequent studies conducted at a higher temperature regime were unable to reproduce the observations [55, 56].

1.1.2 Fish

A masculinization of mosquito fish by paper mill effluent was reported from the U.S. in the 1980s [57]. In the same decade, feminization of teleost fish was first observed in the River Thames [58]. This phenomenon has been termed ‘intersex’ because affected individuals developed hermaphroditic gonads containing both male and female parts [58]. A decade later, the intersex phenomenon was experimentally linked to exposure to sewage treatment effluents [59].[6] As elevated vitellogenin (VTG)[7] levels were observed in exposed male fish, estrogen-like chemicals have been suggested as causation. Since then, numerous field studies have corroborated this connection with several species being affected worldwide (see [60]). For instance, almost all phenotypic male fish caught at polluted sites in England were intersex [61, 62].[8] In a recent study, 98% of fish exposed to sewage treatment effluent were phenotypic females. In a competitive breeding study, sex-reversed males from the effluent did reproduce as females but with a very low success. Moreover, none of the sex-reversed males contributed to the offspring under competitive conditions [63, 64].

Kidd and coworkers approached the issue of ED at population level in a long-term, whole-lake experiment [65]. Over a period of 3 years, an experimental lake was dosed with a low concentration of a synthetic estrogen mimicking the environmentally relevant estrogenic exposure via sewage treatment effluents.[9] Following the second season of exposure, the population of one fish species collapsed almost completely and did not recover in the first 3 years after the removal of the exposure.[10] However, in the spring of the 4th year, adult size-frequency distribution

[6]Given the almost universal contamination of freshwater ecosystems with (treated) wastewater, the issue of feminization became the focal point for research on ED in fish and EDCs in wastewater.

[7]VTG is an egg yolk protein precursor. Synthesized in the female liver, it is transported via the bloodstream for incorporation into oocytes. Its expression is estrogen-dependent [371]. Naturally only produced by mature females, its prevalence in juvenile or male fish is considered a biomarker of exposure to estrogenic substances.

[8]Intersex males had increased VTG and estradiol levels, delayed spermatogenesis, and malformed reproductive ducts as well as reduced milt production, sperm motility, and fertilization rates.

[9]A nominal concentration of 5 ng L^{-1} 17α-ethinylestradiol (EE2), the synthetic estrogen from the birth-control pill, was dosed to a lake in Ontario, Canada [372, 373].

[10]During the exposure period, VTG levels in male fathead minnow (short life cycle) were three orders of magnitude higher than in the reference. In addition, the testicular and ovarian development was arrested [372]. Similar effects have been observed in pearl dace. While in this longer-lived species a clear impact on population size has not yet been observed, population structure was affected as indicated by the loss of the 1-year-old size class [374]. In another species (rainbow trout), fertility was unaffected [375]. This highlights considerable species differences [376].

and abundance had returned to pretreatment levels. Genetic analyses showed that postrecovery fish were descendants of the original EE2-treated population [66]. The lesson learnt from this whole-lake experiment is that chronic exposure to environmentally relevant levels of estrogens clearly impacts the sustainability of wild fish populations.

The intersex phenomenon in fish has long been attributed to estrogen-like compounds. In contrast, recent research provides a strong argument for an additional contribution of antiandrogenic chemicals producing phenotypic effects similar to estrogens [67, 68]. In an innovative approach, Hill et al. [69] analyzed the antiandrogenic activity of tissue extracts from fish exposed to sewage treatment effluents. Combining a fractionation approach with analytical techniques, the authors identified the antimicrobials chlorophene and triclosan as predominant environmental antiandrogens bioavailable to fish [70]. These are excellent examples that research on EDCs in wildlife is shifting from analyzing a few well-established chemicals to the effect-directed identification of unexpected, 'emerging' pollutants [71–75]. Pursuing such approaches provides a more holistic picture of wildlife exposure to EDCs.

1.1.3 Amphibians

The global decline and loss of amphibian biodiversity [76] is of special concern because amphibians appear to be more threatened than either birds or mammals [77]. With a complex causation involved, climate change, pathogens, and habitat loss have been proposed as global drivers [78, 79]. In this picture, the role of chemical pollution is far from clear, but pesticides have been associated with the amphibian decline [17]. Similar to fish (see Sect. 1.1.5), feminization of male frogs, characterized by testicular oocytes and intersex, has been observed in the field as well as in the laboratory (reviewed in [80]). A retrospective analysis of museum specimens suggests an association between intersex and the use of organochlorine chemicals [81].[11] Atrazine, one of the most commonly applied pesticides worldwide, serves as prototypic EDC in amphibians [82]. It induces demasculinization of male frogs at very low, ecologically relevant concentrations (e.g., [83]). The mechanism of atrazine toxicity in vertebrates is well-documented. In brief, the pesticide reduces androgen levels by inhibiting androgen-simulating hormones, enzymes of the androgen biosynthesis, and binding of dihydrotestosterone (DHT) to its target proteins (see [82]). In addition to atrazine, effects of other EDCs (especially pesticides) have been observed in several amphibian species in the field and laboratory [2].

[11] Polychlorinated biphenyls (PCBs) and *p,p*-dichlorodiphenyltrichloroethane (DDT).

1.1.4 Reptiles

Among reptiles, the American alligator is the best-studied species in terms of ED. Here, the case of Lake Apopka in Florida is of special interest because it provided a clear indication of ED in wildlife vertebrates (reviewed in [18]). Experiencing a pesticide spill from a nearby chemical company in 1980, the lake's alligator population subsequently suffered from population decline [84]. Egg viability and post-hatch survival were compromised, most probably because of high concentrations of organochlorine pesticides in the eggs [85, 86].[12] This is supported by laboratory studies in which pesticide exposure caused infertile eggs and increased embryonic mortality (e.g., [87]). Moreover, female hatchlings from Lake Apopka developed polycystic ovaries that resemble symptoms of DES exposure in other vertebrates [88]. Male alligators from Lake Apopka had decreased phallus size and testosterone levels compared to reference populations [85, 89, 90].[13] Mechanistically, several pesticides found in Lake Apopka disrupt steroid signaling and biosynthesis (review in [91, 92]).[14] Recent transcriptomic analyses indicate a loss of sexually dimorphic gene expression in specimen from Lake Apopka, mirroring the morphologic findings. Interestingly, concomitant interference with non-steroidal pathways might also be involved in the disruption of alligator reproduction [93, 94].

1.1.5 Birds

In the middle of the last century, a dramatic decline of raptor populations was observed in Great Britain and North America [95]. For instance, the bald eagle nearly vanished from the Great Lakes during the 1950s through the early 1970s [96, 97]. The phenomenon concurred with eggshell thinning and was supposed to be caused by pesticide (e.g., DDT) exposure [98]. The levels of organochlorine insecticides and PCBs are associated with a number of reproductive outcomes, including eggshell thinning, embryonic malformations, hatchling mortality, and population productivity [99]. Additionally, a number of compounds (e.g., DDT, PBDE, PCBs, TCDD) are experimentally linked to effects observed in the wild [100–103]. Many of these chemicals have been proposed to mediate their toxicity via an estrogenic mechanism [97].[15] However, DDT has a different mechanism for

---

[12]Newer studies indicate that the pesticide levels have not appreciably decreased in the early 2000s [87, 377]. Interestingly, associations between organochlorine pesticides and reproductive performance have also been reported for caimans [378].

[13]Similar effects can be induced by other endocrine disruptors, including 17β-estradiol, atrazine, and Bisphenol A [379].

[14]Several reptilian estrogen receptors have been cloned and display a differential responsiveness to estrogens and pesticides [293, 380].

[15]Interestingly, *in ovo* exposure to estrogenic chemicals feminizes the male gonad in birds [381] as it does in rodent models.

inducing eggshell thinning. Its metabolite *p,p*-DDE inhibits prostaglandin synthesis and, thus, reduces the calcium transport within the eggshell gland [104].

Besides reproduction itself, avian courtship behavior is a target of ED. For instance, exposure to estradiol masculinizes nest building behavior and song structure in zebra finch [105, 106]. Furthermore, early estrogen exposure resulted in a disrupted testicular morphology, leading to male infertility and reproductive failure [107, 108]. Interestingly, bird populations might be affected by EDCs more subtly. When fed on a diet containing environmentally relevant levels of estrogens, male European starlings produced longer and more complex songs.[16] Paradoxically, this rendered the contaminated individuals more attractive to wild-caught females. Because these males' immunocompetence was compromised, consequences at population level appear realistic [109].[17]

1.1.6 Mammals

Apart from rodents used as translational models for human toxicology, not much information is available on ED in wild-living terrestrial mammals. However, there is mounting evidence that marine mammals such as cetaceans and pinnipeds are uniquely at risk from chemical contamination [2]. Because of their long lifespan, high trophic level, and insulating fat tissue (blubber), they are among the species accumulating the highest levels of persistent organic pollutants (POPs [110, 111]). Here, most investigations focus on POPs in blubber biopsied from wild-living specimen and document an extensive contamination (see [2, 112]). More recent studies apply toxicogenomics to investigate the effects of POPs in marine mammals. PCB contamination in Pacific orcas was associated with altered expression levels of several hormone receptors, including AhR, TR, and ER [113]. Analogous indications exist for ringed seals from polluted sites in the Baltic. Here, thyroid and retinoic acid receptor expression was altered when compared to a reference population from Spitsbergen [114]. In accordance with these findings, PCBs and PBDEs affect thyroid hormone levels in white whales and bottlenose dolphins [115, 116].[18] Moreover, dolphins suffered from immune suppression and anemia [115]. Interestingly, similar effects were observed in primates exposed to high doses of PCBs [117].

[16]The exposure scenario was based on the levels of estrogen-like compounds detected in earthworms sampled at wastewater treatment plants. Morphologically, the brain area controlling song complexity was enlarged in exposed compared to control males [109].

[17]In a more recent study, growth and immunocompetence was reduced in nestlings exposed similarly to the previous experiment [382].

[18]Notably, polar bears contaminated with organohalogens had thyroid gland lesions [383].

1.1.7 Conclusions: Endocrine Disruption in Wildlife

Remarkably, the elucidation of ED in wildlife has been triggered by concrete observations in the field (e.g., compromised reproduction). With mollusks (imposex), fish (intersex), reptiles (population decline), and birds (eggshell thinning) these phenomena have been retrospectively linked to a very high presence of man-made compounds. Convincingly, chemical-effect causation has been confirmed for many cases in controlled laboratory experiments. In addition, recent studies focusing on populations rather than on individuals substantiate the hypothesis that EDCs negatively impact ecosystem integrity [6].

The research on ED in wildlife generated considerable insight into the disturbed physiology. In addition, it deepened the understanding of the normal endocrinology of many phyla (e.g., RXR signaling in invertebrates [118]). However, the current state of knowledge on wildlife endocrinology is still fragmentary. Given the extensive crosstalk of endocrine systems and species differences, a comparative approach is needed [8].

Apart from the perspective of conservation, ecotoxicology has provided important stimuli for the examination of the human population [6]. In that sense, BPA-induced super feminization in mollusks is an early example of low dose effects [119] and in ovo exposure to pesticides in birds echoes the concept of the developmental origin of adult disease [120]. Likewise, the impact of EDCs in wastewater treatment plant effluents on fish pioneered mixture toxicology [121]. Most striking are the consistencies in ED observed among different species from all phyla [5]. For instance, testicular oocytes are a sign of feminization of the male gonad in mollusks, fish, amphibians, reptiles, and birds exposed to diverse EDCs (Fig. 1). Notwithstanding the species differences in endocrine systems, the consistency of observations renders ED biologically highly plausible.

1.2 Endocrine-Disrupting Chemicals: Defining a Common Ground

The concept of ED was first specifically addressed in 1991 at a conference held in Racine, Wisconsin. Here, a multidisciplinary expert group formulated the Wingspread Consensus Statement to express their concerns about endocrine-disrupting chemicals (EDCs). They state that "*A large number of man-made chemicals that have been released into the environment [...] have the potential to disrupt the endocrine system of animals, including humans*" and conclude that "*Many wildlife populations are already affected by these compounds*" and "*Humans have been affected by compounds of this nature, too*" [122].[19]

[19]Key aspects of the Wingspread Statement were published in the peer-reviewed literature [384] and formed the cornerstones of Colbourn's 'Our stolen future' [385].

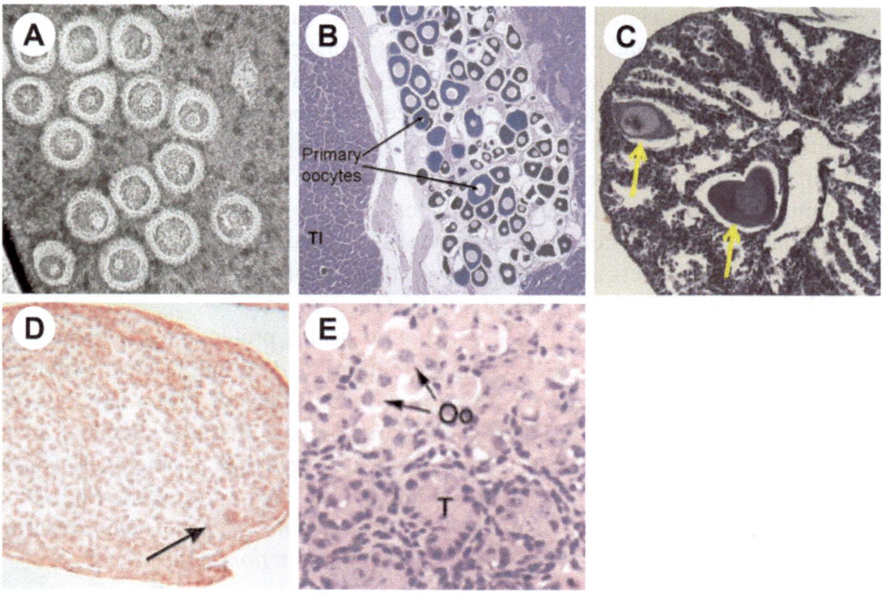

Fig. 1 A comparative view on testicular oocytes in (**a**) mollusks, (**b**) fish, (**c**) amphibians, (**d**) reptiles, and (**e**) birds. Testis histology of male individuals: (**a**) Peppery furrow shell (*Scrobicularia plana*) from a contaminated estuarine area (adopted from [365]), (**b**) Intersex roach (*Rutilus rutilus*) exposed to sewage treatment effluent [58], (**c**) Leopard frog (*Rana pipiens*) exposed to atrazine [82], (**d**) Snapping turtle (*Chelydra serpentine*) exposed to atrazine via soil [366], and (**e**) Japanese quail (*Coturnix japonica*) treated with an estrogen receptor agonist [367]

However, until 1995 a clear definition of what an EDC actually is was lacking. The U.S. Environmental Protection Agency (U.S. EPA) organized a workshop in Raleigh, North Carolina at which the participants defined an EDC as "*an exogenous agent that interferes with the production, release, transport, metabolism, binding, action or elimination of natural hormones in the body responsible for the maintenance of homeostasis and the regulation of developmental processes*" [123].

One year later, a European counterpart workshop entitled 'The Impact of Endocrine Disrupters on Human Health and Wildlife' was held in Weybridge, United Kingdom. It came up with another definition describing an EDC as "*an exogenous substance that causes adverse health effects in an intact organism, or its progeny, secondary to changes in endocrine function*"[20] [124]. Receiving broad acceptance, the Weybridge definition serves as a template for many subsequent attempts to define an EDC. For example, the WHO/IPCS expands the definition: "*An endocrine disrupter is an exogenous substance or mixture that alters function (s) of the endocrine system and consequently causes adverse health effects in an intact organism, or its progeny, or (sub)populations*" [125].

[20]It has been suggested to replace "*secondary*" by "*consequent*" [124].

Whereas the 'Raleigh definition' focuses on biological (sub)systems that might be affected by EDCs and provides a rather mechanistic viewpoint, the 'Weybridge definition' centers on health effects in the organism caused by a disturbance of its endocrine system. Weybridge also introduces the concept of adversity into the definition of an EDC. This is somewhat problematic because a consensual definition of what an 'adverse effect' is missing; there are no unambiguous criteria to discriminate between an impartial biological effect and a toxicological relevant deleterious effect [2]. In contrast to carcinogens and mutagens (which are classified according to their effects in specific assays), the lack of clear-cut criteria generates considerable uncertainty in the debate on EDCs.

In a recent statement on EDCs and public health protection, the Endocrine Society critically reviews the existing definitions and highlights the problems arising from defining an EDC according to specific modes of action (Raleigh) or adversity (Weybridge, WHO/IPCS). The authors propose that any chemical interference with hormone action is a clear predictor of adverse health outcomes. This is in analogy to mutagenicity predicting carcinogenicity. Accordingly, the Endocrine Society defines an endocrine disruptor as "*an exogenous chemical, or mixture of chemicals, that interferes with any aspect of hormone action*" [126].

1.3 High Number of Mechanisms of Action

Besides locally confined signaling[21] mediated by second messengers, transcription factors, neurotransmitters, etc., the 'classical' endocrine systems rely on hormones to communicate their information systemically. The four types of hormones include amino acid derivatives (e.g., adrenaline, thyroid hormones, melatonine), peptides (e.g., growth hormones, insulin, oxytocin), lipid derivatives (e.g., calcitriol, prostaglandins, steroid hormones), and sesquiterpenes (e.g., juvenile hormones). For each hormone, specific receptors exist either at the surface or inside the target cell, e.g., ligand-regulated transporters, G-protein coupled, cytokine, and nuclear receptors (NRs). Concerning EDCs, research has primarily focused on the NR superfamily which consists of 48 members in the human [127]. The 'classical' NRs (e.g., androgen, estrogen, thyroid, glucocorticoid receptors) are DNA-binding transcription factors controlling gene expression via ligand-dependent mechanisms.[22] Basically, ligand binding induces conformational changes within the receptor to enable dimerization,[23] DNA-binding, and subsequent recruitment of transcriptional coregulators to modulate gene expression ultimately [127, 128].

[21]This includes intracrine (intracellular), autocrine (targeting the cell itself), juxtacrine (targeting adjacent cells), and paracrine (targeting cells in the vicinity) signaling.

[22]Half of the human NRs are 'orphan receptors', i.e., its natural ligand is unknown or might not exist [127]. Here, ligand-independent mechanisms might exist.

[23]NRs bind specific hormone response elements (HREs) within the regulatory region(s) of target genes either as monomers, homo-, or heterodimers. In case of cytosolic NRs, nuclear translocation precedes DNA-binding [128, 127].

Because of their importance to reproduction and development, estrogen (ERα/β, NR3A1/2), androgen (AR, NR3C4), aryl hydrocarbon (AhR, dioxin receptor),[24] and, to a lesser extent, progesterone (PR, NR3C3) and thyroid (TRα/β, NR1A1/2) receptors have been studied extensively for their interaction with xenobiotics (see review in [2]). A vast number of chemicals bind directly to hormone receptors, either in an agonistic or antagonistic fashion. For instance, by-products (dioxins), coolants and dielectrics (PCBs), detergents (alkylphenols), flame retardants (PBDEs), food additives (butylated hydroxyanisole), metals (antimony, cadmium), personal care products (parabens, UV filters), pesticides (DDT, vinclozolin), pharmaceuticals (diethylstilbestrol), phyto/mycochemicals (genistein, zearalenone), plastic components (BPA, phthalates), and surfactants (perfluorinated compounds) have been shown to activate or inhibit ERs [2].

Interestingly, many of these (anti)estrogenic chemicals also interact with other NRs. For instance, BPA is a well-known agonist of ERα/β, but also an antagonist of AR, an agonist of AhR and glucocorticoid receptor, and binds to TRα and estrogen-related receptor γ [129, 130]. The list of compounds interacting with NRs is far too long to be covered here, but appropriate examples are provided throughout the text. Inappropriate receptor activation/inhibition is the most extensively studied and probably the most important mode of action inducing ED. However, the complex molecular network of endocrine signaling bears additional potential of xenobiotic interference (Fig. 2) and other mechanisms of action recently came into the spotlight [131].

At the level of endogenous hormones, EDCs are able to intervene with hormone biosynthesis, clearance, and transport, herby changing the concentration of available natural ligands. One classic example is aromatase[25] inhibition by several EDCs, resulting in imbalanced steroid biosynthesis [125].[26] BPA and phthalates activate pregnane X receptor (PXR) [132–134], an NR that induces the xenobiotic metabolism which in turn catabolizes endogenous hormones [131]. A recent study reports a positive association of BPA and sex hormone-binding globulin (SHBG, [135]). Because these proteins bind steroids circulating in the bloodstream, a chemically induced change in SHBG levels might alter hormone bioavailability.

At the receptor level, EDCs can alter receptor expression and degradation. In a primate model, BPA diminished the estradiol-induced expression of PR [136]. Because progesterone normally counteracts 17β-estradiol, this mechanism could be involved in endometrial dysfunction in humans. On the other hand, BPA slows down ERβ degradation in the proteasome [137], thus increasing the receptor level and potentiating its own effect on ERβ [131].[27]

[24]AhR does not belong to the NR superfamily but to the basic helix-loop-helix transcription factors. It is included here because of its relevance to endocrine disruption.

[25]Aromatase is a cytochrome P450 (CYP) enzyme converting C19 androgens into C18 estrogens.

[26]Another example is prostaglandin synthesis. Several EDCs bind to cyclooxygenase enzymes and block the synthesis of prostaglandin precursors [386].

[27]Similarly, phthalic acid blocks PXR degradation in the proteasome [387]. Vice versa, dioxin-activated AhR increases ERα degradation [388].

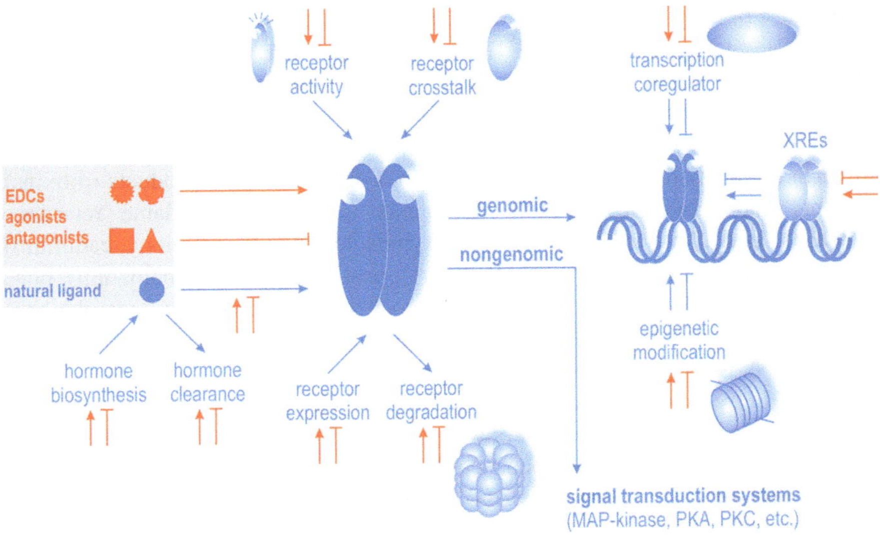

Fig. 2 Potential mechanisms of action by which EDCs interfere with normal hormone signaling (*blue arrows*). *Orange arrows* indicate activation or inhibition of a distinct step of endocrine communication (see text for further details). *MAP* mitogen-activated protein, *PKA/C* protein kinase A/C, *XRE* xenobiotics response element

Another mechanism might involve receptor activity and crosstalk with other NRs. The activity of NRs is modulated by posttranslational modification. Phosphorylation, for example, alters the DNA binding capacity of ERα [138]. Though not well understood, EDCs can interfere with upstream signaling pathways[28] that change receptor phosphorylation and thereby induce ligand-independent activation [139, 140]. In addition, there is extensive crosstalk between the different classes of NRs. For instance, retinoid X receptors (RXR) often heterodimerize with other NRs [128]. Organotin compounds activate RXRα and, by providing more of the active dimerization partner, may interfere with other NRs, such as PPARγ [141].[29]

At transcriptional level, EDCs can target coregulators. One example is the TRAP[30] coactivator complex, which interacts with several NRs. BPA has been shown to enhance the expression of TRAP and its binding to ERβ,[31] thereby increasing the expression of ER-regulated genes [142]. Several NRs also compete for a limited set of coactivators. For instance, the antiestrogenic effect of dioxins can be attributed to a relocation of a coactivator from ERs to AhR [143].[32] AhR

[28]Here, EGF, ERK, MAPK, or Ras signaling might be involved. For example, short-chain fatty acids enhance ER and PR activity through activation of mitogen-activated protein kinase (MAPK) and inhibition of histone deacetylase [389]. Such compounds are termed 'hormone sensitizers'.

[29]Peroxisome proliferator-activated receptor (PPAR).

[30]Thyroid hormone receptor-associated protein.

[31]Interestingly, phthalates increased the binding of TRAP to ERα.

[32]AhR nuclear translocator (ARNT) is the dimerization partner of AhR but also a coactivator of ERs [390]. 2,3,7,8-tetrachlorodibenzo-*p*-dioxin (TCDD) was studied [143].

liganded by xenobiotics can also bind to inhibitory response elements (XREs) close to ER-DNA binding sites and thereby silence the expression of ER target genes [144, 145].

In addition, recently discovered mechanisms of action add to the complexity of ED. Chemicals might induce epigenetic modifications (e.g., [146, 147]) and nongenomic signaling. In the latter case, hormone responses are rapidly communicated via the activation of membrane receptors and subsequent signaling cascades rather than via classical, 'genomic' NRs [148–150]. Interestingly, many EDCs that act as 'weak estrogens' at the genomic pathway affect the nongenomic pathway much more potently [150].

2 In Vitro Bioassays for Endocrine Modes of Action

2.1 Why Do We Use In Vitro Bioassays?

Historically, in vivo models have been used for early detection of pregnancy because of their sensitivity to hormones. An early example is the use of clawed frogs (*Xenopus* spec.) or toad (*Bufo arenarum*) in the Galli–Mainini test between the 1930s and 1960s [151]. Frogs were injected subcutaneously into the dorsal lymph sac with the urine or blood serum of a subject. When the female frog spawned eggs or the male frog released sperm within 24 h, the tested woman was considered pregnant. ED has also been studied using a range of in vivo models. For instance, rodents have been used since the 1930s to study the effects of androgenic and estrogenic chemicals [152, 153].[33] In the same decade, different fish species were used to evaluate the impact of hormones (e.g., [154]).[34] Although in vivo models remain the gold standard in toxicology, the use of animals has ethical, economic, and scientific limitations. First and foremost, ethical objections have resulted in a shrinking societal support for and a stricter control of animal experimentation. Second, in vivo studies are economically demanding, mainly because of their high costs and low throughput.[35] In light of the myriad of chemicals that need to be analyzed, an "in vivo exclusively" approach is not feasible and high-throughput (HTP) approaches are required. Third, the complexity inherent to biological systems often compromises the interpretation of in vivo data.

---

[33]These methods were later standardized by the OECD as Hershberger/Uterotrophic Bioassay in Rats (Guidelines 441 and 440).

[34]Kanter et al. and others [391] developed fish assays for pregnancy testing. They proposed to standardize the animal model because "the results became so confusing that it was felt tests could not be conducted along scientific lines until the fish had been standardized." This is a splendid reminder that the standardization of bioassays has been an issue from the very beginning.

[35]A combined chronic toxicity/carcinogenicity study in rats according to the U.S. EPA's Pesticide Assessment Guidelines costs approximately 2 million dollars and takes approximately 2 years [392].

For instance, low dose and non-monotonic effects of EDCs have long been disregarded as implausible because of an incomplete understanding of the underlying processes [119]. In addition, the mechanism of action of a chemical often cannot be elucidated in vivo.

These three challenges have created strong incentives to develop alternative methods for animal experimentation. In vitro approaches provide such alternatives. By current standards they are ethically sound, economically favorable, offer HTP capability, and provide mechanistic insight. The latter aspect is especially relevant for EDCs. Because of the intricacy of the endocrine system, the response patterns induced by chemical exposures are very complex at the organismal level. Here, the use of in vitro systems reduces the complexity by focusing on one particular mechanism of action (e.g., transactivation of one hormone receptor) and a one-compartment system (mostly to a living cell). Such a reductionist approach has obvious limitations, namely the exclusion of large parts of the toxicokinetic process. However, recent advances in "omics" and in vitro techniques have initiated a paradigm shift in toxicology [155]. Instead of the traditional "black box approach" [156] focusing on apical in vivo endpoint solely, a mechanism-based approach combining in vitro and in silico methods considerably advances our understanding of both the chemically perturbed and the normal function of biological systems.

Taking this idea further, Langley et al. [156] have recently advocated transforming biomedical and toxicological research fundamentally to provide the "big picture." They propose to link environmental sciences and medical research in a systems biology framework to understand intrinsic and extrinsic causes of human disease better. Such a multiscale-pathway-based approach aims at integrating in silico, in vitro, and in vivo data using the adverse outcome pathway concept (AOP).[36] Originally developed to improve environmental risk assessment, the AOP concept provides a framework to organize toxicological knowledge meaningfully in "a sequential series of events that, by definition, span multiple levels of biological organization" [157]. Within this concept, in vitro bioassays can be used to investigate the first anchor of an AOP, the molecular initiating event, and subsequent events at cellular level. A typical, yet simplified, AOP for ED in fish, for instance, would begin with the binding of an EDC to a hormone receptor (studied in ligand-binding assays; see Sect. 2.2.1), consecutive transactivation of the receptor (reporter gene assays; see Sect. 2.2.2), increased VTG synthesis (in fish hepatocytes [158]), feminization of male fish[37] (organismal level), and eventually population decline (mesocosm or ecosystem level). This example highlights that,

[36]An AOP is a conceptual framework that portrays existing knowledge concerning the linkage between two anchor points – THE MOLECULAR INITIATING EVENT, AND AN ADVERSE OUTCOME, CONNECTED BY A CHAIN OF KEY EVENTS and the relationships between them (adopted from AOP-KB, https://aopkb.org).

[37]So far, the cellular processes resulting in feminization are not well understood. VTG induction is here rather used as biomarker for estrogenic effects than causative.

when embedded in the context of AOPs and systems toxicology, the application of in vitro bioassays is a highly valuable contribution to environmental toxicology.

2.2 *Types of In Vitro Bioassays: What Do We Use?*

With regard to their architecture, in vitro bioassays used to investigate ED in ecotoxicology can be broadly classified into three types: cell-free systems, wild-type cell cultures, and reporter-gene assays. This classification somewhat mirrors the history of in vitro bioassays in toxicology. Cell-free systems, for instance ligand-binding assays (LBA), were first used in pharmacological research to study the interaction of chemicals and hormone receptors. Wild-type cell cultures, for instance, immortalized mammary cancer cells, were later used to study apical responses to receptor activation (e.g., proliferation). Genetically modified cell lines, which include reporter genes to study receptor transactivation or inhibition, represent an advancement of wild-type cell lines because they allow for the study of the molecular interaction of an agent with its target.

Traditionally, in vitro bioassays for EDCs are classified according to those methodological considerations (e.g., [159, 160]). However, it is sometimes more meaningful to categorize them according to the biological endpoint under investigation. This is especially important when comparing the results obtained with multiple bioassays. For instance, although commonly referred to as "estrogenic activity," a cell proliferation assay (such as the E-Screen) does not per se deliver the same results as a reporter gene assay for estrogen receptor α (similar to the Yeast Estrogen Screen). The reason for this is that chemically induced proliferation is a response at cellular level whereas receptor transactivation is part of transcription.[38] Because in vitro bioassays operate at different levels of biological organization (Table 1), there are also differences in the complexity of the system under investigation.[39] Accordingly, we discuss bioassays along the level of molecular events resulting from EDCs exposure: the chemical interaction with hormone receptors (Sect. 2.2.1), induced gene expression by the liganded receptor (Sect. 2.2.2), effects of chemicals on protein activity (Sect. 2.2.3), and cellular responses to EDCs (Sect. 2.2.4).

[38]With regard to receptor transactivation, one can argue that the interaction of a chemical with a receptor protein is under investigation (as in the case of LBAs). However, because the induction/repression of the receptor-mediated gene expression is studied, the organizational level under investigation is indeed transcription.

[39]While this difference is obvious, it is worth stressing. In the experience of the authors, the level of biological organization is often neglected when discussing in vitro "endocrine activity."

Table 1 Classification of in vitro bioassays for EDCs according to level of biological organization

Level of response	Type of bioassay	Endpoint	Examples
Receptor	Ligand binding assays	Receptor binding	Several commercially available
Transcription	Reporter gene assays	Receptor transactivation	CALUX, T47D-KBluc, MELN Yeast Estrogen Screen
Translation	Enzyme assays	Enzyme activity	EROD assay using cell lines
Cell	Proliferation assays	Hormone-dependent proliferation	E-Screen (MCF-7) T-Screen (GH3)
	Steroidogenesis	Hormone production	H295R steroidogenesis assay

2.2.1 Receptor Level: Ligand-Binding Assays

Cell-free systems used in ED research mainly include LBA. The basic principle of an LBA is the competition of an analyte and a labeled ligand for binding to the receptor [161]. When incubating homogenates of cells or tissues with a radioactively labeled or fluorescent compound, the chemical-receptor interaction can be determined by the ratio of bound to free ligand concentrations. In an experiment in which the receptor and the ligand concentration is kept constant (competitive LBA), the half-maximal inhibitory concentration (IC_{50}) and the affinity constant K_i of an analyte can be derived. In another variant (saturation LBA), the ligand concentration is varied to determine the maximum receptor density (B_{max}) and the dissociation constant K_d of the receptors in a biological sample (see [162] for review).

Although being instrumental for early pharmaceutical and endocrine research,[40] the application of LBAs in environmental toxicology remained somewhat limited.[41] LBAs have been used mainly to confirm the binding of environmental pollutants to receptors, which were orphan receptors or of non-mammalian origin. For instance, LBAs were first used to confirm the binding of chemicals to AhR [163, 164] or of BPA to a putative mollusk estrogen receptor orthologue [54]. In more recent work, Yost et al. [165] used LBAs to demonstrate that teleost ER subtypes (from *Oryzia slatipes*) have a specific binding pattern to different steroidal estrogens. Medaka ERβ subtype had higher affinities for 17β-estradiol and estriol than the α subtype and may, therefore, play a critical role in mediating estrogenic effects in teleost fish. In environmental health research, LBAs have been used

[40]The first ligand binding assay (a radioimmunoassay for insulin) was presented by Yalow and Berson [393]. Yalow was rewarded with the Nobel prize for her discovery in 1977 [394].

[41]An ISI Web of Science search for "ligand binding assay" returns 20 hits in the research area Environmental Science/Ecology (October 2, 2015).

recently to show that several major flame retardants as well as household dust bind to PPARγ, an NR involved in adipogenesis and potentially obesity [166]. Although LBAs have certain disadvantages, most importantly safety issues when using radioligands and their inability to discern receptor agonists and antagonists, these recent examples demonstrate that they are still useful tools to study receptor interaction.

2.2.2 Transcriptional Responses: Reporter Gene Assays

With the advent of modern molecular biology, genetically engineered cell lines became available to study ED. Today, the majority of in vitro systems with hormonal endpoints are reporter-gene assays. These systems are based on mammalian or yeast cells expressing a certain hormone receptor. Basically, the transactivation of the receptor by a chemical or a sample triggers the expression of a reporter gene. Here, the activated receptor binds a specific DNA response element[42] coupled to the reporter gene. This induces the expression of the reporter protein, in most cases β-galactosidase (lacZ), luciferase, or green fluorescent protein (GFP). The reporter gene activity is quantified biochemically by determining the cleavage of a chromogenic substrate (lacZ), luminescence (luciferase), or fluorescence (GFP). The activity correlates with the degree of receptor activation by the sample.

Traditionally, reporter gene assays have been developed and used for the analysis of estrogen, androgen, and dioxin receptors. Reviewing >200 publications concerned with applying in vitro bioassays to derive bio-equivalents, we found that in the majority of studies reporter gene assays for estrogen (33%) and dioxin receptors (28%) were used [167]. Reporter gene assays for the androgen receptor accounted for only 5%. However, in recent years, bioassays for additional hormone receptors have been established. The ToxCast program, for instance, used high throughput screening to profile 309 environmental chemicals on 25 NRs [168], including other than the classical steroid receptors. This study demonstrated that, beside estrogen receptor α, many environmental compounds activate so far neglected receptors, including PXR, peroxisome proliferator receptors (PPARs), and retinoic acid receptors (RARs). Escher et al. [169] made a similar observation when screening water samples in 103 in vitro bioassays.[43] Here PXR, PPARs, and glucocorticoid receptor (GR) were the most responsive endpoints besides ERs, AR, and AhR. These findings are important because they widen our focus from the traditional steroid-centric view to additional, relevant signaling pathways (see Sect. 1.3).

---

[42]Receptor dimerization or nuclear localization may precede this event. Hormone response elements are regulatory, palindromic DNA sequences. See Ruff et al. [395] for more details.

[43]These included not only endocrine endpoints but also mutagenicity, genotoxicity, stress response, and cytotoxicity.

Yeast-based systems for estrogen and androgen receptors are by far the most commonly used reporter gene assays in environmental toxicology [167]. The templates for these systems are the Yeast Estrogen Screen (YES) and the Yeast Androgen Screen (YAS) developed by the pharmaceutical company Glaxo Group Research in the 1990s [170] and introduced to environmental research by Routledge and Sumpter [171] and Sohoni and Sumpter [172]. In the YES, the hERα gene has been stably integrated to the yeast genome and is expressed under the control of a copper promoter. The expression of the reporter gene lacZ is controlled by multiple copies of the estrogen response element, to which the liganded receptor binds. Upon transactivation, β-galactosidase is expressed and cleaves glycosidic bonds (its actual function in *Escherichia coli* is to hydrolyze galactose to glucose). The β-galactosidase activity is determined using chromogenic, chemiluminescent, or fluorescent substrates.[44] Operating according to the same principle, the transactivation of hAR is studied in the YAS using the lacZ reporter gene coupled to androgen response elements.

As with every bioassay, the YES has undergone a methodological evolution. Several modifications have been proposed to optimize assay sensitivity and efficiency. For instance, Beresford et al. [173] and Dhooge et al. [174] addressed several methodological issues, including the impact of different incubation times, cell densities, and solvents. Murk et al. [175] optimized the sample preparation and application, two factors critical for avoiding false-negative results in in vitro bioassays [176, 177]. A major shortcoming of the original YES protocol was the overlap of the reporter gene induction by the sample and the parallel determination of the reporter gene activity.[45] To overcome this, De Boever et al. [178] used cycloheximide to stop protein synthesis before adding the β-galactosidase substrate. Schultis and Metzger [179] took another approach and used lyticase to lyse the yeast cell walls. This improved the sensitivity of the YES by approximately an order of magnitude and shortened the procedure from >84 h [180] to 7 h. Two other major extensions of the original protocol were to include the ability to test water samples without prior extraction and to detect antagonists of the estrogen as well as the androgen receptor [172]. However, the YES system does not respond to classical antiestrogens, such as ICI 164,384 and fulvestrant (ICI 182,780) as well as tamoxifen [181]. The reasons for that are the impermeability of the cell wall to larger molecules [181] and the lack of metabolic activation. The latter is true for tamoxifen, because its active metabolite 4-hydroxy tamoxifen (OHT) is antiestrogenic in yeast systems [173].[46]

---

[44]Multiple substrates are available. Often the cleavage of *o*-nitrophenyl-β-D-galactopyranosid (ONPG) or chlorophenol red-β-D-galactopyranoside (CPRG) is determined colorimetrically. The fluorescent methylumbelliferyl β-D-galactopyranoside (MUG) has also been used as substrate [396].

[45]The cleavage product of CPRG, chlorophenol red, is estrogenic itself [397], as is phenol red used as pH indicator in cell culture media [398, 399].

[46]Indeed, OHT is a partial agonist inducing a weak estrogenic response in the absence of a full agonist such as 17β-estradiol [173].

Compared to mammalian cell-line, the use of recombinant yeast for screening EDCs offers many advantages, importantly the ease of cultivation and the lack of endogenous hormone production. Accordingly, a plethora of yeast-based reporter gene assays has been developed. The assay with the second widest distribution after the YES by Routledge and Sumpter is an analogous system developed by McDonnell and colleagues [182–184]. In this system, hERα is not integrated in the genome but expressed from a second plasmid. Other assays for estrogenic compounds also rely on a one or two plasmid system [181, 185, 186] but have not been used extensively in environmental research. An interesting example for the commercialization of bioanalytical tools is the so-called A-YES,[47] which, unlike the other yeast-based assays, uses the species *Arxula adeninivorans* instead of *Saccharomyces cerevisiae* and a phytase reporter gene instead of β-galactosidase [187].

Yeast-based reporter gene assays are also available for other human receptors, including ERβ and AR [186, 188–191], AhR [192, 193], GR, MR, and PR [186, 194–197], RARs and RXR [198–201], TRα [202, 203], and VDR [185].[48]

Besides wild-type receptors, an interesting approach is to use mutant receptors to study specific chemicals. Recently, Rajasärkkä et al. applied directed evolution to generate a mutant hERα library and select for receptors specifically binding BPA but not estradiol [204, 205]. Providing a high sensitivity at low costs, this "BPA-receptor" assay is attractive to screen complex samples, such as thermal paper extracts [206]. Taken together, the availability of a broad battery of yeast-based bioassays provides excellent opportunities to investigate the interaction of environmental contaminants with multiple hormone receptors.

As with the yeast-based bioassays, a variety of reporter gene assays have been developed using mammalian cell lines. Arguments for preferring those mainly draw on a closer relationship with humans (regarding toxicokinetics, including membrane permeability and metabolism) and a higher sensitivity. Although the former is not that relevant for environmental toxicology, the latter is advantageous when analyzing environmental samples with low endocrine activity such as surface or drinking water. The Chemical Activated Luciferase gene eXpression (CALUX) assay system has found the widest distribution in environmental sciences and is today commercially available for screening endocrine activity at 13 receptors.[49] The first CALUX assay was developed by Murk et al. [207] to screen for AhR active compounds in sediments and pore waters. In this so-called dioxin-responsive DR-CALUX a rat hepatoma cell line (H4IIE) has been stably transfected with a reporter plasmid containing four dioxin response elements coupled to the firefly

[47]This assay has been developed by the German Leibniz Institute of Plant Genetics and Crop Plant Research and is now marketed by New Diagnostics (www.new-diagnostics.com).

[48]This list is far from comprehensive given that ISI Web of Science lists 13,292 publications under the keywords "yeast" AND "assay" AND "receptor" (on October 22, 2015).

[49]http://www.biodetectionsystems.com/products/bioassays/available-assays.html (last accessed October 23, 2015).

luciferase gene [208, 209]. Shortly after that, Legler et al. [210] presented the ER-CALUX using a human adenocarcinoma cell line (T47D) expressing both estrogen receptor subtypes. Stably transfected with a luciferase reporter plasmid and three estrogen response elements, the ER-CALUX provided excellent specificity and sensitivity. In addition to cell lines expressing AhR and ERs endogenously, human osteoblastic osteosarcoma cell lines (U2-OS) transfected with recombinant human AR, ERα, GR, and PR were developed [211]. Recently, additional CALUX assays for ERβ, PPARC and γ1/2, TRβ, and RAR have been presented [212–214]. As in the case of the YES, the CALUX systems have been constantly optimized [215–222] as well as intensively validated for specific matrices[50] [216, 219–221, 223–230]. Taken together, the high sensitivity and the availability of a broad spectrum of receptors render the CALUX systems a successful example for the commercialization of bioanalytical tools to study ED.[51]

Operating in a similar way to the CALUX, several other reporter gene assays based on mammalian cell lines are available. The US EPA developed the T47D-KBluc cell line [231] which is genetically very similar to the ER-CALUX. Again, ERα/β positive T47D cells are stably transfected with a luciferase reporter plasmid containing three EREs. In the MELN assay, MCF-7 breast cancer cells endogenously expressing ERα were transfected with an ERE-βGlob-Luc reporter construct [232]. The performance of these two non-commercial systems is comparable to the ER-CALUX and they have been successfully used to screen environmental samples (e.g., [233, 234]). The academic community has generated a multitude of other assays using mammalian cells. A recent review lists 34 assays for 11 NRs [235]. This highlights the spectrum of tools available to study EDCs.

2.2.3 Translational Responses: Enzyme Activity

Historically, the activation of AhR by xenobiotics such as dioxins has not been considered a classical endocrine endpoint. Being an orphan receptor, AhR was conceived as a xenobiotics sensor (and key regulator of xenobiotic metabolism) rather than a receptor for endogenous ligands [236]. In recent years, however, our understanding of AhR signaling has considerably increased [237]. For instance, tryptophan derivatives (especially kynurenines) and photoproducts have been proposed as endogenous ligands [238, 239]. Importantly, it became clear that AhR crosstalks with multiple hormone pathways, for instance, estrogen signaling [240, 241]. Accordingly, AhR is involved in multiple endocrine-controlled processes, including reproduction, development, immune response, and carcinogenesis [242].

---

[50] Mainly, the results of the DR-CALUX are compared to chemical analysis, especially of dioxin-like compounds.

[51] BioDetection Systems BV, the company marketing the CALUX assays, is a 2001 spinoff founded by Prof. Abraham Brouwer of the Free University of Amsterdam.

With regard to AhR activation, the EROD assay can be considered the first in vitro bioassay in environmental toxicology. It dates back to the 1980s when it was used to study the activity of ethoxyresorufin-O-deethylase (EROD) in response to benzo(*a*)pyrene [243], nitrosamine [244], dioxins [245], and PAHs [246]. First employed to activate environmental pollutants metabolically, it soon became clear that the EROD induction in vitro (in the rat hepatoma cell line H4IIE) correlates well with in vivo endpoints in rodents, such as body weight loss [245, 247]. Accordingly, it can be argued that the EROD assay is not only the first in vitro assay used in environmental toxicology but also the first case of in vitro–in vivo extrapolation.

Based on the idea that EROD induction is a result of AhR activation by xenobiotics, it has been widely used as a biomarker of exposure when analyzing tissue samples (Eadon, Hanberg). EROD activity in cell cultures (H4IIE, RTL-W1, etc.) was mainly used to characterize the AhR-mediated toxicity of environmental pollutants (biomarker of effect). The former is interesting because the EROD assay for tissues was developed to augment analytical capabilities because back then chemical analysis was unable to detect all congeners of dioxins, furans, and biphenyls [247]. The latter is important because, based on the in vitro and in vivo dioxin-like activity of environmental chemicals, the concept of toxic/TCDD equivalency factor was developed and adopted by WHO [248, 249].

2.2.4 Cellular Response: Proliferation and Steroidogenesis Assays

The cellular responses to EDCs can be investigated in cultured wild-type cell lines, especially by looking at more apical endpoints of hormone receptor activation. One of the earliest and most prominent bioassays applying that principle is the E-Screen. Here, an immortal line of human mammary carcinoma cells (MCF-7) is used to study estrogenic effects [250–252]. The basic idea is that the proliferation of MCF-7 cells is induced via the activation of ERα by an estrogenic chemical or complex sample. The estrogen-dependent proliferation can be determined by various means, including nuclei counting (original method), staining DNA or proteins (using bromodeoxyuridine or sulphorhodamine B), and metabolic activity (conversion of resazurin or similar dyes). Having been in use for over two decades, the E-Screen has been constantly improved, namely by identifying sensitive sub-lines [253], optimization and miniaturization to 96-well plate format [254–256], and the use of flow-cytometry to determine proliferating S-phase cells [257].

The superior sensitivity has led to a broad application of the E-Screen in (eco) toxicological research. Since 1994, it has been used in almost 250 studies[52] to determine the estrogenicity of chemicals and complex samples. Despite the wide distribution, the E-Screen has certain disadvantages, including the inability to discern cytotoxicity from antagonistic effects and the need to confirm the receptor-mediated mechanism of action of an enhanced proliferation. With regard

[52]According to an ISI Web of Science search (October 5, 2015).

to the latter, a known ER antagonist (e.g., fulvestrant or tamoxifen) can be co-administered to determine whether the proliferative effect of a sample is suppressed by receptor inhibition. Although multiple signaling pathways converge to induce proliferation of MCF-7 cells, some of them hormone-independent (e.g., by ER phosphorylation [258]), the E-Screen is a valuable tool to study estrogenicity (Leusch et al. [323]).

Besides the estrogen receptors, inadequate chemical interference with the thyroid system is of specific interest because it can lead to an impaired (neuro) development (especially in amphibians [8]). In analogy to the E-Screen, the T-Screen has been developed to study TR-dependent cell proliferation in the GH3 cell line derived from rat pituitary [259]. Out of the wide spectrum of thyroid-disrupting chemicals, certain PCBs [260], PAHs [261], parabens [262], plasticizers [263], phytoestrogens [264], perfluoroalkyl acids [265], and pesticides [266] are active in the T-Screen.

Besides inducing cell proliferation, EDCs can also affect steroid biosynthesis [267]. A range of in vitro models for steroidogenesis is available, including rodent Leydig cell lines and human H295R adrenocortical carcinoma cells [268]. The H295R assay has been validated by OECD [269] and is part of the US EPA Endocrine Disruptor Screening Program [270]. The H295R cell line expresses the full set of enzymes needed for converting cholesterol to the key steroids such as estrogens and androgens as well as progesterone, glucocorticoids, and aldosterone [271, 272]. To investigate the impact of EDCs on steroidogenesis, H295R cells are exposed to the test compound/sample for 48–72 h and the secreted hormones are extracted from the medium. Hormone concentrations, mainly those of 17β-estradiol and testosterone, are determined using radio immunoassays, ELISA (enzyme-linked immunosorbent assay), or chemical analysis [273]. In addition, quantitative real-time PCR can be used to study the expression of steroidogenic key enzymes [272, 274]. The H295R cell line has been used to study the effect of numerous EDCs and environmental samples on steroid production. Beyond this, Maglich et al. [275] have demonstrated that the system is not only useful to study steroidogenesis but is a sensitive and specific predictor for reproductive toxicity in vivo. It has been argued that the adrenal gland is the most neglected organ in endocrine toxicology [271]. In that sense, the H295R system provides a versatile tool to study the effects of EDCs at the start of the vertebrate hypothalamus-pituitary-adrenal axis.

3 In Vitro Endpoints: What Are We Looking At?

Although the spectrum of available in vitro assays for studying EDCs is very broad (see Sect. 2.1), it becomes clear that the investigated endpoints or, more specifically. mechanisms of action are quite narrow. Notwithstanding the different levels of biological organization, most bioassays focus on chemical interactions with hormone receptors as the key initiating event. In LBAs the binding of chemicals to receptors is studied directly, and in reporter gene and cell proliferation assays the immediate downstream effects of receptor (ant)agonism are investigated. The reason for this is that receptor-mediated toxicity is at the center of ED research. However, in the light of the diversity of potential molecular sites and processes with which EDCs can interfere (see Fig. 2), such receptor-centric view is too simplistic. Unfortunately, the development of in vitro systems for other mechanisms/modes of action (besides hormone synthesis, H295R) has not made much progress in the field of environmental toxicology. The main aspect constraining the development of novel bioassays is our considerable lack of knowledge on the mechanistic basis of ED and the endocrine systems of the non-mammalian models used in environmental toxicology. Here, recent advances in (eco)toxicogenomics and metabolomics provide valuable insight [276–278] which in turn might result in the extension of in vitro approaches beyond receptor-interactions.

Within the realm of hormone receptors,[53] the spectrum of targets we are looking at is again quite narrow. Most available in vitro systems target estrogen, androgen, and, to a lesser extent, thyroid pathways (EAT). The focus on estrogens and androgens evolved because of the critical role steroids and their environmental mimics play in the sexual development and reproduction of vertebrates (see Sect. 1.1). Furthermore, the initial focus on estrogens and androgens, as well as their environmental mimics, was very much driven by intersex observations in the 1980s involving masculinization and feminization of fish. However, many recent studies indicate that other pathways beside steroid signaling are relevant both from an effect as well as an exposure perspective. One example of the former is the family of retinoid receptors (RXRs and RARs) which, inadvertently activated by environmental compounds such as TBT, mediate obesity in vertebrates [279] and potentially in invertebrates [280], imposex in gastropods (see Sect. 1.2), and teratogenicity in amphibians [281]. On the exposure side, retinoid-like pollutants appear to be quite common in the aquatic environment as recent in vitro studies demonstrate [169, 282–285]. This highlights the need for our focus to widen beyond the classical steroid receptors.

Another limitation of the currently available in vitro systems is that they almost exclusively rely on human receptors and cell lines. Although EAT signaling appears to be conserved in the vertebrate lineage, considerable interspecies differences exist with regard to the specificity and sensitivity of hormone receptors to EDCs. For instance, the work of Iguchi and Katsu (see [118] for review) demonstrates that although ERs from six different fish species are similarly sensitive to estradiol in vitro, this is not the case for DDT and its metabolites. Recently, Miyagawa et al. [286] used ERα reporter gene assays for nine fish species and demonstrated that the latter is also true for other common EDCs. They conclude that the interspecies differences “indicate environmental risk assessment for estrogens cannot necessarily be predicted for all fish by simply examining receptor activation for a few model fish species.” This is especially problematic because

[53]Humans possess 48 nuclear receptors, *Caenorhabditis elegans* 284 [400].

most of the in vitro studies for estrogen-like compounds have not been conducted using fish but human ERα. Several studies (e.g., [287–289]) have shown that the sensitivity of fish and human ERs to EDCs differ considerably.[54] Accordingly, Ihara et al. [290] demonstrated that a reporter gene assay with medaka ERα was better in predicting VTG expression in vivo than an assay with human ERα. To complicate the matter further, other ER subtypes, such as ERβ, might be more relevant for estrogenic effects in fish [165].

Besides mammals and (teleost) fish, many other phyla and classes are affected by EDCs, including reptiles, amphibians, and mollusks (see Sect. 1.2). Although relevant from an ecological perspective, in vitro bioassays are scarce for these groups. Again, the main reason is our limited understanding of their endocrinology [291]. One example is the role of steroid receptors in sexual differentiation and reproduction. Even though it is clear that ERs control both processes in vertebrates, including non-mammalian species, their role in the reproductive biology of invertebrates remains far from clear. For instance, it is commonly accepted that estrogen-like compounds affect diverse mollusk species [15]. Although mollusks possess orthologues of the vertebrate estrogen receptors [292–296], their function is not well understood. Here, reporter gene assays have helped to demonstrate that molluskan ERs are not responsive to vertebrate steroids[55] and selected EDCs (e.g., [297]) but are constitutively active instead (e.g., [298]).[56] These examples demonstrate that in vitro methods for non-model species, although in an early stage of development, have been instrumental in understanding not only mechanistic aspects of toxicity but also basic biology.

4 Applications for In Vitro Bioassays: What Are We Using It For?

The range of application for in vitro bioassays for endocrine endpoints is very diverse. However, the term "screening" can be used to describe the common features of the majority of studies. In contrast to mechanistic studies applying in vitro techniques to understand the molecular basis of ED in (non)model species, screening studies are rather descriptive: Bioassays for well-established mechanisms of actions (e.g., transactivation of human ERs) are used to quantify the endocrine effects ("activity") of chemicals or environmental samples. With regard to the

---

[54] A meta-analysis by Dang et al. [401] demonstrated that the EC_{50}s of several EDCs at human and fish ERs correlate and concluded that "transactivation of ERs in one vertebrate species or one subtype of ERs could be extrapolated to other species or subtypes of ERs for the purpose of chemical screening." However, the EC_{50}s can differ by two orders of magnitude for many EDCs.

[55] An interesting question remains as to why mollusks possess steroid hormones anyway [402, 403].

[56] Notably, other mollusk receptors (e.g., RAR) also appear to be constitutively active [404]. If so, the mechanism of action mediating endocrine disruption in mollusks remains unknown.

former, in vitro bioassays have played a critical role in identifying novel EDCs. For instance, the estrogenic activity of today's well-established EDCs, including BPA, alkylphenols, parabens, and some UV filters, was first demonstrated in vitro. More recently, in vitro approaches have been used to identify so-called obesogens [48, 299, 300] and EDCs acting via nongenomic pathways [301]. In vitro chemical screening was also important to understand mixture toxicity. Although the theoretical basis was established in the 1920s, Silva and others [302, 303] published their landmark "something from nothing" studies, demonstrating that combinations of EDCs produce significant effects when mixed at concentrations that are ineffective individually. Subsequent in vitro studies have corroborated mathematical models that can be used to predict the toxicity of complex chemical mixtures (reviewed in [121, 304]). Interestingly, this stimulated investigations into the in vivo effects of EDC mixtures arriving at basically the same conclusions, namely that the concentration addition model can be used to predict mixture effects [305–307].

The in vitro screening of environmental samples reverses the chemical mixture approach. Instead of combining known compounds, samples likely containing mixtures of EDCs are analyzed for their total endocrine activity. This is advantageous because in vitro (as well as in vivo) bioassays integrate the joint effects of all EDCs present in complex samples, no matter whether the causative chemicals and the mixture composition are known or not.[57] This, together with the increased throughput capacity, has led to an explosion in the number of in vitro studies screening environmental samples for diverse endocrine effects. In our meta-analysis we found that most of the studies applying bio-equivalents to quantify the effects of complex samples [167] screened for endocrine activity in soils and sediments, surface water, and wastewater (in almost equal parts). To a lesser extent, the endocrine effects in biological matrices (e.g., in tissues), foodstuff, and herbal medicine (e.g., in milk) as well as consumer products (e.g., leaching from plastic toys) were investigated. Again, effect-based in vitro screening has been important in demonstrating that EDCs are omnipresent in the aquatic environment, especially in wastewater (see, e.g., [308]), foodstuff and bottled drinking water [176, 309, 310], and plastic products [311].

Although in vitro screening is instrumental in providing information on the presence of EDCs in the environment (i.e., demonstrating a potential exposure), two challenges remain. First, many of the EDCs causing in vitro effects in complex samples remain unknown. Second, it remains unclear whether and how in vitro effects translate to ED in vivo (see Sect. 5.2). Regarding the first aspect, bioassays can be coupled to other techniques to identify the causative EDCs. This approach is often referred to as effect-directed analysis (EDA).[58] EDA essentially relies on the physical separation of the active from the inactive compounds present in a complex sample and the subsequent bioassay-guided analytical identification of the active

[57]Despite tremendous advances in analytical chemistry, they remain unknown in most cases.

[58]In the U.S., the term toxicity identification evaluation (TIE) is used. However, TIE rather refers to in vivo effects [405].

compound(s). The chemicals can be separated according to different molecular properties, including polarity and size [312]. Accordingly, a sample fractionation can be achieved either by liquid chromatography or by size exclusion. The resulting fractions are subsequently analyzed for their toxicity in vitro and the active fractions are subjected to chemical analysis. With each step of EDA being separate and time-consuming, recent advances have been made in terms of parallelization. Buchinger et al. [313] combined the fractionation and the bioassay analysis by directly coupling thin-layer chromatography to the YES. Another approach is the parallel analysis of fractionated samples in the bioassay and high-resolution mass spectrometry [314].

Although early attempts have been "disappointing" [315], recent EDA studies were successful in identifying previously unknown EDCs as well as known EDCs in unexpected places. For instance, chlorophene [316], di-*tert*-butylphenol, and di-*tert*-butyl-methoxy-*p*-cresol [317] have been identified as novel ER ligands in river sediments. The latter study also found that di-iso-octylphthalate is an agonist of PXR. Chlorinated biphenyls in the blood plasma of polar bear cubs were found to be thyroid disrupting [318], as were triclosan and nonylphenols in sediments [319]. PAHs were found to cause in vitro antiandrogenic activity in river sediments [74] and in extracts of the clam *Scrobicularia plana* [320]. In fish bile, antiandrogenic effects were caused by chlorophene, triclosan, and a range of other phenols [70]. In our own work, we have shown that di(2-ethylhexyl) fumarate (DEHF) is a novel antiestrogen in bottled drinking water [177] and that parabens unexpectedly caused estrogenic and antiandrogenic effects leaching from plastic baby toys [321].

Admittedly, there are probably at least as many unsuccessful attempts in EDA as the main challenge remains the analytical identification of truly unknown compounds (i.e., those that have never been measured before). Here, the capability for structural elucidation by mass spectrometry is limited because mass spectra cannot be readily translated to one specific chemical entity in most cases. An alternative approach is nuclear magnetic resonance (NMR) spectroscopy, the gold standard for structural elucidation. However, NMR requires relatively high amounts of a pure compound, which can rarely be attained from fractionated environmental samples.[59] These limitations notwithstanding, EDA studies have not only extended the list of EDCs. More importantly, they have demonstrated that it is possible to bring together chemical and biological methods to identify novel environmental pollutants that are toxicologically relevant. In the light of the complexity of chemical exposures, such prioritization is urgently needed.

[59]For instance, had we attempted to identify DEHF by NMR, we would have needed to extract, fractionate, and test at least 1,000 L of bottled drinking water.

5 Translational Aspects: How Can We Put In Vitro Data into Context?

Effect-directed analysis represents a good example of how in vitro bioassays can be applied translationally. In the case of EDA, in vitro toxicity is translated to the chemical(s) causing the effect. Other aspects are not that straightforward. The question of how endocrine effects determined in one type of in vitro bioassay relate to the results obtained with another (in vitro–in vitro extrapolation) is far from answered. The matter becomes even more complicated when attempting to make reliable in vitro–in vivo extrapolations (IVIVE), which are indeed the major goal of in vitro toxicology.

5.1 In Vitro–In Vitro Comparison

With regard to the comparisons of different in vitro bioassays, multiple studies focusing on estrogenic activities are available. For instance, Fang et al. [322] compared LBA, YES, and E-Screen data for 29 compounds from 5 different laboratories. The authors arrived at the conclusion that the assays "generally provided consistent information on quantitatively determining estrogenic activity." Interestingly, LBAs were especially predictive for the two other assays when antiestrogens were excluded. The same was true when correlating the YES and E-Screen results. Moreover, Fang et al. [322] emphasize the advantages of comparing the results of different assay types to learn about assay characteristics and activity patterns of EDCs (especially receptor antagonists). Along the same line, Scrimshaw and Lester [159] compared literature data from ER binding assays, yeast screens, and the E-Screen and concluded that "no single test is suitable to determine whether any compound (or mixture) is or is not (likely to be) an endocrine disrupter."

Moving from pure chemicals to environmental samples, Leusch et al. [323] compared the estrogenic effects of ground, waste, and surface water samples in the YES, ER-CALUX, MELN, T47D-KBluc, and E-Screen, tests being performed in seven laboratories. Overall, all bioassays generated comparable results and were in good agreement with chemical analysis. However, the Leusch study highlights the individual advantages and disadvantages of each bioassay. For instance, the YES is less sensitive compared to the assays based on mammalian cell lines but has lower costs and is technically more mature. The ER-CALUX offered the highest performance but requires more training and resources. Taking the bioanalytical approach further, Escher et al. [169] applied a battery of 103 unique in vitro bioassays at 20 laboratories to provide a toxicity profile of 10 water samples. This massive battery included, among others, several bioassays for endocrine endpoints. For instance, 12 assays for ER-mediated effects were applied. Here, the overall "effect pattern across the different samples was similar for all bioassays." Moreover,

ER-CALUX, E-Screen, and MCF7-ERE were most responsive, corroborating the idea that bioassays with mammalian cell lines are more sensitive than those based on yeast strains.

Taking assay-specific differences into account (e.g., level of biological organization, metabolic activity, etc.), there appears to be a reasonable agreement in the results of different in vitro bioassays for estrogenic effects. In addition, applying a battery of bioassays for multiple molecular initiating events is preferable to provide a more complete picture of the toxicity of a chemical or sample [324]. However, it has to be taken into account that comparative assessments of assays for other endocrine endpoints is currently lacking. In summary, we and others advocate a "fit-for-purpose" approach, that is, the assay type should be selected based on the nature of the research question. Leusch et al. [323] conclude that "any of the five bioassays may be suitable for testing estrogenic activity in environmental samples as long as their limitations and technical requirements are clearly understood by the researcher."

5.2 In Vitro–In Vivo Extrapolations

Translating endocrine effects observed in vitro to the in vivo situation remains the major challenge in human and environmental toxicology. With regard to the former, in vitro assays appear to be predictive when testing chemicals for estrogenic effects in vivo in the uterotrophic assay [325, 326]. In a recent study, Rotroff et al. [327] used HTP data from the Tox21 program to derive ER Interaction Scores for 1814 chemicals screened in 13 assays. Comparing the results to 45 compounds, for which in vivo data were available, resulted in a high specificity and sensitivity of the score to predict in vivo effects. Accordingly, it has been argued that in vitro data can be used to prioritize chemicals for in vivo screening [328] and human risk assessment [329, 330].

In environmental toxicology, less progress regarding IVIVE has been made and relatively few comparative studies are available, mainly focusing on single EDCs. For instance, Villeneuve et al. [331] compared the effects of six chemicals on steroidogenesis in the H295R assay and in ex vivo fathead minnow ovary explants. Although both assays displayed different sensitivities to the compounds, the ex vivo data were more predictive for in vivo effects. In another example, Puy-Azurmendi et al. [332] investigated the estrogenic effects of different octyl- and nonylphenol isomers in reporter gene assays and in vivo with early life stages of zebra fish. Here, octylphenols were estrogenic in vitro and in vivo and nonylphenols induced effects in vitro but not in *Danio rerio*. Comparing the in vitro, ex vivo, and in vivo effects of benzothiazoles on the thyroid axis, Hornung et al. [333] found that the in vitro inhibition of thyroid peroxidase activity was a good indicator of thyroid disrupting effects in vivo in *Xenopus laevis*.[60]

---

[60]Interestingly, ex vivo data were not predictive.

Only a few studies compare the in vitro/in vivo effects of complex environmental samples. Leusch et al. [334] investigated the endocrine effects of two wastewater treatment plant effluents in reporter gene assays and in situ with mosquito fish (*Gambusia holbrooki*). They found low estrogenicity in the E-Screen, which corresponded to a slightly increased VTG expression in male mosquito fish. Because chemical, in vitro, and in vivo data were well in line, they promote the "utility of an ecotoxicity toolbox combining multiple lines of evidence, including chemical analysis, a battery of in vitro bioassays, and in situ monitoring in the assessment of water quality." In a study of male brown trout in Swiss rivers, Vermeirssen et al. [335] found a positive but not significant correlation of estrogenic activity in vitro and VTG levels in vivo. Using a similar approach, Henneberg et al. [336] investigated the estrogenicity of river water samples in the E-Screen and compared that to the reproduction of the mudsnail *Potamopyrgus antipodarum* and the VTG induction in brown trout exposed in situ. From a qualitative point of view, in vitro data "reasonably reflect reproduction and ED observed in snails and field-exposed fish." In a more quantitative assessment, Ihara et al. [290] demonstrated for the first time that the in vitro estrogenicity of wastewater was an excellent predictor for VTG induction in male medaka. Interestingly, the in vitro bioassay with medaka ER outperformed chemical analysis, indicating that the co-occurrence of ER agonists and antagonists in wastewater determines the joint in vivo effects.

Although these examples are promising,[61] one needs to take into account that the studies mentioned above investigate a well-known mode to toxicity: VTG expression in fish is under tight control of estrogens and mediated via ERs, accordingly. However, it remains unknown how effects of EDCs on gene expression translate to more apical endpoints, for instance, intersex. In that sense, we are dealing with a relatively simple AOP, whose activation may be well predicted in vitro. Once we move toward higher levels of biological organization and non-model species with less well-researched endocrinology, the complexity increases considerably. Accordingly, IVIVE is complicated. The challenges inherent to this are caused by differences in toxicokinetics and toxicodynamics. With regard to the latter, in vitro testing simplifies the situation because one mechanism of action is investigated at a time. In vivo, however, many EDCs often affect multiple pathways, resulting in much more complex response patterns (see the example of BPA in Sect. 1.3). In addition, the considerable lack of knowledge with regard to mechanisms of (endocrine) action in non-model species further complicates the matter.

Besides in vitro–in vivo differences in toxicodynamics, toxicokinetics is likely the main factor complicating a translation of in vitro data to in vivo effects. This is because in vitro bioassays exclude the larger part of the toxicokinetic process. First of all, the bioavailability of a compound is very different when administered in a multi-well plate compared to in vivo exposure via water, sediments, food, etc. Obviously, chemicals – depending on their physico-chemical characteristics –bind not only to the cells but also to the medium and the plastic materials used in vitro. In

[61]This might be because of publication bias.

addition, compounds may degrade or evaporate during the test duration [337]. Accordingly, it is important to consider dose metrics other than nominal concentrations when quantifying in vitro effects [337].[62] In addition, the rest of the adsorption, distribution, metabolism, and elimination (ADME) cascade differ in vitro and in vivo. Although the distribution and elimination aspects are largely neglected in vitro, metabolism might or might not occur. For instance, heterocyclic aromatic hydrocarbons are metabolically activated and thus estrogenic in the ER-CALUX but not in the YES [338]. Differences in ADME may be accounted for by applying physiologically-based pharmacokinetic (PBPK) models. Although successfully applied in human toxicology [339, 340], such an approach is still rarely used in environmental toxicology.

Taken together, IVIVE remains the main challenge for in vitro toxicology, especially with regard to the effects of EDCs in wildlife. To advance environmental toxicology in this regard, we first need to elucidate relevant mechanisms of action for ED in non-model species. Then, we need to develop in vitro bioassays specific to these mechanisms. Here, adopting methods from other disciplines, for instance from drug discovery and omics, has already advanced our capabilities and insights and will continue to do so in the future. Most importantly, we need suitable frameworks to put the in vitro data into context. Here, promising concepts are already available from quantitative IVIVE [341, 342], AOPs [157, 343, 344], and systems toxicology [345–349].

6 What Is the Practical Relevance of In Vitro Data?

6.1 Potential of In Vitro Bioassays Compared to Chemical Analysis

In vitro bioassays can, on the one hand, serve as screening tools for chemicals with a specific mechanism of action (see Sect. 4), for example, that can be followed up upon with chemical or further effect-based investigations. They can also serve as a tool for the prioritization of chemicals. For instance, in vitro HTP assays are incorporated as Tier 1 in the U.S. EPA's Endocrine Disruptor Screening Program for the twenty-first century (EDSP21).[63] The results of the chemical screening inform the in vivo assays performed at Tier 2.

In contrast to chemical analysis, in vitro bioassays are unable to discern single chemicals present in a complex sample. However, they are selective for specific endpoints and can measure the effect of chemicals with the same mechanism of action. In vitro bioassays can determine mixture effects of all bioactive chemicals

[62]For instance, the total cell concentration or the dose at the molecular target (biologically effective dose).

[63]More information at http://www.epa.gov/endocrine-disruption.

present in the sample, whereas, with chemical analysis, such effects can only be determined for detectable substances and by applying respective mathematical models (see Sect. 2.2). By measuring the mixture effect of a sample, effects of transformation products generated during wastewater treatment with the same mechanism of action are also taken into account. With regard to sensitivity, in vitro bioassays are often less sensitive for the detection of single chemicals than chemical analysis. However, they enable a sensitive detection of all single chemicals with a common mode of action [350]. In the case of estrogenic substances, where proposed environmental quality standard (EQS) values (e.g., for 17α-ethinylestradiol of 0.035 ng/L) are below the limits of quantification of most of the currently available routine chemical analytical methods, in vitro bioassays can represent a useful tool to detect such substances (e.g., [351, 352]).

6.2 Current Regulatory Application

These potentials of in vitro bioassays make them valuable for a consideration in the regulatory assessment of ED. With regard to chemical regulation, such bioassays are already implemented, for example, in the Endocrine Disrupter Screening and Assessment (EDSA) Programme of the Organization for Economic Co-operation and Development [353]. In research, many in vitro bioassays for ED are also regularly used with several of them being potentially suitable for water quality assessment and monitoring (for review see [350]). Although routine applications of in vitro bioassays for water quality testing are lacking, it has to be kept in mind that routine applications for analyzing foodstuffs for dioxin-like activity are accepted in a regulatory context.[64]

Regulations and water quality guidelines, such as the EU water framework directive [354], are rather focused on the chemical monitoring of a list of priority substances. Based on such a chemical monitoring, mixture toxicity is also addressed in the assessment of fresh and marine water quality [355], the assessment of combination effects from exposure to multiple chemicals (EU [356]), and the investigation of cumulative effects of pesticides [357]. Only one in vitro bioassay for estrogenic activity (the YES) is applied in Sweden to monitor wastewater and surface water quality [358].

[64]See COMMISSION REGULATION (EU) No 589/2014 of 2 June 2014 laying down methods of sampling and analysis for the control of levels of dioxins, dioxin-like PCBs, and non-dioxin-like PCBs in certain foodstuffs and repealing Regulation (EU) No 252/2012. Official Journal of the European Union L 164/18.

6.3 Standardization of In Vitro Bioassays

One crucial point for the acceptance of in vitro bioassays in regulation is their standardization. Bioassay standardization for single chemicals is performed by the OECD, whereas the validation and standardization of bioassays for the assessment of environmental samples is part of the work program of several technical committees of the International Organization for Standardization (ISO). Prerequisites for standardization are (1) reliability, indicating the extent of reproducibility of results from a test within and among laboratories over time, when using the same standardized protocol and (2) relevance, describing the extent to which the test method correctly measures or predicts the (biological) effect of interest. Regulatory need, usefulness, and limitations of the test method are aspects of its relevance [359].

Albeit many in vitro bioassays for ED are in discussion for the OECD's EDSA programme [353], currently only two of them are fully standardized: an in vitro reporter gene assay for the detection of estrogenic chemicals (OECD guideline 455) and the H295R steroidogenesis assay (OECD guideline 456). Validation and standardization of the ER-CALUX assay for the detection of estrogenic chemicals (for inclusion in OECD guideline 455) is currently underway. For water quality assessment, currently several estrogen receptor transactivation assays (ER-TAs) to determine the estrogenic potential of water and wastewater have been accepted as new work item proposals within the ISO technical committee 147 (currently: ISO/CD 19040–1, −2, −3).[65] Two of them are yeast-based assays and the third is based on human cell lines. This work should result in three new international standards within 2016.

6.4 Risk Assessment: Trigger Values

To date, there are no formal guideline values available in water quality regulation that are based on in vitro bioassay data [360]. Recently, two common EDCs with pharmaceutical use and of emerging concern (17α-ethinylestradiol (EE2) and 17-β-estradiol) have been included in the European monitoring list ("watch list") of the European Directive 2013/39/EU [361] dealing with the review of priority substances in surface water bodies. These substances have a very high biological activity in the environment, resulting in chronic environmental quality criteria of less than 1 ng/L.[66] However, detecting the substances at such low levels requires

[65]ISO Standards Catalogue. ISO/TC 147 – Water quality. Standards under development (see www.iso.org).

[66]Annual-Average Environmental Quality Standards (AA-EQS) for EE2: 0.035 ng/L and for E2 of 0.4 ng/L. These concentrations should not be exceeded in order to protect the aquatic environment and human health.

great effort in analytical chemistry. In this respect, in vitro bioassays have the potential to support or replace chemical analytical methods as has recently been suggested by Jarošová et al. [308] and Kunz et al. [351]. Using bioassays as replacements for chemical analytical methods requires a paradigm shift, however. Under the current scheme, such as the water framework directive (WFD), maximal protective concentrations of individual priority pollutants (environmental quality standards, EQS) are monitored. As bioassays determine the joint effect of an unknown mixture, they may indicate significant effects in a sample whilst the target EQS compound is only a minor contributor to this effect or may even be absent in the mixture altogether. In this worst-case scenario, assessing water quality based on bioassay data alone, bioassays produce results that cannot be interpreted within the WFD paradigm/concept.

To evaluate water quality based on results of in vitro bioassays, suitable trigger values have to be established. Concepts for such trigger values are suggested by Brand et al. [362] with regard to human toxicology, and by Jarošová et al. [363], Escher et al. [364], and Kunz et al. [351] regarding environmental effects.

Deriving trigger values for human toxicology is suggested by using (1) acceptable or tolerable daily intake (ADI/TDI) values of specific compounds, (2) pharmacokinetic factors defining their bioavailability, (3) estimations of the bioavailability of unknown compounds with equivalent hormonal activity, (4) relative endocrine potencies, and (5) physiological and drinking water allocation factors [362]. The trigger values define a level above which human health or environmental risk cannot be waived a priori. If these values are exceeded, additional examination of specific endocrine activity may be necessary. The authors pointed out that the limits should be sufficiently conservative to serve as a warning signal, but not too conservative to avoid unnecessary and costly additional mitigation measures. This approach resulted in trigger values of 3.8 ng E2 equivalents (EEQ)/L (for estrogenic effects), 11 ng DHT-EQ/L (for androgenic effects), 21 ng dexamethasone (DEX)-EQ/L (for progesterone-like effects), and 333 ng Org2058-eq/L (for glucocorticoid-like effects) [362].

Suggested trigger values for environmental risk assessment are considerably lower, for example, Escher et al. [364] suggested a trigger value of 0.2–12 ng/L for estrogenic effects, and Jarošová et al. [363] and Kunz et al. [351] suggested trigger values of 0.1–0.4 ng EEQ/L (partly depending on the bioassay used for derivation of the trigger value). The concepts for deriving these trigger values differ. Escher et al. [364] applied a statistical method to be able to read directly across from chemical guideline values to the trigger values using bioanalytical equivalents. The authors proposed statistical distribution methods to derive a specific effect-based trigger bioanalytical equivalent concentration (EBT-BEQ) for bioassays targeting receptor-mediated toxicity. The values were derived by matching effect concentrations with existing chemical guideline values, followed by the selection of appropriate chemicals giving a response in the respective bioassays at concentrations in the range of the guideline values [364]. The differing sensitivity of the applied bioassays was also taken into account. However, currently, this concept depends on the availability of experimental bioassay data on individual regulated chemicals. Therefore, the trigger values are provisional.

Another concept suggests the application of the chronic environmental quality standard for 17β-estradiol (0.4 ng/L) as EEQ-based as trigger value. This value was suggested for the following reasons [351]:

- EEQ are commonly used in bioanalytics and biomonitoring.
- The in vitro to in vivo potency of the natural steroid hormone E2 is between those of estrone (E1) and EE2, and therefore the authors consider EEQs as assessing mixture effects most accurately.
- The main contributors to estrogenic activity in surface water are likely E1 and E2 for aquatic vertebrates (as well as Estriol (E3) with regard to in vitro bioassays), and, to a lesser extent, EE2, which makes E2 more representative of estrogenic substances than EE2.
- The in vitro potency of EE2 is only slightly (1.4–1.9 times) higher than that of E2, whereas the in vivo potency in vertebrate species is 10–25 times higher. Therefore, the risk might be overestimated by using EE2 equivalents and false positives might occur, as both E2 and E1 also bind to the receptor.

Jarošová et al. [363] arrived at a very similar value [0.3 ng/L, named EEQ Safe regarding Steroid Estrogens (EEQs-SSE)] for municipal WWTP effluents based on the assumption that the steroid estrogens are responsible for more than 90% of in vitro estrogenicity. The authors derived EEQs-SSEs by applying bioassay and testing protocol-specific in vitro potencies of steroid estrogens, and in vivo predicted no effect concentration (PNECs) of these compounds plus their relative contributions to the overall estrogenicity detected in municipal WWTP effluents. Resulting EEQs-SSE ranged from 0.1 to 0.4 ng/L depending on the applied bioassay and protocol. For all these concepts, the trigger values can be compared with the measured bioanalytical equivalent concentrations as routinely done in environmental risk assessment to calculate risk quotients (RQ = BEQ/trigger value). If the RQ is higher than 1, a risk for aquatic organisms cannot be excluded; if the RQ is lower than 1, no risk is anticipated.

7 Summary

EDCs adversely affect a wide range of wildlife. Their investigation is therefore a main focus of ecotoxicological research. Although a plethora of molecular mechanisms of action exist, a chemical interaction with hormone receptors is central for inducing an endocrine disrupting effect. Accordingly, a broad battery of in vitro bioassays has been developed to study the receptor-mediated effects of EDCs. Besides ethical and economical advantages, in vitro techniques considerably reduce the toxicokinetic and toxicodynamic complexity inherent in in vivo systems. Such a reductionist approach has largely expanded our knowledge of EDCs.

In vitro bioassays can be classified according to the different levels of biological organization under investigation. Systems are available to study chemical interactions at the level of the receptor (ligand binding assays), transcription (reporter gene assays), translation (enzyme assays), and cell (proliferation and steroidogenesis assays). Independent of the assay type, ecotoxicological research concentrates on few in vitro endpoints, namely estrogenic, androgenic, and – to a lesser extent – thyroid effects. This focus has historical reasons but needs to be widened, given the growing spectrum of EDCs acting via other receptors and mechanisms. Another limitation is that most of the in vitro bioassays currently in use are based on human hormone receptors. This anthropocentric view is problematic because emerging evidence implies that non-human receptor orthologs have different sensitivities toward EDCs. Understanding these interspecific differences and elucidating the mechanisms of EDC action in non-model species remains a future challenge.

In ecotoxicology, in vitro bioassays are used to provide mechanistic insight. However, in the majority of studies they are applied to screen for endocrine activity of chemicals and environmental samples. Here, in vitro techniques have been instrumental for identifying new EDCs as well as for demonstrating that EDCs are omnipresent in the environment. Such in vitro screening has the advantage of providing an integrated response, which includes the effects of so far unknown EDCs and mixtures. Accordingly, in vitro bioassays are used for environmental monitoring, and protective levels for estrogenic activity in surface waters, so-called trigger values, have been proposed. Along the same line, the international standardization of certain assay systems is currently ongoing and needed to integrate in vitro data in environmental legislation.

One specific area of application is effect-directed analysis (EDA). Here, information from in vitro and chemical analyses is combined to identify the EDCs causing the observed effect. Although EDA has been successfully applied to find novel EDCs, the main challenge remains the structural elucidation of truly unknown chemicals.

In vitro data can not only be translated to the causative chemicals but needs to be put into context with the results of other in vitro and in vivo bioassays. With regard to in vitro–in vitro extrapolation, there appears to be general agreement between the different assay types for estrogenic endpoints. In vitro–in vivo extrapolation as the main goal of in vitro toxicology is complicated by differences in toxicokinetics and dynamics. Here, reasonable agreement can be established for estrogenic activity of water samples and vitellogenin induction in male fish. However, this represents a well-described and simple AOP, and predicting more complex in vivo responses to EDCs based on in vitro data remains challenging. To tackle this, we need to improve our mechanistic understanding of ED, especially in invertebrate species. Here, emerging techniques such as ecotoxicogenomics as well as conceptual frameworks, including AOP and systems toxicology, will greatly contribute to advance the field of in vitro ecotoxicology.

References

1. LaFont R (2000) The endocrinology of invertebrates. Ecotoxicology 9(1–2):41–57
2. Kortenkamp A, Martin O, Faust M, Evans R, McKinlay R, Orton F, Rosivatz E (2011) State of the art assessment of endocrine disruptors, final report. European Commission, The Directorate-General for the Environment, p 433
3. Thornton JW (2001) Evolution of vertebrate steroid receptors from an ancestral estrogen receptor by ligand exploitation and serial genome expansions. Proc Natl Acad Sci U S A 98(10):5671–5676
4. Thornton JW, Need E, Crews D (2003) Resurrecting the ancestral steroid receptor: ancient origin of estrogen signaling. Science 301(5640):1714–1717
5. Edwards TM, Moore BC, Guillette LJ (2006) Reproductive dysgenesis in wildlife: a comparative view. Int J Androl 29(1):109–120
6. Guillette LJ (2000) Contaminant-induced endocrine disruption in wildlife. Growth Horm IGF Res 10:45–50
7. Guillette LJ Jr (2006) Endocrine disrupting contaminants: beyond the dogma. Environ Health Perspect 114:9–12
8. Kloas W, Urbatzka R, Opitz R, Wuertz S, Behrends T, Hermelink B, Hofmann F, Jagnytsch O, Kroupova H, Lorenz C, Neumann N, Pietsch C, Trubiroha A, Van Ballegooy C, Wiedemann C, Lutz I (2009) Endocrine disruption in aquatic vertebrates. Trends Comp Endocrinol Neurobiol 1163:187–200
9. McLachlan JA (2001) Environmental signaling: what embryos and evolution teach us about endocrine disrupting chemicals. Endocr Rev 22(3):319–341
10. Milnes MR, Bermudez DS, Bryan TA, Edwards TM, Gunderson MP, Larkin ILV, Moore BC, Guillette LJ (2006) Contaminant-induced feminization and demasculinization of nonmammalian vertebrate males in aquatic environments. Environ Res 100(1):3–17
11. Oehlmann J, Schulte-Oehlmann U, Kloas W, Jagnytsch O, Lutz I, Kusk KO, Wollenberger L, Santos EM, Paull GC, Van Look KJW, Tyler CR (2009) A critical analysis of the biological impacts of plasticizers on wildlife. Philos Trans R Soc B-Biol Sci 364(1526):2047–2062
12. Sumpter JP (2005) Endocrine disrupters in the aquatic environment: an overview. Acta Hydrochim Hydrobiol 33(1):9–16
13. Tyler CR, Jobling S, Sumpter JP (1998) Endocrine disruption in wildlife: a critical review of the evidence. Crit Rev Toxicol 28(4):319–361
14. Vos JG, Dybing E, Greim HA, Ladefoged O, Lambre C, Tarazona JV, Brandt I, Vethaak AD (2000) Health effects of endocrine-disrupting chemicals on wildlife, with special reference to the European situation. Crit Rev Toxicol 30(1):71–133
15. Oehlmann J, Di Benedetto P, Tillmann M, Duft M, Oetken M, Schulte-Oehlmann U (2007) Endocrine disruption in prosobranch molluscs: evidence and ecological relevance. Ecotoxicology 16(1):29–43
16. Mills LJ, Chichester C (2005) Review of evidence: are endocrine-disrupting chemicals in the aquatic environment impacting fish populations? Sci Total Environ 343(1–3):1–34
17. Hayes TB, Case P, Chui S, Chung D, Haeffele C, Haston K, Lee M, Mai VP, Marjuoa Y, Parker J, Tsui M (2006) Pesticide mixtures, endocrine disruption, and amphibian declines: are we underestimating the impact? Environ Health Perspect 114:40–50
18. Milnes MR, Guillette LJ Jr (2008) Alligator tales: new lessons about environmental contaminants from a sentinel species. Bioscience 58(11):1027–1036
19. Ottinger MA, Lauoie E, Thompson N, Barton A, Whitehouse K, Barton M, Abdelnabi M, Quinn M Jr, Panzica G, Viglietti-Panzica C (2008) Neuroendocrine and behavioral effects of embryonic exposure to endocrine disrupting chemicals in birds. Brain Res Rev 57(2): 376–385
20. Iguchi T, Watanabe H, Katsu Y (2006) Application of ecotoxicogenomics for studying endocrine disruption in vertebrates and invertebrates. Environ Health Perspect 114:101–105

21. Hamlin HJ, Guillette LJ Jr (2011) Embryos as targets of endocrine disrupting contaminants in wildlife. Birth Defects Res C: Embryo Today 93(1):19–33
22. Rhind SM (2009) Anthropogenic pollutants: a threat to ecosystem sustainability? Philos Trans R Soc B-Biol Sci 364(1534):3391–3401
23. Acevedo-Whitehouse K, Duffus ALJ (2009) Effects of environmental change on wildlife health. Philos Trans R Soc B-Biol Sci 364(1534):3429–3438
24. Jenssen BM (2006) Endocrine-disrupting chemicals and climate change: a worst-case combination for arctic marine mammals and seabirds? Environ Health Perspect 114:76–80
25. Matthiessen P, Gibbs PE (1998) Critical appraisal of the evidence for tributyltin-mediated endocrine disruption in mollusks. Environ Toxicol Chem 17(1):37–43
26. Blaber SJM (1970) The occurrence of a penis-like out-growth behind the right tentacle in spent females of *Nucella lapillus* (L.). Proc Malacol Soc Lond 39:231–232
27. Smith BS (1971) Sexuality in the American mud snail, *Nassarius obsoletus* SAY. Proc Malacol Soc Lond 39:377–378
28. Smith BS (1981) Male characteristics on female mud snails caused by antifouling bottom paints. J Appl Toxicol 1(1):22–25
29. Smith BS (1981) Reproductive anomalies in stenoglossan snails related to pollution from marinas. J Appl Toxicol 1(1):15–21
30. Titley-O'Neal CP, Munkittrick KR, MacDonald BA (2011) The effects of organotin on female gastropods. J Environ Monit 13(9):2360–2388
31. Oetken M, Bachmann J, Schulte-Oehlmann U, Oehlmann J (2004) Evidence for endocrine disruption in invertebrates. Int Rev Cytol 236:1–44
32. Galante-Oliveira S, Oliveira I, Ferreira N, Santos JA, Pacheco M, Barroso C (2011) Nucella lapillus L. imposex levels after legislation prohibiting TBT antifoulants: temporal trends from 2003 to 2008 along the Portuguese coast. J Environ Monit 13(2):304–312
33. Guomundsdottir LO, Ho KKY, Lam JCW, Svavarsson J, Leung KMY (2011) Long-term temporal trends (1992–2008) of imposex status associated with organotin contamination in the dogwhelk *Nucella lapillus* along the Icelandic coast. Mar Pollut Bull 63(5–12):500–507
34. Nicolaus EE, Barry J (2015) Imposex in the dogwhelk (*Nucella lapillus*): 22-year monitoring around England and Wales. Environ Monit Assess 187(12):736
35. Santos JA, Galante-Oliveira S, Barroso C (2011) An innovative statistical approach for analysing non-continuous variables in environmental monitoring: assessing temporal trends of TBT pollution. J Environ Monit 13(3):673–680
36. Titley-O'Neal CP, MacDonald BA, Pelletier E, Saint-Louis R (2011) Using Nucella lapillus (L.) as a bioindicator of tributyltin (TBT) pollution in eastern Canada: a historical perspective. Water Qual Res J Can 46(1):74–86
37. Sternberg RM, Gooding MP, Hotchkiss AK, LeBlanc GA (2010) Environmental-endocrine control of reproductive maturation in gastropods: implications for the mechanism of tributyltin-induced imposex in prosobranchs. Ecotoxicology 19(1):4–23
38. Bettin C, Oehlmann J, Stroben E (1996) TBT-induced imposex in marine neogastropods is mediated by an increasing androgen level. Helgoländer Meeresun 50(3):299–317
39. Oehlmann J, Stroben E, Schulte-Oehlmann U, Bauer B, Fioroni P, Markert B (1996) Tributyltin biomonitoring using prosobranchs as sentinel organisms. Fresenius J Anal Chem 354(5–6):540–545
40. Spooner N, Gibbs PE, Bryan GW, Goad LJ (1991) The effect of tributyltin upon steroid titers in the female dogwhelk, *Nucella lapillus*, and the development of imposex. Mar Environ Res 32(1–4):37–49
41. Schulte-Oehlmann U, Tillmann M, Markert B, Oehlmann J, Watermann B, Scherf S (2000) Effects of endocrine disruptors on prosobranch snails (Mollusca: Gastropoda) in the laboratory. Part II: Triphenyltin as a xeno-androgen. Ecotoxicology 9(6):399–412
42. Abidli S, Santos MM, Lahbib Y, Costa Castro LF, Reis-Henriques MA, El Menif NT (2012) Tributyltin (TBT) effects on *Hexaplex trunculus* and *Bolinus brandaris* (Gastropoda: Muricidae): imposex induction and sex hormone levels insights. Ecol Indic 13(1):13–21

43. Castro LFC, Lima D, Machado A, Melo C, Hiromori Y, Nishikawa J, Nakanishi T, Reis-Henriques MA, Santos MM (2007) Imposex induction is mediated through the retinoid X receptor signalling pathway in the neogastropod *Nucella lapillus*. Aquat Toxicol 85(1):57–66
44. Horiguchi T, Nishikawa T, Ohta Y, Shiraishi H, Morita M (2010) Time course of expression of the retinoid X receptor gene and induction of imposex in the rock shell, *Thais clavigera*, exposed to triphenyltin chloride. Anal Bioanal Chem 396(2):597–607
45. Horiguchi T, Ohta Y, Nishikawa T, Shiraishi F, Shiraishi H, Morita M (2008) Exposure to 9-cis retinoic acid induces penis and vas deferens development in the female rock shell, Thais clavigera. Cell Biol Toxicol 24(6):553–562
46. Nakanishi T (2008) Endocrine disruption induced by organotin compounds; organotins function as a powerful agonist for nuclear receptors rather than an aromatase inhibitor. J Toxicol Sci 33(3):269–276
47. Nishikawa J (2006) Imposex in marine gastropods may be caused by binding of organotins to retinoid X receptor. Mar Biol 149(1):117–124
48. Nishikawa J, Mamiya S, Kanayama T, Nishikawa T, Shiraishi F, Horiguchi T (2004) Involvement of the retinoid X receptor in the development of imposex caused by organotins in gastropods. Environ Sci Technol 38(23):6271–6276
49. Urushitani H, Katsu Y, Ohta Y, Shiraishi H, Iguchi T, Horiguchi T (2011) Cloning and characterization of retinoid X receptor (RXR) isoforms in the rock shell, *Thais clavigera*. Aquat Toxicol 103(1–2):101–111
50. Stange D, Sieratowicz A, Oehlmann J (2012) Imposex development in *Nucella lapillus*: evidence for the involvement of retinoid X receptor and androgen signalling pathways *in vivo*. Aquat Toxicol 106–107:20–24
51. Oehlmann J, Schulte-Oehlmann U, Tillmann M, Markert B (2000) Effects of endocrine disruptors on prosobranch snails (Mollusca: Gastropoda) in the laboratory. Part I: Bisphenol A and octylphenol as xeno-estrogens. Ecotoxicology 9(6):383–397
52. Schulte-Oehlmann U, Tillmann M, Casey D, Duft M, Markert B, Oehlmann J (2001) Östrogenartige Wirkungen von Bisphenol A auf Vorderkiemenschnecken (Mollusca: Gastropoda: Prosobranchia). Umweltwiss Schadst Forsch 13(6):319–333
53. Staples CA, Woodburn K, Caspers N, Hall AT, Klecka GM (2002) A weight of evidence approach to the aquatic hazard assessment of bisphenol A. Hum Ecol Risk Assess 8(5): 1083–1105
54. Oehlmann J, Schulte-Oehlmann U, Bachmann J, Oetken M, Lutz I, Kloas W, Ternes TA (2006) Bisphenol A induces superfeminization in the ramshorn snail *Marisa cornuarietis* (Gastropoda: Prosobranchia) at environmentally relevant concentrations. Environ Health Perspect 114:127–133
55. Forbes VE, Aufderheide J, Warbritton R, van der Hoevene N, Caspers N (2007) Does bisphenol A induce superfeminization in Marisa cornuarietis? Part II: Toxicity test results and requirements for statistical power analyses. Ecotoxicol Environ Saf 66(3):319–325
56. Forbes VE, Selck H, Palmqvist A, Aufderheide J, Warbritton R, Pounds N, Thompson R, van der Hoeven N, Caspers N (2007) Does bisphenol A induce superfeminization in *Marisa cornuarietis*? Part I: Intra- and inter-laboratory variability in test endpoints. Ecotoxicol Environ Saf 66(3):309–318
57. Bortone SA, Davis WP (1994) Fish intersexuality as indicator of environmental-stress. Bioscience 44(3):165–172
58. Tyler CR, Jobling S (2008) Roach, sex, and gender-bending chemicals: the feminization of wild fish in English rivers. Bioscience 58(11):1051–1059
59. Purdom CE (1994) Estrogenic effects of effluents from sewage treatment works. Chem Ecol 8:275–285
60. Desforges J-PW, Peachey BDL, Sanderson PM, White PA, Blais JM (2010) Plasma vitellogenin in male teleost fish from 43 rivers worldwide is correlated with upstream human population size. Environ Pollut 158(10):3279–3284

61. Jobling S, Beresford N, Nolan M, Rodgers-Gray T, Brighty GC, Sumpter JP, Tyler CR (2002) Altered sexual maturation and gamete production in wild roach (*Rutilus rutilus*) living in rivers that receive treated sewage effluents. Biol Reprod 66(2):272–281
62. Jobling S, Coey S, Whitmore JG, Kime DE, Van Look KJW, McAllister BG, Beresford N, Henshaw AC, Brighty G, Tyler CR, Sumpter JP (2002) Wild intersex roach (*Rutilus rutilus*) have reduced fertility. Biol Reprod 67(2):515–524
63. Harris CA, Hamilton PB, Runnalls TJ, Vinciotti V, Henshaw A, Hodgson D, Coe TS, Jobling S, Tyler CR, Sumpter JP (2011) The consequences of feminization in breeding groups of wild fish. Environ Health Perspect 119(3):306–311
64. Lange A, Paull GC, Hamilton PB, Iguchi T, Tyler CR (2011) Implications of persistent exposure to treated wastewater effluent for breeding in wild roach (*Rutilus rutilus*) populations. Environ Sci Technol 45(4):1673–1679
65. Kidd KA, Blanchfield PJ, Mills KH, Palace VP, Evans RE, Lazorchak JM, Flick RW (2007) Collapse of a fish population after exposure to a synthetic estrogen. Proc Natl Acad Sci U S A 104(21):8897–8901
66. Blanchfield PJ, Kidd KA, Docker MF, Palace VP, Park BJ, Postma LD (2015) Recovery of a wild fish population from whole-lake additions of a synthetic estrogen. Environ Sci Technol 49(5):3136–3144
67. Filby AL, Thorpe KL, Maack G, Tyler CR (2007) Gene expression profiles revealing the mechanisms of anti-androgen-and estrogen-induced feminization in fish. Aquat Toxicol 81(2):219–231
68. Jobling S, Burn RW, Thorpe K, Williams R, Tyler C (2009) Statistical modeling suggests that antiandrogens in effluents from wastewater treatment works contribute to widespread sexual disruption in fish living in English rivers. Environ Health Perspect 117(5):797–802
69. Hill EM, Evans KL, Horwood J, Rostkowski P, Oladapo FO, Gibson R, Shears JA, Tyler CR (2010) Profiles and some initial identifications of (anti)androgenic compounds in fish exposed to wastewater treatment works effluents. Environ Sci Technol 44(3):1137–1143
70. Rostkowski P, Horwood J, Shears JA, Lange A, Oladapo FO, Besselink HT, Tyler CR, Hill EM (2011) Bioassay-directed identification of novel antiandrogenic compounds in bile of fish exposed to wastewater effluents. Environ Sci Technol 45(24):10660–10667
71. Schriks M, van Leerdam JA, van der Linden SC, van der Burg B, van Wezel AP, de Voogt P (2010) High-resolution mass spectrometric identification and quantification of glucocorticoid compounds in various wastewaters in the Netherlands. Environ Sci Technol 44(12): 4766–4774
72. Suzuki G, Tue NM, van der Linden S, Brouwer A, van der Burg B, van Velzen M, Lamoree M, Someya M, Takahashi S, Isobe T, Tajima Y, Yamada TK, Takigami H, Tanabe S (2011) Identification of major dioxin-like compounds and androgen receptor antagonist in acid-treated tissue extracts of high trophic-level animals. Environ Sci Technol 45(23): 10203–10211
73. Thomas KV, Langford K, Petersen K, Smith AJ, Tollefsen KE (2009) Effect-directed identification of naphthenic acids as important *in vitro* xeno-estrogens and anti-androgens in North sea offshore produced water discharges. Environ Sci Technol 43(21):8066–8071
74. Weiss JM, Hamers T, Thomas KV, van der Linden S, Leonards PEG, Lamoree MH (2009) Masking effect of anti-androgens on androgenic activity in European river sediment unveiled by effect-directed analysis. Anal Bioanal Chem 394(5):1385–1397
75. Weiss JM, Simon E, Stroomberg GJ, de Boer R, de Boer J, van der Linden SC, Leonards PEG, Lamoree MH (2011) Identification strategy for unknown pollutants using high-resolution mass spectrometry: androgen-disrupting compounds identified through effect-directed analysis. Anal Bioanal Chem 400(9):3141–3149
76. Alford RA, Richards SJ (1999) Global amphibian declines: a problem in applied ecology. Annu Rev Ecol Syst 30:133–165

77. Stuart SN, Chanson JS, Cox NA, Young BE, Rodrigues ASL, Fischman DL, Waller RW (2004) Status and trends of amphibian declines and extinctions worldwide. Science 306(5702):1783–1786
78. Pounds JA, Bustamante MR, Coloma LA, Consuegra JA, Fogden MPL, Foster PN, La Marca E, Masters KL, Merino-Viteri A, Puschendorf R, Ron SR, Sanchez-Azofeifa GA, Still CJ, Young BE (2006) Widespread amphibian extinctions from epidemic disease driven by global warming. Nature 439(7073):161–167
79. Sodhi NS, Bickford D, Diesmos AC, Lee TM, Koh LP, Brook BW, Sekercioglu CH, Bradshaw CJA (2008) Measuring the meltdown: drivers of global amphibian extinction and decline. PLoS One 3(2)
80. Mann RM, Hyne RV, Choung CB, Wilson SP (2009) Amphibians and agricultural chemicals: review of the risks in a complex environment. Environ Pollut 157(11):2903–2927
81. Reeder AL, Ruiz MO, Pessier A, Brown LE, Levengood JM, Phillips CA, Wheeler MB, Warner RE, Beasley VR (2005) Intersexuality and the cricket frog decline: historic and geographic trends. Environ Health Perspect 113(3):261–265
82. Hayes TB, Anderson LL, Beasley VR, de Solla SR, Iguchi T, Ingraham H, Kestemont P, Kniewald J, Kniewald Z, Langlois VS, Luque EH, McCoy KA, Munoz-de-Toro M, Oka T, Oliveira CA, Orton F, Ruby S, Suzawa M, Tavera-Mendoza LE, Trudeau VL, Victor-Costa AB, Willingham E (2011) Demasculinization and feminization of male gonads by atrazine: consistent effects across vertebrate classes. J Steroid Biochem Mol Biol 127(1–2):64–73
83. Hayes TB, Khoury V, Narayan A, Nazir M, Park A, Brown T, Adame L, Chan E, Buchholz D, Stueve T, Gallipeau S (2010) Atrazine induces complete feminization and chemical castration in male African clawed frogs (*Xenopus laevis*). Proc Natl Acad Sci U S A 107(10): 4612–4617
84. Woodward AR, Percival HF, Jennings ML, Clinton TM (1993) Low clutch viability of American alligators. Fla Sci 56(1):52–63
85. Guillette LJ, Brock JW, Rooney AA, Woodward AR (1999) Serum concentrations of various environmental contaminants and their relationship to sex steroid concentrations and phallus size in juvenile American alligators. Arch Environ Contam Toxicol 36(4):447–455
86. Heinz GH, Percival HF, Jennings ML (1991) Contaminants in American alligator eggs from Lake Apopka, Lake Griffin, and Lake Okeechobee, Florida. Environ Monit Assess 16(3): 277–285
87. Rauschenberger RH, Wiebe JJ, Sepulveda MS, Scarborough JE, Gross TS (2007) Parental exposure to pesticides and poor clutch viability in American alligators. Environ Sci Technol 41(15):5559–5563
88. Guillette LJ, Gross TS, Masson GR, Matter JM, Percival HF, Woodward AR (1994) Developmental abnormalities of the gonad and abnormal sex-hormone concentrations in juvenile alligators from contaminated and control lakes in Florida. Environ Health Perspect 102(8): 680–688
89. Crain DA, Guillette LJ, Pickford DB, Percival HF, Woodward AR (1998) Sex-steroid and thyroid hormone concentrations in juvenile alligators (*Alligator mississippiensis*) from contaminated and reference lakes in Florida, USA. Environ Toxicol Chem 17(3):446–452
90. Guillette LJ, Pickford DB, Crain DA, Rooney AA, Percival HF (1996) Reduction in penis size and plasma testosterone concentrations in juvenile alligators living in a contaminated environment. Gen Comp Endocrinol 101(1):32–42
91. Crain DA, Guillette LJ (1998) Reptiles as models of contaminant-induced endocrine disruption. Anim Reprod Sci 53(1–4):77–86
92. Guillette LJ Jr, Edwards TM, Moore BC (2007) Alligators, contaminants and steroid hormones. Environ Sci 14(6):331–347
93. Moore BC, Hamlin HJ, Botteri NL, Guillette LJ Jr (2010) Gonadal mRNA expression levels of TGF beta superfamily signaling factors correspond with post-hatching morphological development in American alligators. Sex Dev 4(1–2):62–72

94. Moore BC, Milnes MR, Kohno S, Katsu Y, Iguchi T, Woodruff TK, Guillette LJ Jr (2011) Altered gonadal expression of TGF-beta superfamily signaling factors in environmental contaminant-exposed juvenile alligators. J Steroid Biochem Mol Biol 127(1–2):58–63
95. Ratcliffe DA (1967) Decrease in eggshell weight in certain birds of prey. Nature 215(5097):208–210
96. Bowerman WW, Giesy JP, Best DA, Kramer VJ (1995) A review of factors affecting productivity of bald eagles in the Great-lakes region - implications for recovery. Environ Health Perspect 103:51–59
97. Fry DM (1995) Reproductive effects in birds exposed to pesticides and industrial-chemicals. Environ Health Perspect 103:165–171
98. Ratcliffe DA (1970) Changes attributable to pesticides in egg breakage frequency and eggshell thickness in some British birds. J Appl Ecol 7(1):67–107
99. Best DA, Elliott KH, Bowerman WW, Shieldcastle M, Postupalsky S, Kubiak TJ, Tillitt DE, Elliott JE (2010) Productivity, embryo and eggshell characteristics, and contaminants in bald eagles from the Great Lakes, USA, 1986 to 2000. Environ Toxicol Chem 29(7):1581–1592
100. Bruggeman V, Swennen Q, De Ketelaere B, Onagbesan O, Tona K, Decuypere E (2003) Embryonic exposure to 2,3,7,8-tetrachlorodibenzo-p-dioxin in chickens: effects of dose and embryonic stage on hatchability and growth. Comp Biochem Physiol Part C: Toxicol Pharmacol 136(1):17–28
101. Fernie KJ, Shutt JL, Mayne G, Hoffman D, Letcher RJ, Drouillard KG, Ritchie IJ (2005) Exposure to polybrominated diphenyl ethers (PBDEs): changes in thyroid, vitamin A, glutathione homeostasis, and oxidative stress in American kestrels (*Falco sparverius*). Toxicol Sci 88(2):375–383
102. Holm L, Blomqvist A, Brandt I, Brunstrom B, Ridderstrale Y, Berg C (2006) Embryonic exposure to o, p'-DDT causes eggshell thinning and altered shell gland carbonic anhydrase expression in the domestic hen. Environ Toxicol Chem 25(10):2787–2793
103. Hoogesteijn AL, DeVoogd TJ, Quimby FW, De Caprio T, Kollias GV (2005) Reproductive impairment in zebra finches (*Taeniopygia guttata*). Environ Toxicol Chem 24(1):219–223
104. Lundholm CE (1997) DDE-induced eggshell thinning in birds: effects of p, p'-DDE on the calcium and prostaglandin metabolism of the eggshell gland. Comp Biochem Physiol C Pharmacol Toxicol Endocrinol 118(2):113–128
105. Quaglino AE, Craig-Veit CB, Viant MR, Erichsen AL, Fry DM, Millam JR (2002) Oral estrogen masculinizes female zebra finch song system. Horm Behav 41(2):236–241
106. Rochester JR, Heiblum R, Rozenboim I, Millam JR (2008) Post-hatch oral estrogen exposure reduces oviduct and egg mass and alters nest-building behavior in adult zebra finches (*Taeniopygia guttata*). Physiol Behav 95(3):370–380
107. Millam JR, Craig-Veit CB, Quaglino AE, Erichsen AL, Famula TR, Fry DM (2001) Posthatch oral estrogen exposure impairs adult reproductive performance of zebra finch in a sex-specific manner. Horm Behav 40(4):542–549
108. Rochester JR, Forstmeier W, Millam JR (2010) Post-hatch oral estrogen in zebra finches (Taeniopygia guttata): is infertility due to disrupted testes morphology or reduced copulatory behavior? Physiol Behav 101(1):13–21
109. Markman S, Leitner S, Catchpole C, Barnsley S, Mueller CT, Pascoe D, Buchanan KL (2008) Pollutants increase song complexity and the volume of the brain area HVC in a songbird. PLoS One 3(2)
110. Braune BM, Outridge PM, Fisk AT, Muir DCG, Helm PA, Hobbs K, Hoekstra PF, Kuzyk ZA, Kwan M, Letcher RJ, Lockhart WL, Norstrom RJ, Stern GA, Stirling I (2005) Persistent organic pollutants and mercury in marine biota of the Canadian Arctic: an overview of spatial and temporal trends. Sci Total Environ 351:4–56
111. Letcher RJ, Bustnes JO, Dietz R, Jenssen BM, Jorgensen EH, Sonne C, Verreault J, Vijayan MM, Gabrielsen GW (2010) Exposure and effects assessment of persistent organohalogen contaminants in arctic wildlife and fish. Sci Total Environ 408(15):2995–3043

112. Tanabe S (2002) Contamination and toxic effects of persistent endocrine disrupters in marine mammals and birds. Mar Pollut Bull 45(1–12):69–77
113. Buckman AH, Veldhoen N, Ellis G, Ford JKB, Helbing CC, Ross PS (2011) PCB-associated changes in mRNA expression in killer whales (*Orcinus orca*) from the NE Pacific Ocean. Environ Sci Technol 45(23):10194–10202
114. Routti H, Arukwe A, Jenssen BM, Letcher RJ, Nyman M, Backman C, Gabrielsen GW (2010) Comparative endocrine disruptive effects of contaminants in ringed seals (*Phoca hispida*) from Svalbard and the Baltic Sea. Comp Biochem Physiol Part C: Toxicol Pharmacol 152(3):306–312
115. Schwacke LH, Zolman ES, Balmer BC, De Guise S, George RC, Hoguet J, Hohn AA, Kucklick JR, Lamb S, Levin M, Litz JA, McFee WE, Place NJ, Townsend FI, Wells RS, Rowles TK (2012) Anaemia, hypothyroidism and immune suppression associated with polychlorinated biphenyl exposure in bottlenose dolphins (*Tursiops truncatus*). Proc R Soc B Biol Sci 279(1726):48–57
116. Villanger GD, Lydersen C, Kovacs KM, Lie E, Skaare JU, Jenssen BM (2011) Disruptive effects of persistent organohalogen contaminants on thyroid function in white whales (*Delphinapterus leucas*) from Svalbard. Sci Total Environ 409(13):2511–2524
117. Van den Berg KJ, Zurcher C, Brouwer A, Vanbekkum DW (1988) Chronic toxicity of 3,4,3',4'-tetrachlorobiphenyl in the marmoset monkey (*Callithrix jacchus*). Toxicology 48(2):209–224
118. Iguchi T, Katsu Y (2008) Commonality in signaling of endocrine disruption from snail to human. Bioscience 58(11):1061–1067
119. Vandenberg LN, Colborn T, Hayes TB, Heindel JJ, Jacobs DR Jr, Lee D-H, Shioda T, Soto AM, vom Saal FS, Welshons WV, Zoeller RT, Myers JP (2012) Hormones and endocrine-disrupting chemicals: low-dose effects and nonmonotonic dose responses. Endoc Rev 33(3):378–455
120. Barouki R, Gluckman PD, Grandjean P, Hanson M, Heindel JJ (2012) Developmental origins of non-communicable disease: implications for research and public health. Environ Health 11:1–9, http://www.ehjournal.net/content/11/1/42
121. Kortenkamp A (2007) Ten years of mixing cocktails: a review of combination effects of endocrine-disrupting chemicals. Environ Health Perspect 115:98–105
122. Bern HA, Blair P, Brasseur S, Colborn T, Cunha GR, Davis W, Döhler KD, Fox G, Fry M, Gray LE, Green R, Hines MJKT, McLachlan J, Myers JP, Peterson RW, Reijnders PJH, Soto AM, Van Der Kraak G, vom Saal FS, Whitten P (1992) Statement from the work session on chemically-induced alterations in sexual development: the wildlife/human connection. In: Colborn T, Clement C (eds) Chemically-induced alterations in sexual and functional development: the wildlife/human connection. Advances in modern environmental toxicology, vol. xxi. Princeton Scientific Pub, Princeton, NJ, p 403
123. Kavlock RJ, Daston GP, DeRosa C, Fenner-Crisp P, Gray LE, Kaattari S, Lucier G, Luster M, Mac MJ, Maczka C, Miller R, Moore J, Rolland R, Scott G, Sheehan DM, Sinks T, Tilson HA (1996) Research needs for the risk assessment of health and environmental effects of endocrine disruptors: a report of the U.S. EPA-sponsored workshop. Environ Health Perspect 104(Suppl 4):715–740
124. European Commission (1996) Report of Proceedings: European Workshop on the Impact of Endocrine Disrupters on Human Health and Wildlife, Weybridge, UK, EUR 17549, European Commission, Environment and Climate Research Programme of DG XII, 2–4 December 1996
125. WHO/IPCS (2002) Global assessment of the state-of-the-science of endocrine disruptors. WHO publication no. WHO/PCS/EDC/02.2. International Programme on Chemical Safety, World Health Organisation, Geneva, Switzerland
126. Zoeller RT, Brown TR, Doan LL, Gore AC, Skakkebaek NE, Soto AM, Woodruff TJ, Vom Saal FS (2012) Endocrine-disrupting chemicals and public health protection: a statement of principles from the Endocrine Society. Endocrinology
127. Germain P, Staels B, Dacquet C, Spedding M, Laudet V (2006) Overview of nomenclature of nuclear receptors. Pharmacol Rev 58(4):685–704

128. Gronemeyer H, Gustafsson JA, Laudet V (2004) Principles for modulation of the nuclear receptor superfamily. Nat Rev Drug Discov 3(11):950–964
129. Rubin BS (2011) Bisphenol A: an endocrine disruptor with widespread exposure and multiple effects. J Steroid Biochem Mol Biol 127(1–2):27–34
130. Vandenberg LN, Maffini MV, Sonnenschein C, Rubin BS, Soto AM (2009) Bisphenol-A and the great divide: a review of controversies in the field of endocrine disruption. Endocr Rev 30(1):75–95
131. Tabb MM, Blumberg B (2006) New modes of action for endocrine-disrupting chemicals. Mol Endocrinol 20(3):475–482
132. Masuyama H, Hiramatsu Y, Kunitomi M, Kudo T, MacDonald PN (2000) Endocrine disrupting chemicals, phthalic acid and nonylphenol, activate Pregnane X receptor-mediated transcription. Mol Endocrinol 14(3):421–428
133. Takeshita A, Igarashi-Migitaka J, Nishiyama K, Takahashi H, Takeuchi Y, Koibuchi N (2011) Acetyl tributyl citrate, the most widely used phthalate substitute plasticizer, induces cytochrome P450 3A through steroid and xenobiotic receptor. Toxicol Sci 123(2):460–470
134. Takeshita A, Koibuchi N, Oka J, Taguchi M, Shishiba Y, Ozawa Y (2001) Bisphenol-A, an environmental estrogen, activates the human orphan nuclear receptor, steroid and xenobiotic receptor-mediated transcription. Eur J Endocrinol 145(4):513–517
135. Mendiola J, Jorgensen N, Andersson A-M, Calafat AM, Ye X, Redmon JB, Drobnis EZ, Wang C, Sparks A, Thurston SW, Liu F, Swan SH (2010) Are environmental levels of bisphenol A associated with reproductive function in fertile men? Environ Health Perspect 118(9):1286–1291
136. Aldad TS, Rahmani N, Leranth C, Taylor HS (2011) Bisphenol-A exposure alters endometrial progesterone receptor expression in the nonhuman primate. Fertil Steril 96(1): 175–179
137. Masuyama H, Hiramatsu Y (2004) Involvement of suppressor for Gal 1 in the ubiquitin/proteasome-mediated degradation of estrogen receptors. J Biol Chem 279(13):12020–12026
138. Weigel NL, Moore NL (2007) Steroid receptor phosphorylation: a key modulator of multiple receptor functions. Mol Endocrinol 21(10):2311–2319
139. Gillesby BE, Zacharewski TR (1998) Exoestrogens: mechanisms of action and strategies for identification and assessment. Environ Toxicol Chem 17(1):3–14
140. Zsarnovszky A, Le HH, Wang HS, Belcher SM (2005) Ontogeny of rapid estrogen-mediated extracellular signal-regulated kinase signaling in the rat cerebellar cortex: potent nongenomic agonist and endocrine disrupting activity of the xenoestrogen bisphenol A. Endocrinology 146(12):5388–5396
141. le Maire A, Grimaldi M, Roecklin D, Dagnino S, Vivat-Hannah V, Balaguer P, Bourguet W (2009) Activation of RXR-PPAR heterodimers by organotin environmental endocrine disruptors. EMBO Rep 10(4):367–373
142. Inoshita H, Masuyama H, Hiramatsu Y (2003) The different effects of endocrine-disrupting chemicals on estrogen receptor-mediated transcription through interaction with coactivator TRAP220 in uterine tissue. J Mol Endocrinol 31(3):551–561
143. Rüegg J, Swedenborg E, Wahlstroem D, Escande A, Balaguer P, Pettersson K, Pongratz I (2008) The transcription factor aryl hydrocarbon receptor nuclear translocator functions as an estrogen receptor beta-selective coactivator, and its recruitment to alternative pathways mediates antiestrogenic effects of dioxin. Mol Endocrinol 22(2):304–316
144. Safe S, Wormke M (2003) Inhibitory aryl hydrocarbon receptor-estrogen receptor a cross-talk and mechanisms of action. Chem Res Toxicol 16(7):807–816
145. Swedenborg E, Ruegg J, Makela S, Pongratz I (2009) Endocrine disruptive chemicals: mechanisms of action and involvement in metabolic disorders. J Mol Endocrinol 43(1–2): 1–10
146. Crews D, Gillette R, Scarpino SV, Manikkam M, Savenkova MI, Skinner MK (2012) Epigenetic transgenerational inheritance of altered stress responses. Proc Natl Acad Sci U S A 109(23):9143–9148

147. Manikkam M, Tracey R, Guerrero-Bosagna C, Skinner MK (2013) Plastics derived endocrine disruptors (BPA, DEHP and DBP) induce epigenetic transgenerational inheritance of obesity, reproductive disease and sperm epimutations. PLoS One 8(1):e55387
148. Bjornstrom L, Sjoberg M (2005) Mechanisms of estrogen receptor signaling: convergence of genomic and nongenomic actions on target genes. Mol Endocrinol 19(4):833–842
149. Falkenstein E, Tillmann HC, Christ M, Feuring M, Wehling M (2000) Multiple actions of steroid hormones: a focus on rapid, nongenomic effects. Pharmacol Rev 52(4):513–555
150. Watson CS, Jeng Y-J, Guptarak J (2011) Endocrine disruption via estrogen receptors that participate in nongenomic signaling pathways. J Steroid Biochem Mol Biol 127(1–2):44–50
151. Mainini CG (1947) Pregnancy test using the male toad. J Clin Endocrinol Metab 7(9): 653–658
152. Bulbring E, Burn JH (1935) The estimation of oestrin and of male hormone in oily solution. J Physiol-Lond 85(3):320–333
153. Korenchevsky V (1932) The assay of testicular hormone preparations. Biochem J 26: 413–422
154. Kanter AE, Bauer CP, Klawans AH (1934) A new biologic test for hormones in pregnancy urine: preliminary report. JAMA 103(26):2026–2027
155. National Research Council (2007) Toxicity testing in the 21st century: a vision and a strategy, vol xvii. National Academies Press, Washington, DC, p 196
156. Langley G, Austin CP, Balapure AK, Birnbaum LS, Bucher JR, Fentem J, Fitzpatrick SC, Fowle JR III, Kavlock RJ, Kitano H, Lidbury BA, Muotri AR, Peng SQ, Sakharov D, Seidle T, Trez T, Tonevitsky A, van de Stolpe A, Whelan M, Willett C (2015) Lessons from toxicology: developing a 21st-century paradigm for medical research. Environ Health Perspect 123:A268–A272
157. Ankley GT, Bennett RS, Erickson RJ, Hoff DJ, Hornung MW, Johnson RD, Mount DR, Nichols JW, Russom CL, Schmieder PK, Serrrano JA, Tietge JE, Villeneuve DL (2010) Adverse outcome pathways: a conceptual framework to support ecotoxicology research and risk assessment. Environ Toxicol Chem 29(3):730–741
158. Hultman MT, Rundberget JT, Tollefsen KE (2015) Evaluation of the sensitivity, responsiveness and reproducibility of primary rainbow trout hepatocyte vitellogenin expression as a screening assay for estrogen mimics. Aquat Toxicol 159:233–244
159. Scrimshaw MD, Lester JN (2004) In-vitro assays for determination of oestrogenic activity. Anal Bioanal Chem 378(3):576–581
160. Zacharewski T (1997) *In vitro* bioassays for assessing estrogenic substances. Environ Sci Technol 31(3):613–623
161. de Jong LAA, Uges DRA, Franke JP, Bischoff R (2005) Receptor-ligand binding assays: technologies and applications. J Chromatogr B Analyt Technol Biomed Life Sci 829(1–2): 1–25
162. Hulme EC, Trevethick MA (2010) Ligand binding assays at equilibrium: validation and interpretation. Br J Pharmacol 161(6):1219–1237
163. Chae K, Albro PW, Luster MI, Mckinney JD (1984) A screening assay for the tetrachlorodibenzo-para-dioxin receptor using the [iodovaleramide-i-125 derivative of trichlorodibenzo-para-dioxin as the binding ligand. Int J Environ Anal Chem 17(3–4):267–274
164. Poland A, Glover E, Kende AS (1976) Stereospecific, high affinity binding of 2,3,7,8-tetrachlorodibenzo-para-dioxin by hepatic cytosol: evidence that binding species is receptor for induction of aryl-hydrocarbon hydroxylase. J Biol Chem 251(16):4936–4946
165. Yost EE, Pow CL, Hawkins MB, Kullman SW (2014) Bridging the gap from screening assays to estrogenic effects in fish: potential roles of multiple estrogen receptor subtypes. Environ Sci Technol 48(9):5211–5219
166. Fang ML, Webster TF, Ferguson PL, Stapleton HM (2015) Characterizing the peroxisome proliferator-activated receptor (PPAR gamma) ligand binding potential of several major

flame retardants, their metabolites, and chemical mixtures in house dust. Environ Health Perspect 123(2):166–172

167. Wagner M, Vermeirssen ELM, Buchinger S, Behr M, Magdeburg A, Oehlmann J (2013) Deriving bio-equivalents from *in vitro* bioassays: assessment of existing uncertainties and strategies to improve accuracy and reporting. Environ Toxicol Chem 32(8):1906–1917
168. Martin MT, Dix DJ, Judson RS, Kavlock RJ, Reif DM, Richard AM, Rotroff DM, Romanov S, Medvedev A, Poltoratskaya N, Gambarian M, Moeser M, Makarov SS, Houck KA (2010) Impact of environmental chemicals on key transcription regulators and correlation to toxicity end points within EPA's ToxCast program. Chem Res Toxicol 23(3):578–590
169. Escher BI, Allinson M, Altenburger R, Bain PA, Balaguer P, Busch W, Crago J, Denslow ND, Dopp E, Hilscherova K, Humpage AR, Kumar A, Grimaldi M, Jayasinghe BS, Jarosova B, Jia A, Makarov S, Maruya KA, Medvedev A, Mehinto AC, Mendez JE, Poulsen A, Prochazka E, Richard J, Schifferli A, Schlenk D, Scholz S, Shiraish F, Snyder S, Su GY, Tang JYM, van der Burg B, van der Linden SC, Werner I, Westerheide SD, Wong CKC, Yang M, Yeung BHY, Zhang XW, Leusch FDL (2014) Benchmarking organic micropollutants in wastewater, recycled water and drinking water with *in vitro* bioassays. Environ Sci Technol 48(3):1940–1956
170. Purvis IJ, Chotai D, Dykes CW, Lubahn DB, French FS, Wilson EM, Hobden AN (1991) An androgen-inducible expression system for *Saccharomyces cerevisiae*. Gene 106(1):35–42
171. Routledge EJ, Sumpter JP (1996) Estrogenic activity of surfactants and some of their degradation products assessed using a recombinant yeast screen. Environ Toxicol Chem 15(3):241–248
172. Sohoni P, Sumpter JP (1998) Several environmental oestrogens are also anti-androgens. J Endocrinol 158(3):327–339
173. Beresford N, Routledge EJ, Harris CA, Sumpter JP (2000) Issues arising when interpreting results from an *in vitro* assay for estrogenic activity. Toxicol Appl Pharmacol 162(1):22–33
174. Dhooge W, Arijs K, D'Haese I, Stuyvaert S, Versonnen B, Janssen C, Verstraete W, Comhaire F (2006) Experimental parameters affecting sensitivity and specificity of a yeast assay for estrogenic compounds: results of an interlaboratory validation exercise. Anal Bioanal Chem 386(5):1419–1428
175. Murk AJ, Legler J, van Lipzig MMH, Meerman JHN, Belfroid AC, Spenkelink A, van der Burg B, Rijs GBJ, Vethaak D (2002) Detection of estrogenic potency in wastewater and surface water with three *in vitro* bioassays. Environ Toxicol Chem 21(1):16–23
176. Wagner M, Oehlmann J (2011) Endocrine disruptors in bottled mineral water: estrogenic activity in the E-Screen. J Steroid Biochem Mol Biol 127(1–2):128–135
177. Wagner M, Schlusener MP, Ternes TA, Oehlmann J (2013) Identification of putative steroid receptor antagonists in bottled water: combining bioassays and high-resolution mass spectrometry. PLoS One 8(8), e72472
178. De Boever P, Demare W, Vanderperren E, Cooreman K, Bossier P, Verstraete W (2001) Optimization of a yeast estrogen screen and its applicability to study the release of estrogenic isoflavones from a soygerm powder. Environ Health Perspect 109(7):691–697
179. Schultis T, Metzger JW (2004) Determination of estrogenic activity by LYES-assay (yeast estrogen screen-assay assisted by enzymatic digestion with lyticase). Chemosphere 57(11): 1649–1655
180. Routledge EJ, Parker J, Odum J, Ashby J, Sumpter JP (1998) Some alkyl hydroxy benzoate preservatives (parabens) are estrogenic. Toxicol Appl Pharmacol 153(1):12–19
181. Lyttle CR, Damianmatsumura P, Juul H, Butt TR (1992) Human estrogen-receptor regulation in a yeast model system and studies on receptor agonists and antagonists. J Steroid Biochem Mol Biol 42(7):677–685
182. Fuqua SAW, Fitzgerald SD, Chamness GC, Tandon AK, McDonnell DP, Nawaz Z, Omalley BW, McGuire WL (1991) Variant human breast-tumor estrogen-receptor with constitutive transcriptional activity. Cancer Res 51(1):105–109

183. McDonnell DP, Nawaz Z, Densmore C, Weigel NL, Pham TA, Clark JH, Omalley BW (1991) High-level expression of biologically-active estrogen-receptor in *Saccharomyces cerevisiae*. J Steroid Biochem Mol Biol 39(3):291–297
184. McDonnell DP, Nawaz Z, Omalley BW (1991) *In situ* distinction between steroid-receptor binding and transactivation at a target gene. Mol Cell Biol 11(9):4350–4355
185. Inoue D, Nakama K, Matsui H, Sei K, Ike M (2009) Detection of agonistic activities against five human nuclear receptors in river environments of Japan using a yeast two-hybrid assay. Bull Environ Contam Toxicol 82(4):399–404
186. Miller CA, Tan XB, Wilson M, Bhattacharyya S, Ludwig S (2010) Single plasmids expressing human steroid hormone receptors and a reporter gene for use in yeast signaling assays. Plasmid 63(2):73–78
187. Hahn T, Tag K, Riedel K, Uhlig S, Baronian K, Gellissen G, Kunze G (2006) A novel estrogen sensor based on recombinant *Arxula adeninivorans* cells. Biosens Bioelectron 21(11):2078–2085
188. Bovee TFH, Helsdingen RJR, Hamers ARM, van Duursen MBM, Nielen MWF, Hoogenboom RLAP (2007) A new highly specific and robust yeast androgen bioassay for the detection of agonists and antagonists. Anal Bioanal Chem 389(5):1549–1558
189. Kamata R, Shiraishi F, Nakajima D, Takigami H, Shiraishi H (2009) Mono-hydroxylated polychlorinated biphenyls are potent aryl hydrocarbon receptor ligands in recombinant yeast cells. Toxicol in Vitro 23(4):736–743
190. Lee HJ, Lee YS, Kwon HB, Lee K (2003) Novel yeast bioassay system for detection of androgenic and antiandrogenic compounds. Toxicol in Vitro 17(2):237–244
191. Leskinen P, Michelini E, Picard D, Karp M, Virta M (2005) Bioluminescent yeast assays for detecting estrogenic and androgenic activity in different matrices. Chemosphere 61(2): 259–266
192. Miller CA (1997) Expression of the human aryl hydrocarbon receptor complex in yeast: activation of transcription by indole compounds. J Biol Chem 272(52):32824–32829
193. Miller CA, Martinat MA, Hyman LE (1998) Assessment of aryl hydrocarbon receptor complex interactions using pBEVY plasmids: expression vectors with bi-directional promoters for use in *Saccharomyces cerevisiae*. Nucleic Acids Res 26(15):3577–3583
194. Bovee TFH, Helsdingen RJR, Hamers ARM, Brouwer BA, Nielen MWF (2011) Recombinant cell bioassays for the detection of (gluco)corticosteroids and endocrine-disrupting potencies of several environmental PCB contaminants. Anal Bioanal Chem 401(3):873–882
195. Gaido KW, Leonard LS, Lovell S, Gould JC, Babai D, Portier CJ, McDonnell DP (1997) Evaluation of chemicals with endocrine modulating activity in a yeast-based steroid hormone receptor gene transcription assay. Toxicol Appl Pharmacol 143(1):205–212
196. Garcia-Reyero N, Grau E, Castillo M, De Alda MJL, Barcelo D, Pina B (2001) Monitoring of endocrine disruptors in surface waters by the yeast recombinant assay. Environ Toxicol Chem 20(6):1152–1158
197. Ito-Harashima S, Shiizaki K, Kawanishi M, Kakiuchi K, Onishi K, Yamaji R, Yagi T (2015) Construction of sensitive reporter assay yeasts for comprehensive detection of ligand activities of human corticosteroid receptors through inactivation of CWP and PDR genes. J Pharmacol Toxicol Methods 74:41–52
198. Inoue D, Nakama K, Sawada K, Watanabe T, Takagi M, Sei K, Yang M, Hirotsuji J, Hu J, Nishikawa J-I, Nakanishi T, Ike M (2010) Contamination with retinoic acid receptor agonists in two rivers in the Kinki region of Japan. Water Res 44(8):2409–2418
199. Kamata R, Shiraishi F, Nishikawa J-I, Yonemoto J, Shiraishi H (2008) Screening and detection of the *in vitro* agonistic activity of xenobiotics on the retinoic acid receptor. Toxicol in Vitro 22(4):1050–1061
200. Li J, Ma M, Wang ZJ (2008) A two-hybrid yeast assay to quantify the effects of xenobiotics on retinoid X receptor-mediated gene expression. Toxicol Lett 176(3):198–206

201. Zhen H, Wu X, Hu J, Xiao Y, Yang M, Hirotsuji J, Nishikawa J-I, Nakanishi T, Ike M (2009) Identification of retinoic acid receptor agonists in sewage treatment plants. Environ Sci Technol 43(17):6611–6616
202. Kitagawa Y, Takatori S, Oda H, Nishikawa J, Nishihara T, Nakazawa H, Hori S (2003) Detection of thyroid hormone receptor-binding activities of chemicals using a yeast two-hybrid assay. J Health Sci 49(2):99–104
203. Shiraishi F, Okumura T, Nomachi M, Serizawa S, Nishikawa J, Edmonds JS, Shiraishi H, Morita M (2003) Estrogenic and thyroid hormone activity of a series of hydroxy-polychlorinated biphenyls. Chemosphere 52(1):33–42
204. Rajasarkka J, Hakkila K, Virta M (2011) Developing a compound-specific receptor for bisphenol A by directed evolution of human estrogen receptor alpha. Biotechnol Bioeng 108(11):2526–2534
205. Rajasarkka J, Virta M (2013) Characterization of a bisphenol A specific yeast bioreporter utilizing the bisphenol A-targeted receptor. Anal Chem 85(21):10067–10074
206. Rajasarkka J, Koponen J, Airaksinen R, Kiviranta H, Virta M (2014) Monitoring bisphenol A and estrogenic chemicals in thermal paper with yeast-based bioreporter assay. Anal Bioanal Chem 406(23):5695–5702
207. Murk AJ, Legler J, Denison MS, Giesy JP, van de Guchte C, Brouwer A (1996) Chemical-activated luciferase gene expression (CALUX): a novel *in vitro* bioassay for Ah receptor active compounds in sediments and pore water. Fundam Appl Toxicol 33(1):149–160
208. Aarts JMMJG, Denison MS, Cox MA, Schalk MAC, Garrison PM, Tullis K, Dehaan LHJ, Brouwer A (1995) Species-specific antagonism of Ah receptor action by 2,2',5,5'--tetrachlorobiphenyl and 2,2',3,3',4,4'-hexachlorobiphenyl. Eur J Pharmacol Environ Toxicol Pharmacol 293(4):463–474
209. Garrison PM, Tullis K, Aarts JMMJG, Brouwer A, Giesy JP, Denison MS (1996) Species-specific recombinant cell lines as bioassay systems for the detection of 2,3,7,8-tetrachlorodibenzo-p-dioxin-like chemicals. Fundam Appl Toxicol 30(2):194–203
210. Legler J, van den Brink CE, Brouwer A, Murk AJ, van der Saag PT, Vethaak AD, van der Burg P (1999) Development of a stably transfected estrogen receptor-mediated luciferase reporter gene assay in the human T47D breast cancer cell line. Toxicol Sci 48(1):55–66
211. Sonneveld E, Jansen HJ, Riteco JAC, Brouwer A, van der Burg B (2005) Development of androgen- and estrogen-responsive bioassays, members of a panel of human cell line-based highly selective steroid-responsive bioassays. Toxicol Sci 83(1):136–148
212. Gijsbers L, Man H-Y, Kloet SK, de Haan LHJ, Keijer J, Rietjens IMCM, van der Burg B, Aarts JMMJG (2011) Stable reporter cell lines for peroxisome proliferator-activated receptor gamma (PPAR gamma)-mediated modulation of gene expression. Anal Biochem 414(1): 77–83
213. Piersma AH, Bosgra S, van Duursen MBM, Hermsen SAB, Jonker LRA, Kroese ED, van der Linden SC, Man H, Roelofs MJE, Schulpen SHW, Schwarz M, Uibel F, van Vugt-Lussenburg BMA, Westerhout J, Wolterbeek APM, van der Burg B (2013) Evaluation of an alternative *in vitro* test battery for detecting reproductive toxicants. Reprod Toxicol 38:53–64
214. van der Burg B, van der Linden S, Man H-Y, Winter R, Jonker L, van Vugt-Lussenburg B, Brouwer A (2013) A panel of quantitative calux® reporter gene assays for reliable high-throughput toxicity screening of chemicals and complex mixtures, p 532
215. Brennan JC, He G, Tsutsumi T, Zhao J, Wirth E, Fulton MH, Denison MS (2015) Development of species-specific Ah receptor-responsive third generation CALUX cell lines with enhanced responsiveness and improved detection limits. Environ Sci Technol 49(19): 11903–11912
216. Gizzi G, Hoogenboom LAP, Von Holst C, Rose M, Anklam E (2005) Determination of dioxins (PCDDs/PCDFs) and PCBs in food and feed using the DR CALUX (R) bioassay: results of an international validation study. Food Addit Contam 22(5):472–481

217. He G, Tsutsumi T, Zhao B, Baston DS, Zhao J, Heath-Pagliuso S, Denison MS (2011) Third-generation Ah receptor-responsive luciferase reporter plasmids: amplification of dioxin-responsive elements dramatically increases CALUX bioassay sensitivity and responsiveness. Toxicol Sci 123(2):511–522
218. Pieterse B, Felzel E, Winter R, van der Burg B, Brouwer A (2013) PAH-CALUX, an optimized bioassay for AhR-mediated hazard identification of polycyclic aromatic hydrocarbons (PAHs) as individual compounds and in complex mixtures. Environ Sci Technol 47(20):11651–11659
219. van der Burg B, Winter R, Man H-Y, Vangenechten C, Berckmans P, Weimer M, Witters H, van der Linden S (2010) Optimization and prevalidation of the *in vitro* AR CALUX method to test androgenic and antiandrogenic activity of compounds. Reproductive Toxicology 30(1):18–24
220. van der Burg B, Winter R, Weimer M, Berckmans P, Suzuki G, Gijsbers L, Jonas A, van der Linden S, Witters H, Aarts J, Legler J, Kopp-Schneider A, Bremer S (2010) Optimization and prevalidation of the *in vitro* ER alpha CALUX method to test estrogenic and antiestrogenic activity of compounds. Reprod Toxicol 30(1):73–80
221. Van Loco J, Van Leeuwen SPJ, Roos P, Carbonnelle S, de Boer J, Goeyens L, Beernaert H (2004) The international validation of bio- and chemical-analytical screening methods for dioxins and dioxin-like PCBs: the DIFFERENCE project rounds 1 and 2. Talanta 63(5): 1169–1182
222. Windal I, Denison MS, Birnbaum LS, Van Wouwe N, Baeyens W, Goeyens L (2005) Chemically activated luciferase gene expression (CALUX) cell bioassay analysis for the estimation of dioxin-like activity: critical parameters of the CALUX procedure that impact assay results. Environ Sci Technol 39(19):7357–7364
223. Anezaki K, Yamaguchi K, Takeuchi S, Iida M, Jin K, Kojima H (2009) Application of a bioassay using DR-Ecoscreen cells to the determination of dioxins in ambient air: a comparative study with HRGC-HRMS analysis. Environ Sci Technol 43(19):7478–7483
224. Besselink HT, Schipper C, Klamer H, Leonards P, Verhaar H, Felzel E, Murk AJ, Thain J, Hosoe K, Schoeters G, Legler J, Brouwer B (2004) Intra- and interlaboratory calibration of the DR Calux (R) bioassay for the analysis of dioxins and dioxin-like chemicals in sediments. Environ Toxicol Chem 23(12):2781–2789
225. Bovee TFH, Heskamp HH, Helsdingen RJR, Hamers ARM, Brouwer BA, Nielen MWF (2013) Validation of a recombinant cell bioassay for the detection of (gluco)corticosteroids in feed. Food Addit Contam Part A Chem Anal Control Expo Risk Assess 30(2):264–271
226. Bovee TFH, Hoogenboom LAP, Hamers ARM, Traag WA, Zuidema T, Aarts J, Brouwer A, Kuiper HA (1998) Validation and use of the CALUX-bioassay for the determination of dioxins and PCBs in bovine milk. Food Addit Contam 15(8):863–875
227. Chou IC, Lee W-J, Wang L-C, Chang-Chien G-P, Lee W-S, Lee H (2008) Validation of the CALUX bioassay as a screening and semi-quantitative method for PCDD/F levels in cow's milk. J Hazard Mater 154(1–3):1166–1172
228. Sonneveld E, Pieterse B, Schoonen WG, van der Burg B (2011) Validation of *in vitro* screening models for progestagenic activities: inter-assay comparison and correlation with *in vivo* activity in rabbits. Toxicol in Vitro 25(2):545–554
229. Tsutsumi T, Amakura Y, Nakamura M, Brown DJ, Clark GC, Sasaki K, Toyoda M, Maitani T (2003) Validation of the CALUX bioassay for the screening of PCDD/Fs and dioxin-like PCBs in retail fish. Analyst 128(5):486–492
230. Windal I, Van Wouwe N, Eppe G, Xhrouet C, Debacker V, Baeyens W, De Pauw E, Goeyens L (2005) Validation and interpretation of CALUX as a tool for the estimation of dioxin-like activity in marine biological matrixes. Environ Sci Technol 39(6):1741–1748
231. Wilson VS, Bobseine K, Gray LE (2004) Development and characterization of a cell line that stably expresses an estrogen-responsive luciferase reporter for the detection of estrogen receptor agonist and antagonists. Toxicol Sci 81(1):69–77

232. Balaguer P, Francois F, Comunale F, Fenet H, Boussioux AM, Pons M, Nicolas JC, Casellas C (1999) Reporter cell lines to study the estrogenic effects of xenoestrogens. Sci Total Environ 233(1–3):47–56
233. Jugan ML, Oziol L, Bimbot M, Huteau V, Tamisier-Karolak S, Blondeau JP, Levi Y (2009) *In vitro* assessment of thyroid and estrogenic endocrine disruptors in wastewater treatment plants, rivers and drinking water supplies in the greater Paris area (France). Sci Total Environ 407(11):3579–3587
234. Wehmas LC, Cavallin JE, Durhan EJ, Kahl MD, Martinovic D, Mayasich J, Tuominen T, Villeneuve DL, Ankley GT (2011) Screening complex effluents for estrogenic activity with the T47D-kbluc cell bioassay: assay optimization and comparison with *in vivo* responses in fish. Environ Toxicol Chem 30(2):439–445
235. Grimaldi M, Boulahtouf A, Delfosse V, Thouennon E, Bourguet W, Balaguer P (2015) Reporter cell lines for the characterization of the interactions between human nuclear receptors and endocrine disruptors. Front Endocrinol 6:62–62
236. Nguyen LP, Bradfield CA (2008) The search for endogenous activators of the aryl hydrocarbon receptor. Chem Res Toxicol 21(1):102–116
237. Tian J, Feng Y, Fu H, Xie HQ, Jiang JX, Zhao B (2015) The aryl hydrocarbon receptor: a key bridging molecule of external and internal chemical signals. Environ Sci Technol 49(16): 9518–9531
238. Fritsche E, Schaefer C, Calles C, Bernsmann T, Bernshausen T, Wurm M, Huebenthal U, Cline JE, Hajimiragha H, Schroeder P, Klotz L-O, Rannug A, Fuerst P, Hanenberg H, Abel J, Krutmann J (2007) Lightening up the UV response by identification of the arylhydrocarbon receptor as a cytoplasmatic target for ultraviolet B radiation. Proc Natl Acad Sci U S A 104(21):8851–8856
239. Opitz CA, Litzenburger UM, Sahm F, Ott M, Tritschler I, Trump S, Schumacher T, Jestaedt L, Schrenk D, Weller M, Jugold M, Guillemin GJ, Miller CL, Lutz C, Radlwimmer B, Lehmann I, von Deimling A, Wick W, Platten M (2011) An endogenous tumour-promoting ligand of the human aryl hydrocarbon receptor. Nature 478(7368): 197–203
240. Beischlag TV, Morales JL, Hollingshead BD, Perdew GH (2008) The aryl hydrocarbon receptor complex and the control of gene expression. Crit Rev Eukaryot Gene Expr 18(3): 207–250
241. Tijet N, Boutros PC, Moffat ID, Okey AB, Tuomisto J, Pohjanvirta R (2006) Aryl hydrocarbon receptor regulates distinct dioxin-dependent and dioxin-independent gene batteries. Mol Pharmacol 69(1):140–153
242. Mandal PK (2005) Dioxin: a review of its environmental effects and its aryl hydrocarbon receptor biology. J Comp Physiol B 175(4):221–230
243. Doolittle DJ, Newton JF, Goodman JI (1983) Quantitative studies on metabolic-activation in vitro employing benzo[a]pyrene as the test compound. Mutat Res 108(1–3):29–44
244. Doolittle DJ, Goodman JI (1984) Quantitative studies on the *in vitro* metabolic-activation of dimethylnitrosamine by rat-liver postmitochondrial supernatant. Environ Health Perspect 57(Aug):327–332
245. Mason G, Farrell K, Keys B, Piskorskapliszczynska J, Safe L, Safe S (1986) Polychlorinated dibenzo-para-dioxins: quantitative *in vitro* and *in vivo* structure-activity-relationships. Toxicology 41(1):21–31
246. Piskorska-Pliszczynska J, Keys B, Safe S, Newman MS (1986) The cytosolic receptor-binding affinities and AHH induction potencies of 29 polynuclear aromatic-hydrocarbons. Toxicol Lett 34(1):67–74
247. Safe S (1987) Determination of 2,3,7,8-Tcdd toxic equivalent factors (TEFs): support for the use of the in vitro AHH induction assay. Chemosphere 16(4):791–802
248. Van den Berg M, Birnbaum L, Bosveld ATC, Brunstrom B, Cook P, Feeley M, Giesy JP, Hanberg A, Hasegawa R, Kennedy SW, Kubiak T, Larsen JC, van Leeuwen FXR, Liem AKD, Nolt C, Peterson RE, Poellinger L, Safe S, Schrenk D, Tillitt D, Tysklind M,

Younes M, Waern F, Zacharewski T (1998) Toxic equivalency factors (TEFs) for PCBs, PCDDs, PCDFs for humans and wildlife. Environ Health Perspect 106(12):775–792
249. Van den Berg M, Birnbaum LS, Denison M, De Vito M, Farland W, Feeley M, Fiedler H, Hakansson H, Hanberg A, Haws L, Rose M, Safe S, Schrenk D, Tohyama C, Tritscher A, Tuomisto J, Tysklind M, Walker N, Peterson RE (2006) The 2005 World Health Organization reevaluation of human and mammalian toxic equivalency factors for dioxins and dioxin-like compounds. Toxicol Sci 93(2):223–241
250. Soto AM, Chung KL, Sonnenschein C (1994) The pesticides endosulfan, toxaphene, and dieldrin have estrogenic effects on human estrogen-sensitive cells. Environ Health Perspect 102(4):380–383
251. Soto AM, Sonnenschein C (1985) The role of estrogens on the proliferation of human breast tumor cells (MCF-7). J Steroid Biochem Mol Biol 23(1):87–94
252. Soto AM, Sonnenschein C, Chung KL, Fernandez MF, Olea N, Serrano FO (1995) The E-Screen assay as a tool to identify estrogens: an update on estrogenic environmental-pollutants. Environ Health Perspect 103:113–122
253. Villalobos M, Olea N, Brotons JA, Oleaserrano MF, Dealmodovar JMR, Pedraza V (1995) The E-Screen assay: a comparison of different MCF7 cell stocks. Environ Health Perspect 103(9):844–850
254. Desaulniers D, Leingartner K, Zacharewski T, Foster WG (1998) Optimization of an MCF7-E3 cell proliferation assay and effects of environmental pollutants and industrial chemicals. Toxicol in Vitro 12(4):409–422
255. Korner W, Hanf V, Schuller W, Kempter C, Metzger J, Hagenmaier H (1999) Development of a sensitive E-screen assay for quantitative analysis of estrogenic activity in municipal sewage plant effluents. Sci Total Environ 225(1–2):33–48
256. Rasmussen TH, Nielsen JB (2002) Critical parameters in the MCF-7 cell proliferation bioassay (E-Screen). Biomarkers 7(4):322–336
257. Vanparys C, Maras M, Lenjou M, Robbens J, Van Bockstaele D, Blust R, De Coen W (2006) Flow cytometric cell cycle analysis allows for rapid screening of estrogenicity in MCF-7 breast cancer cells. Toxicol in Vitro 20(7):1238–1248
258. Riverso M, Kortenkamp A, Silva E (2014) Non-tumorigenic epithelial cells secrete MCP-1 and other cytokines that promote cell division in breast cancer cells by activating ER alpha via PI3K/Akt/mTOR signaling. Int J Biochem Cell Biol 53:281–294
259. Gutleb AC, Meerts I, Bergsma JH, Schriks M, Murk AJ (2005) T-Screen as a tool to identify thyroid hormone receptor active compounds. Environ Toxicol Pharmacol 19(2):231–238
260. Schriks M, Vrabie CM, Gutleb AC, Faassen EJ, Rietjens I, Murk AJ (2006) T-screen to quantify functional potentiating, antagonistic and thyroid hormone-like activities of poly halogenated aromatic hydrocarbons (PHAHs). Toxicol in Vitro 20(4):490–498
261. Gutleb AC, Mossink L, Schriks M, van den Berg HJH, Murk AJ (2007) Delayed effects of environmentally relevant concentrations of 3,3',4,4'-tetrachlorobiphenyl (PCB-77) and non-polar sediment extracts detected in the prolonged-FETAX. Sci Total Environ 381(1–3):307–315
262. Taxvig C, Vinggaard AM, Hass U, Axelstad M, Boberg J, Hansen PR, Frederiksen H, Nellemann C (2008) Do parabens have the ability to interfere with steroidogenesis? Toxicol Sci 106(1):206–213
263. Ghisari M, Bonefeld-Jorgensen EC (2009) Effects of plasticizers and their mixtures on estrogen receptor and thyroid hormone functions. Toxicol Lett 189(1):67–77
264. Long M, Kruger T, Ghisari M, Bonefeld-Jorgensen EC (2012) Effects of selected phytoestrogens and their mixtures on the function of the thyroid hormone and the aryl hydrocarbon receptor. Nutr Cancer 64(7):1008–1019
265. Long M, Ghisari M, Bonefeld-Jorgensen EC (2013) Effects of perfluoroalkyl acids on the function of the thyroid hormone and the aryl hydrocarbon receptor. Environ Sci Pollut Res 20(11):8045–8056

266. Ghisari M, Long M, Tabbo A, Bonefeld-Jorgensen EC (2015) Effects of currently used pesticides and their mixtures on the function of thyroid hormone and aryl hydrocarbon receptor in cell culture. Toxicol Appl Pharmacol 284(3):292–303
267. Sanderson JT (2006) The steroid hormone biosynthesis pathway as a target for endocrine-disrupting chemicals. Toxicol Sci 94(1):3–21
268. Connolly L, Ropstad E, Verhaegen S (2011) *In vitro* bioassays for the study of endocrine-disrupting food additives and contaminants. TrAC Trends Anal Chem 30(2):227–238
269. OECD (2011) Test no. 456: H295R steroidogenesis assay. OECD, Paris. doi: 10.1787/9789264122642-en
270. US EPA (2009) Endocrine disruptor screening program test guidelines. OPPTS 890.1550: steroidogenesis (human cell line – H295R). Office of Prevention, Pesticides and Toxic Substances, United States Environmental Protection Agency
271. Harvey PW, Everett DJ, Springall CJ (2007) Adrenal toxicology: a strategy for assessment of functional toxicity to the adrenal cortex and steroidogenesis. J Appl Toxicol 27(2):103–115
272. Zhang XW, Yu RMK, Jones PD, Lam GKW, Newsted JL, Gracia T, Hecker M, Hilscherova K, Sanderson JT, Wu RSS, Giesy JP (2005) Quantitative RT-PCR methods for evaluating toxicant-induced effects on steroidogenesis using the H295R cell line. Environ Sci Technol 39(8):2777–2785
273. Hecker M, Hollert H, Cooper R, Vinggaard AM, Akahori Y, Murphy M, Nellemann C, Higley E, Newsted J, Laskey J, Buckalew A, Grund S, Maletz S, Giesy J, Timm G (2011) The OECD validation program of the H295R steroidogenesis assay: phase 3. Final inter-laboratory validation study. Environ Sci Pollut Res 18(3):503–515
274. Hilscherova K, Jones PD, Gracia T, Newsted JL, Zhang XW, Sanderson JT, Yu RMK, Wu RSS, Giesy JP (2004) Assessment of the effects of chemicals on the expression of ten steroidogenic genes in the H295R cell line using real-time PCR. Toxicol Sci 81(1):78–89
275. Maglich JM, Kuhn M, Chapin RE, Pletcher MT (2014) More than just hormones: H295R cells as predictors of reproductive toxicity. Reprod Toxicol 45:77–86
276. Antczak P, Jo HJ, Woo S, Scanlan L, Poynton H, Loguinov A, Chan S, Falciani F, Vulpe C (2013) Molecular toxicity identification evaluation (mTIE) approach predicts chemical exposure in *Daphnia magna*. Environ Sci Technol 47(20):11747–11756
277. Bundy JG, Davey MP, Viant MR (2009) Environmental metabolomics: a critical review and future perspectives. Metabolomics 5(1):3–21
278. Iguchi T, Watanabe H, Katsu Y (2007) Toxicogenomics and ecotoxicogenomics for studying endocrine disruption and basic biology. Gen Comp Endocrinol 153(1–3):25–29
279. Grun F, Blumberg B (2006) Environmental obesogens: organotins and endocrine disruption via nuclear receptor signaling. Endocrinology 147(6, Suppl. S):S50–S55
280. Jordao R, Casas J, Fabrias G, Campos B, Pina B, Lemos MFL, Soares AMVM, Tauler R, Barata C (2015) Obesogens beyond vertebrates: lipid perturbation by tributyltin in the crustacean *Daphnia magna*. Environ Health Perspect 123(8):813–819
281. Alsop DH, Brown SB, van der Kraak GJ (2004) Dietary retinoic acid induces hindlimb and eye deformities in Xenopus laevis. Environ Sci Technol 38(23):6290–6299
282. Chen C-H, Chou P-H, Kawanishi M, Yagi T (2014) Occurrence of xenobiotic ligands for retinoid X receptors and thyroid hormone receptors in the aquatic environment of Taiwan. Mar Pollut Bull 85(2):613–618
283. Inoue D, Nakama K, Sawada K, Watanabe T, Matsui H, Sei K, Nakanishi T, Ike M (2011) Screening of agonistic activities against four nuclear receptors in wastewater treatment plants in Japan using a yeast two-hybrid assay. J Environ Sci (China) 23(1):125–132
284. Inoue D, Sawada K, Wada Y, Sei K, Ike M (2013) Removal characteristics of retinoic acids and 4-oxo-retinoic acids in wastewater by activated sludge treatment. Water Sci Technol 67(12):2868–2874
285. Sawada K, Inoue D, Wada Y, Sei K, Nakanishi T, Ike M (2012) Detection of retinoic acid receptor agonistic activity and identification of causative compounds in municipal wastewater treatment plants in Japan. Environ Toxicol Chem 31(2):307–315

286. Miyagawa S, Lange A, Hirakawa I, Tohyama S, Ogino Y, Mizutani T, Kagami Y, Kusano T, Ihara M, Tanaka H, Tatarazako N, Ohta Y, Katsu Y, Tyler CR, Iguchi T (2014) Differing species responsiveness of estrogenic contaminants in fish is conferred by the ligand binding domain of the estrogen receptor. Environ Sci Technol 48(9):5254–5263
287. Kunz PY, Fent K (2006) Estrogenic activity of UV filter mixtures. Toxicol Appl Pharmacol 217(1):86–99
288. Legler J, Zeinstra LM, Schuitemaker F, Lanser PH, Bogerd J, Brouwer A, Vethaak AD, De Voogt P, Murk AJ, Van der Burg B (2002) Comparison of *in vivo* and *in vitro* reporter gene assays for short-term screening of estrogenic activity. Environ Sci Technol 36(20): 4410–4415
289. Simmons DB, Marlatt VL, Trudeau VL, Sherry JP, Metcalfe CD (2010) Interaction of Galaxolide(R) with the human and trout estrogen receptor-alpha. Sci Total Environ 408(24):6158–6164
290. Ihara M, Kitamura T, Kumar V, Park C-B, Ihara MO, Lee S-J, Yamashita N, Miyagawa S, Iguchi T, Okamoto S, Suzuki Y, Tanaka H (2015) Evaluation of estrogenic activity of wastewater: comparison among *in vitro* er alpha reporter gene assay, *in vivo* vitellogenin induction, and chemical analysis. Environ Sci Technol 49(10):6319–6326
291. Kohno S, Katsu Y, Iguchi T, Guillette LJ Jr (2008) Novel approaches for the study of vertebrate steroid hormone receptors. Integr Comp Biol 48(4):527–534
292. Hultin CL, Hallgren P, Persson A, Hansson MC (2014) Identification of an estrogen receptor gene in the natural freshwater snail *Bithynia tentaculata*. Gene 540(1):26–31
293. Katsu Y, Matsubara K, Kohno S, Matsuda Y, Toriba M, Oka K, Guillette LJ Jr, Ohta Y, Iguchi T (2010) Molecular cloning, characterization, and chromosome mapping of reptilian estrogen receptors. Endocrinology 151(12):5710–5720
294. Kaur S, Jobling S, Jones CS, Noble LR, Routledge EJ, Lockyer AE (2015) The nuclear receptors of *Biomphalaria glabrata* and *Lottia gigantea*: implications for developing new model organisms. PLoS One 10(4)
295. Nagasawa K, Treen N, Kondo R, Otoki Y, Itoh N, Rotchell JM, Osada M (2015) Molecular characterization of an estrogen receptor and estrogen-related receptor and their autoregulatory capabilities in two Mytilus species. Gene 564(2):153–159
296. Vogeler S, Galloway TS, Lyons BP, Bean TP (2014) The nuclear receptor gene family in the Pacific oyster, *Crassostrea gigas*, contains a novel subfamily group. BMC Genomics 15
297. Bannister R, Beresford N, Granger DW, Pounds NA, Rand-Weaver M, White R, Jobling S, Routledge EJ (2013) No substantial changes in estrogen receptor and estrogen-related receptor orthologue gene transcription in *Marisa cornuarietis* exposed to estrogenic chemicals. Aquat Toxicol 140:19–26
298. Bridgham JT, Keay J, Ortlund EA, Thornton JW (2014) Vestigialization of an allosteric switch: genetic and structural mechanisms for the evolution of constitutive activity in a steroid hormone receptor. PLoS Genet 10(1)
299. Grun F, Watanabe H, Zamanian Z, Maeda L, Arima K, Cubacha R, Gardiner DM, Kanno J, Iguchi T, Blumberg B (2006) Endocrine-disrupting organotin compounds are potent inducers of adipogenesis in vertebrates. Mol Endocrinol 20(9):2141–2155
300. Kanayama T, Kobayashi N, Mamiya S, Nakanishi T, Nishikawa J (2005) Organotin compounds promote adipocyte differentiation as agonists of the peroxisome proliferator-activated receptor gamma/retinoid X receptor pathway. Mol Pharmacol 67(3):766–774
301. Watson CS, Bulayeva NN, Wozniak AL, Alyea RA (2007) Xenoestrogens are potent activators of nongenomic estrogenic responses. Steroids 72(2):124–134
302. Rajapakse N, Silva E, Kortenkamp A (2002) Combining xenoestrogens at levels below individual no-observed-effect concentrations dramatically enhances steroid hormone action. Environ Health Perspect 110(9):917–921
303. Silva E, Rajapakse N, Kortenkamp A (2002) Something from "nothing": eight weak estrogenic chemicals combined at concentrations below NOECs produce significant mixture effects. Environ Sci Technol 36(8):1751–1756

304. Kortenkamp A (2014) Low dose mixture effects of endocrine disrupters and their implications for regulatory thresholds in chemical risk assessment. Curr Opin Pharmacol 19: 105–111
305. Christiansen S, Kortenkamp A, Axelstad M, Boberg J, Scholze M, Jacobsen PR, Faust M, Lichtensteiger W, Schlumpf M, Burdorf A, Hass U (2012) Mixtures of endocrine disrupting contaminants modelled on human high end exposures: an exploratory study in rats. Int J Androl 35(3):303–316
306. Christiansen S, Scholze M, Dalgaard M, Vinggaard AM, Axelstad M, Kortenkamp A, Hass U (2009) Synergistic disruption of external male sex organ development by a mixture of four antiandrogens. Environ Health Perspect 117(12):1839–1846
307. Hass U, Scholze M, Christiansen S, Dalgaard M, Vinggaard AM, Axelstad M, Metzdorff SB, Kortenkamp A (2007) Combined exposure to anti-androgens exacerbates disruption of sexual differentiation in the rat. Environ Health Perspect 115:122–128
308. Jarošová B, Erseková A, Hilscherová K, Loos R, Gawlik BM, Giesy JP, Bláha L (2014) Europe-wide survey of estrogenicity in wastewater treatment plant effluents: the need for the effect-based monitoring. Environ Sci Pollut Res 21(18):10970–10982
309. Behr M, Oehlmann J, Wagner M (2011) Estrogens in the daily diet: *in vitro* analysis indicates that estrogenic activity is omnipresent in foodstuff and infant formula. Food Chem Toxicol 49(10):2681–2688
310. Wagner M, Oehlmann J (2009) Endocrine disruptors in bottled mineral water: total estrogenic burden and migration from plastic bottles. Environ Sci Pollut Res 16(3):278–286
311. Yang CZ, Yaniger SI, Jordan VC, Klein DJ, Bittner GD (2011) Most plastic products release estrogenic chemicals: a potential health problem that can be solved. Environ Health Perspect 119(7):989–996
312. Brack W (2003) Effect-directed analysis: a promising tool for the identification of organic toxicants in complex mixtures? Anal Bioanal Chem 377(3):397–407
313. Buchinger S, Spira D, Broder K, Schlusener M, Ternes T, Reifferscheid G (2013) Direct coupling of thin-layer chromatography with a bioassay for the detection of estrogenic compounds: applications for effect-directed analysis. Anal Chem 85(15):7248–7256
314. Jonker W, Lamoree MH, Houtman CJ, Hamers T, Somsen GW, Kool J (2015) Rapid activity-directed screening of estrogens by parallel coupling of liquid chromatography with a functional gene reporter assay and mass spectrometry. J Chromatogr A 1406:165–174
315. Reemtsma T (2001) Prospects of toxicity-directed wastewater analysis. Anal Chim Acta 426(2):279–287
316. Schmitt S, Reifferscheid G, Claus E, Schluesener M, Buchinger S (2012) Effect directed analysis and mixture effects of estrogenic compounds in a sediment of the river Elbe. Environ Sci Pollut Res 19(8):3350–3361
317. Creusot N, Budzinski H, Balaguer P, Kinani S, Porcher J-M, Ait-Aissa S (2013) Effect-directed analysis of endocrine-disrupting compounds in multi-contaminated sediment: identification of novel ligands of estrogen and pregnane X receptors. Anal Bioanal Chem 405(8):2553–2566
318. Simon E, van Velzen M, Brandsma SH, Lie E, Loken K, de Boer J, Bytingsvik J, Jenssen BM, Aars J, Hamers T, Lamoree MH (2013) Effect-directed analysis to explore the polar bear exposome: identification of thyroid hormone disrupting compounds in plasma. Environ Sci Technol 47(15):8902–8912
319. Weiss JM, Andersson PL, Zhang J, Simon E, Leonards PEG, Hamers T, Lamoree MH (2015) Tracing thyroid hormone-disrupting compounds: database compilation and structure-activity evaluation for an effect-directed analysis of sediment. Anal Bioanal Chem 407(19): 5625–5634
320. Alvarez-Munoz D, Indiveri P, Rostkowski P, Horwood J, Greer E, Minier C, Pope N, Langston WJ, Hill EM (2015) Widespread contamination of coastal sediments in the Transmanche Channel with anti-androgenic compounds. Mar Pollut Bull 95(2):590–597

321. Berger E, Potouridis T, Haeger A, Puettmann W, Wagner M (2015) Effect-directed identification of endocrine disruptors in plastic baby teethers. J Appl Toxicol 35(11):1254–1261
322. Fang H, Tong W, Perkins R, Soto AM, Prechtl NV, Sheehan DM (2000) Quantitative comparisons of *in vitro* assays for estrogenic activities. Environ Health Perspect 108(8): 723–729
323. Leusch FD, de Jager C, Levi Y, Lim R, Puijker L, Sacher F, Tremblay LA, Wilson VS, Chapman HF (2010) Comparison of five *in vitro* bioassays to measure estrogenic activity in environmental waters. Environ Sci Technol 44(10):3853–3860
324. Dreier DA, Connors KA, Brooks BW (2015) Comparative endpoint sensitivity of *in vitro* estrogen agonist assays. Regul Toxicol Pharmacol 72(2):185–193
325. Akahori Y, Nakai M, Yamasaki K, Takatsuki M, Shimohigashi Y, Ohtaki M (2008) Relationship between the results of *in vitro* receptor binding assay to human estrogen receptor alpha and *in vivo* uterotrophic assay: comparative study with 65 selected chemicals. Toxicol in Vitro 22(1):225–231
326. Coldham NG, Dave M, Sivapathasundaram S, McDonnell DP, Connor C, Sauer MJ (1997) Evaluation of a recombinant yeast cell estrogen screening assay. Environ Health Perspect 105(7):734–742
327. Rotroff DM, Martin MT, Dix DJ, Filer DL, Houck KA, Knudsen TB, Sipes NS, Reif DM, Xia M, Huang R, Judson RS (2014) Predictive endocrine testing in the 21st century using *in vitro* assays of estrogen receptor signaling responses. Environ Sci Technol 48(15): 8706–8716
328. Rotroff DM, Dix DJ, Houck KA, Knudsen TB, Martin MT, McLaurin KW, Reif DM, Crofton KM, Singh AV, Xia M, Huang R, Judson RS (2013) Using *in vitro* high throughput screening assays to identify potential endocrine-disrupting chemicals. Environ Health Perspect 121(1): 7–14
329. Blaauboer BJ (2015) The long and winding road of progress in the use of *in vitro* data for risk assessment purposes: from "carnation test" to integrated testing strategies. Toxicology 332: 4–7
330. Meek ME, Lipscomb JC (2015) Gaining acceptance for the use of *in vitro* toxicity assays and QIVIVE in regulatory risk assessment. Toxicology 332:112–123
331. Villeneuve DL, Ankley GT, Makynen EA, Blake LS, Greene KJ, Higley EB, Newsted JL, Giesy JP, Hecker M (2007) Comparison of fathead minnow ovary explant and H295R cell-based steroidogenesis assays for identifying endocrine-active chemicals. Ecotoxicol Environ Saf 68(1):20–32
332. Puy-Azurmendi E, Olivares A, Vallejo A, Ortiz-Zarragoitia M, Pina B, Zuloaga O, Cajaraville MP (2014) Estrogenic effects of nonylphenol and octylphenol isomers *in vitro* by recombinant yeast assay (RYA) and *in vivo* with early life stages of zebrafish. Sci Total Environ 466:1–10
333. Hornung MW, Kosian PA, Haselman JT, Korte JJ, Challis K, Macherla C, Nevalainen E, Degitz SJ (2015) *In vitro*, ex vivo, and *in vivo* determination of thyroid hormone modulating activity of benzothiazoles. Toxicol Sci 146(2):254–264
334. Leusch FDL, Khan SJ, Gagnon MM, Quayle P, Trinh T, Coleman H, Rawson C, Chapman HF, Blair P, Nice H, Reitsema T (2014) Assessment of wastewater and recycled water quality: a comparison of lines of evidence from *in vitro*, *in vivo* and chemical analyses. Water Res 50:420–431
335. Vermeirssen ELM, Burki R, Joris C, Peter A, Segner H, Suter MJF, Burkhardt-Holm P (2005) Characterization of the estrogenicity of swiss midland rivers using a recombinant yeast bioassay and plasma vitellogenin concentrations in feral male brown trout. Environ Toxicol Chem 24(9):2226–2233
336. Henneberg A, Bender K, Blaha L, Giebner S, Kuch B, Koehler H-R, Maier D, Oehlmann J, Richter D, Scheurer M, Schulte-Oehlmann U, Sieratowicz A, Ziebart S, Triebskorn R (2014) Are *in vitro* methods for the detection of endocrine potentials in the aquatic environment

predictive for *in vivo* effects? Outcomes of the projects SchussenAktiv and SchussenAktiv plus in the Lake Constance area, Germany. PLoS One 9(6)
337. Groothuis FA, Heringa MB, Nicol B, Hermens JL, Blaauboer BJ, Kramer NI (2015) Dose metric considerations in *in vitro* assays to improve quantitative *in vitro-in vivo* dose extrapolations. Toxicology 332:30–40
338. Brinkmann M, Maletz S, Krauss M, Bluhm K, Schiwy S, Kuckelkorn J, Tiehm A, Brack W, Hollert H (2014) Heterocyclic aromatic hydrocarbons show estrogenic activity upon metabolization in a recombinant transactivation assay. Environ Sci Technol 48(10):5892–5901
339. Louisse J, Bosgra S, Blaauboer BJ, Rietjens IM, Verwei M (2015) Prediction of *in vivo* developmental toxicity of all-trans-retinoic acid based on *in vitro* toxicity data and in silico physiologically based kinetic modeling. Arch Toxicol 89(7):1135–1148
340. Yoon M, Campbell JL, Andersen ME, Clewell HJ (2012) Quantitative *in vitro* to *in vivo* extrapolation of cell-based toxicity assay results. Crit Rev Toxicol 42(8):633–652
341. Wetmore BA (2015) Quantitative *in vitro*-to-*in vivo* extrapolation in a high-throughput environment. Toxicology 332:94–101
342. Wilk-Zasadna I, Bernasconi C, Pelkonen O, Coecke S (2015) Biotransformation *in vitro*: an essential consideration in the quantitative *in vitro*-to-*in vivo* extrapolation (QIVIVE) of toxicity data. Toxicology 332:8–19
343. Burden N, Sewell F, Andersen ME, Boobis A, Chipman JK, Cronin MTD, Hutchinson TH, Kimber I, Whelan M (2015) Adverse outcome pathways can drive non-animal approaches for safety assessment. J Appl Toxicol 35(9):971–975
344. Lee JW, Won E-J, Raisuddin S, Lee J-S (2015) Significance of adverse outcome pathways in biomarker-based environmental risk assessment in aquatic organisms. J Environ Sci 35: 115–127
345. Plant NJ (2015) An introduction to systems toxicology. Toxicol Res 4(1):9–22
346. Simmons DBD, Benskin JP, Cosgrove JR, Duncker BP, Ekman DR, Martyniuk CJ, Sherry JP (2015) Omics for aquatic ecotoxicology: control of extraneous variability to enhance the analysis of environmental effects. Environ Toxicol Chem 34(8):1693–1704
347. Spurgeon DJ, Jones OAH, Dorne J-LCM, Svendsen C, Swain S, Stuerzenbaum SR (2010) Systems toxicology approaches for understanding the joint effects of environmental chemical mixtures. Sci Total Environ 408(18):3725–3734
348. Waters MD, Fostel JM (2004) Toxicogenomics and systems toxicology: aims and prospects. Nat Rev Genet 5(12):936–948
349. Yoon K, Kwack SJ, Kim HS, Lee B-M (2014) Estrogenic endocrine-disrupting chemicals: molecular mechanisms of actions on putative human diseases. J Toxicol Environ Health B Crit Rev 17(3):127–174
350. Escher B, Leusch F (2012) Bioanalytical tools in water quality assessment. International Water Association, London
351. Kunz PY, Kienle C, Carere M, Homazava N, Kase R (2015) *In vitro* bioassays to screen for endocrine active pharmaceuticals in surface and waste waters. J Pharm Biomed Anal 106: 107–115
352. Loos R (2012) Analytical methods relevant to the European Commission's 2012 proposal on priority substances under the water framework directive. Publications Office of the European Union
353. OECD (2012) OECD Series on Testing and Assessment Number 178. Detailed review paper on the state of the science on novel *in vitro* and *in vivo* screening and testing methods and endpoints for evaluating endocrine disruptors. ENV/JM/MONO(2012)23
354. EU (2008) DIRECTIVE 2008/105/EC OF THE EUROPEAN PARLIAMENT AND OF THE COUNCIL of 16 December 2008 on environmental quality standards in the field of water policy, amending and subsequently repealing Council Directives 82/176/EEC, 83/513/EEC, 84/156/EEC, 84/491/EEC, 86/280/EEC and amending Directive 2000/60/EC of the European Parliament and of the Council. Off J Eur Union L 348/84

355. ANZECC, ARMCANZ (2000) Australian and New Zealand guidelines for fresh and marine water quality. Aquatic ecosystems: Rationale and background information, vol. 2. Australian Water Association, Artarmon, AU
356. EU Council (2009) Council conclusions on combination effects of chemicals Brussels, Belgium. 2988th Environment Council meeting, Council of the European Union, 22 Dec 2009
357. Epa U (2002) Guidance on cumulative risk assessment of pesticide chemicals that have a common mechanism of toxicity, federal register. U.S. Environmental Protection Agency. Office of Pesticide Programs, Washington, DC
358. OSPAR Commission (2007) Practical guidance document on whole effluent assessment, No. 316/2007. OSPAR Commission, London
359. OECD (2005) OECD Series on Testing and Assessment Number 34. Guidance document on the validation and international acceptance of new or updated test methods for hazard assessment. ENV/JM/MONO(2005)14
360. Tang JYM, McCarty S, Glenn E, Neale PA, Warne MSJ, Escher BI (2013) Mixture effects of organic micropollutants present in water: towards the development of effect-based water quality trigger values for baseline toxicity. Water Res 47(10):3300–3314
361. EU (2013) DIRECTIVE 2013/39/EU OF THE EUROPEAN PARLIAMENT AND OF THE COUNCIL of 12 August 2013 amending Directives 2000/60/EC and 2008/105/EC as regards priority substances in the field of water policy. Off J Eur Union L226/1
362. Brand W, de Jongh CM, van der Linden SC, Mennes W, Puijker LM, van Leeuwen CJ, van Wezel AP, Schriks M, Heringa MB (2013) Trigger values for investigation of hormonal activity in drinking water and its sources using CALUX bioassays. Environ Int 55:109–118
363. Jarošová B, Bláha L, Giesy JP, Hilscherová K (2014) What level of estrogenic activity determined by *in vitro* assays in municipal waste waters can be considered as safe? Environ Int 64:98–109
364. Escher BI, Neale PA, Leusch FDL (2015) Effect-based trigger values for *in vitro* bioassays: reading across from existing water quality guideline values. Water Res 81:137–148
365. Langston WJ, Burt GR, Chesman BS (2007) Feminisation of male clams Scrobicularia plana from estuaries in Southwest UK and its induction by endocrine-disrupting chemicals. Mar Ecol Prog Ser 333:173–184
366. De Solla SR, Martin PA, Fernie KJ, Park BJ, Mayne G (2006) Effects of environmentally relevant concentrations of atrazine on gonadal development of snapping turtles (*Chelydra serpentina*). Environ Toxicol Chem 25(2):520–526
367. Mattsson A, Olsson JA, Brunstrom B (2008) Selective estrogen receptor alpha activation disrupts sex organ differentiation and induces expression of vitellogenin II and very low-density apolipoprotein II in Japanese quail embryos. Reproduction 136(2):175–186
368. Wagner M (2012) Endocrine disruptors in bottled water and foodstuff: a bioassay-based approach. Dr. Phil. Nat., Faculty of Biological Sciences, Goethe University, Frankfurt, p 206
369. Fernandes D, Loi B, Porte C (2011) Biosynthesis and metabolism of steroids in molluscs. J Steroid Biochem Mol Biol 127(3–5):189–195
370. Dodds EC, Lawson W (1936) Synthetic oestrogenic agents without the phenanthrene nucleus. Nature 137:996
371. Vanbohemen CG, Lambert JGD, Goos HJT, Vanoordt P (1982) Estrone and estradiol participation during exogenous vitallogenesis in the female rainbow-trout, *Salmo gairdneri*. Gen Comp Endocrinol 46(1):81–92
372. Palace VP, Evans RE, Wautier K, Baron C, Vandenbyllardt L, Vandersteen W, Kidd K (2002) Induction of vitellogenin and histological effects in wild fathead minnows from a lake experimentally treated with the synthetic estrogen, ethynylestradiol. Water Qual Res J Can 37(3):637–650
373. Park BJ, Kidd K (2005) Effects of the synthetic estrogen ethinylestradiol on early life stages of mink frogs and green frogs in the wild and *in situ*. Environ Toxicol Chem 24(8):2027–2036

374. Palace VP, Wautier KG, Evans RE, Blanchfield PJ, Mills KH, Chalanchuk SM, Godard D, McMaster ME, Tetreault GR, Peters LE, Vandenbyllaardt L, Kidd KA (2006) Biochemical and histopathological effects in pearl dace (*Margariscus margarita*) chronically exposed to a synthetic estrogen in a whole lake experiment. Environ Toxicol Chem 25(4):1114–1125
375. Werner J, Palace VP, Wautier KG, Mills KH, Chalanchuk SM, Kidd KA (2006) Reproductive fitness of lake trout (*Salvelinus namaycush*) exposed to environmentally relevant concentrations of the potent estrogen ethynylestradiol (EE2) in a whole lake exposure experiment. Sci Mar 70:59–66
376. Palace VP, Evans RE, Wautier KG, Mills KH, Blanchfield PJ, Park BJ, Baron CL, Kidd KA (2009) Interspecies differences in biochemical, histopathological, and population responses in four wild fish species exposed to ethynylestradiol added to a whole lake. Can J Fish Aquat Sci 66(11):1920–1935
377. Milnes MR, Bryan TA, Katsu Y, Kohno S, Moore BC, Iguchi T, Guillette LJ Jr (2008) Increased posthatching mortality and loss of sexually dimorphic gene expression in alligators (*Alligator mississippiensis*) from a contaminated environment. Biol Reprod 78(5):932–938
378. Stoker C, Repetti MR, Garcia SR, Zayas MA, Galoppo GH, Beldomenico HR, Loque EH, Munoz-de-Toro M (2011) Organochlorine compound residues in the eggs of broad-snouted caimans (*Caiman latirostris*) and correlation with measures of reproductive performance. Chemosphere 84(3):311–317
379. Stoker C, Beldomenico PM, Bosquiazzo VL, Zayas MA, Rey F, Rodriguez H, Munoz-de-Toro M, Luque EH (2008) Developmental exposure to endocrine disruptor chemicals alters follicular dynamics and steroid levels in Caiman latirostris. Gen Comp Endocrinol 156(3): 603–612
380. Rider CV, Hartig PC, Cardon MC, Lambright CR, Bobseine KL, Guillette LJ Jr, Gray LE Jr, Wilson VS (2010) Differences in sensitivity but not selectivity of xenoestrogen binding to alligator versus human estrogen receptor alpha. Environ Toxicol Chem 29(9):2064–2071
381. Fry DM, Toone CK (1981) DDT-induced feminization of gull embryos. Science 213(4510): 922–924
382. Markman S, Mueller CT, Pascoe D, Dawson A, Buchanan KL (2011) Pollutants affect development in nestling starlings *Sturnus vulgaris*. J Appl Ecol 48(2):391–397
383. Sonne C, Iburg T, Leifsson PS, Born EW, Letcher RJ, Dietz R (2011) Thyroid gland lesions in organohalogen contaminated East Greenland polar bears (*Ursus maritimus*). Toxicol Environ Chem 93(4):789–805
384. Colborn T, vom Saal FS, Soto AM (1993) Developmental effects of endocrine-disrupting chemicals in wildlife and humans. Environ Health Perspect 101(5):378–384
385. Colborn T, Dumanoski D, Myers JP (1996) Our stolen future: are we threatening our fertility, intelligence, and survival? A scientific detective story. Dutton, New York
386. Kristensen DM, Skalkam ML, Audouze K, Lesne L, Desdoits-Lethimonier C, Frederiksen H, Brunak S, Skakkebaek NE, Jegou B, Hansen JB, Junker S, Leffers H (2011) Many putative endocrine disruptors inhibit prostaglandin synthesis. Environ Health Perspect 119(4): 534–541
387. Masuyama H, Inoshita H, Hiramatsu Y, Kudo T (2002) Ligands have various potential effects on the degradation of pregnane X receptor by proteasome. Endocrinology 143(1):55–61
388. Wormke M, Stoner M, Saville B, Walker K, Abdelrahim M, Burghardt R, Safe S (2003) The aryl hydrocarbon receptor mediates degradation of estrogen receptor alpha through activation of proteasomes. Mol Cell Biol 23(6):1843–1855
389. Jansen MS, Nagel SC, Miranda PJ, Lobenhofer EK, Afshari CA, McDonnell DP (2004) Short-chain fatty acids enhance nuclear receptor activity through mitogen-activated protein kinase activation and histone deacetylase inhibition. Proc Natl Acad Sci U S A 101(18): 7199–7204
390. Brunnberg S, Pettersson K, Rydin E, Matthews J, Hanberg A, Pongratz I (2003) The basic helix-loop-helix-PAS protein ARNT functions as a potent coactivator of estrogen receptor-dependent transcription. Proc Natl Acad Sci U S A 100(11):6517–6522

391. Kleiner IS, Weisman AI, Barowsky H (1935) An investigation of the new biologic test for hormones in pregnancy urine: preliminary report. JAMA 104:1318–1319
392. Klaassen CD, Casarett LJ, Doull J (2013) Casarett and Doull's toxicology: the basic science of poisons. McGraw-Hill, New York
393. Yalow RS, Berson SA (1960) Immunoassay of endogenous plasma insulin in man. J Clin Invest 39(7):1157–1175
394. Khan MN, Findlay JWA (2010) Ligand-binding assays: development, validation, and implementation in the drug development arena. Wiley, Hoboken, NJ
395. Ruff M, Gangloff M, Wurtz JM, Moras D (2000) Estrogen receptor transcription and transactivation: structure-function relationship in DNA- and ligand-binding domains of estrogen receptors. Breast Cancer Res 2(5):353–359
396. Muller MB, Dausend C, Weins C, Frimmel FH (2004) A new bioautographic screening method for the detection of estrogenic compounds. Chromatographia 60(3–4):207–211
397. Vanderperren E, Demare W, Blust R, Cooreman K, Bossier P (2001) Oestrogenic activity of CPRG (chlorophenol red-beta-D-galactopyranoside), a beta-galactosidase substrate commonly used in recombinant yeast oestrogenic assays. Biomarkers 6(5):375–380
398. Berthois Y, Katzenellenbogen JA, Katzenellenbogen BS (1986) Phenol red in tissue-culture media is a weak estrogen: implications concerning the study of estrogen-responsive cells in culture. Proc Natl Acad Sci U S A 83(8):2496–2500
399. Welshons WV, Wolf MF, Murphy CS, Jordan VC (1988) Estrogenic activity of phenol red. Mol Cell Endocrinol 57(3):169–178
400. Antebi A (2006) Nuclear hormone receptors in *C. elegans*. In: WormBook (ed). The *C. elegans* Research Community, WormBook. doi:10.1895/wormbook.1.64.1. http://www.wormbook.org
401. Dang Z, Ru S, Wang W, Rorije E, Hakkert B, Vermeire T (2011) Comparison of chemical-induced transcriptional activation of fish and human estrogen receptors: regulatory implications. Toxicol Lett 201(2):152–175
402. Scott AP (2012) Do mollusks use vertebrate sex steroids as reproductive hormones? Part I: Critical appraisal of the evidence for the presence, biosynthesis and uptake of steroids. Steroids 77(13):1450–1468
403. Scott AP (2013) Do mollusks use vertebrate sex steroids as reproductive hormones? II. Critical review of the evidence that steroids have biological effects. Steroids 78(2): 268–281
404. Gutierrez-Mazariegos J, Nadendla EK, Lima D, Pierzchalski K, Jones JW, Kane M, Nishikawa J-I, Hiromori Y, Nakanishi T, Santos MM, Castro LFC, Bourguet W, Schubert M, Laudet V (2014) A mollusk retinoic acid receptor (RAR) ortholog sheds light on the evolution of ligand binding. Endocrinology 155(11):4275–4286
405. Burgess RM, Ho KT, Brack W, Lamoree M (2013) Effects-directed analysis (EDA) and toxicity identification evaluation (TIE): complementary but different approaches for diagnosing causes of environmental toxicity. Environ Toxicol Chem 32(9):1935–1945

Adv Biochem Eng Biotechnol (2017) 157: 59–80
DOI: 10.1007/10_2015_5018

Published online: 9 September 2016

In Vitro Genotoxicity Testing: Significance and Use in Environmental Monitoring

Gisela De Aragao Umbuzeiro, Minne Heringa, and Errol Zeiger

Abstract There is ongoing concern about the consequences of mutations in humans and biota arising from environmental exposures to industrial and other chemicals. Genetic toxicity tests have been used to analyze chemicals, foods, drugs, and environmental matrices such as air, water, soil, and wastewaters. This is because the mutagenicity of a substance is highly correlated with its carcinogenicity. However, no less important are the germ cell mutations, because the adverse outcome is related not only to an individual but also to population levels. For environmental analysis the most common choices are in vitro assays, and among them the most widely used is the Ames test (*Salmonella*/microsome assay). There are several protocols and methodological approaches to be applied when environmental samples are tested and these are discussed in this chapter, along with the meaning and relevance of the obtained responses. Two case studies illustrate the utility of in vitro mutagenicity tests such as the Ames test. It is clear that, although it is not possible to use the outcome of the test directly in risk assessment, the application of the assays provides a great opportunity to monitor the exposure of humans and biota to mutagenic substances for the purpose of reducing or quantifying that exposure.

Keywords Ames test, Cancer, Environmental testing, Germ cell mutations, Mutagenicity

G.D.A. Umbuzeiro (✉)
School of Technology, UNICAMP, Rua Dr Paschoal Marmo 1888, 13484-332 Limeira, SP, Brazil
e-mail: giselau@ft.unicamp.br

M. Heringa
National Institute of Public Health and the Environment (RIVM), PO Box 1, 3720 BA Bilthoven, The Netherlands

E. Zeiger
Errol Zeiger Consulting, 800 Indian Springs Road, Chapel Hill, NC 27514, USA

Contents

1 Introduction

There is ongoing concern about the consequences of mutations in humans and biota arising from environmental exposures. If somatic cells are mutated, cancer is a possible outcome in the affected individuals, and this has been the primary emphasis of mutation testing. In contrast, germ cell mutations [1], which have had less attention, can have effects that go beyond the individual level and affect future generations, and also lead to population level alterations such as reduction in fitness, reproduction impairment, higher rates of offspring carrying deleterious mutations, and population decline.

Mutagenicity is one aspect of genotoxicity[1] and can be measured in tests that detect alterations at the molecular level (e.g., DNA base modifications) or at the chromosome level (e.g., structural damage, micronuclei, or aneuploidy). Genotoxicity is a broader term and refers to any type of damage to the DNA, chromosome, or nuclear material whether or not it leads to mutation. Short-term, in vitro genetic toxicology tests measure damage or changes to the cell's DNA, chromosomes, or replication machinery, and are widely used to screen for potential mutagenic and carcinogenic substances, or for contamination in environmental samples.

The widespread use of genetic toxicity tests has been based on the finding that the mutagenicity of a substance is highly correlated with its carcinogenicity. Germ cell mutations are also a concern because of the potential effects on future generations, and

[1] The terms mutagenicity, genetic toxicity, and genotoxicity are often used interchangeably. Genetic toxicity and genotoxicity generally refer to any damage or effect on the cell's DNA, chromosomes, or replication machinery, regardless of the outcome to the cell. In contrast, mutagenicity refers only to specific base-pair changes, additions, or deletions at the DNA level, and can include structural chromosome damage, which is potentially heritable.

less information on the consequences of germ cell mutation is available, especially at population level [2]. For these reasons, genetic toxicity tests are widely used as an early screen in drug, food additive, pesticide, and chemical development.

Another advantage from the use of in vitro genetic toxicity tests is that they can efficiently screen environmental samples for hazardous substances. Although these tests do not identify the specific chemicals, they are able to signal the presence, and possibly levels, of mutagenic, and potentially carcinogenic, substances in environmental samples, e.g., air, water, sediment, soil, and effluents [3–7]. Once a mutagenic sample has been identified and its potency quantified, the same test can be used to monitor changes with time or to evaluate intervention techniques [8]. As a follow-up, the samples testing positive can be subjected to further testing or analysis to identify the specific mutagens present and to provide information for risk assessment [9] and possible mitigation. This can be done using Effect Directed Analysis (EDA), first described by Brack [10]. Umbuzeiro et al. [11] provide key information on how to proceed with an EDA when using in vitro mutagenicity assays.

Therefore, genotoxicity testing of environmental samples has many aspects: detection and identification of potential carcinogens and germ cell mutagens in the environmental compartments, tracking the source(s) of genotoxin contamination, mitigating the presence of genotoxic chemicals through changing manufacturing, engineering or waste disposal processes, and providing monitoring quality information. However, evaluation of the health or ecological risks of a mutagenic environmental sample to humans and biota remains a challenge.

Although the primary impetus to the development and use of these assays has been directed towards testing chemicals for hazard to humans, this aspect is not addressed here. Instead, this chapter presents a brief history of genetic toxicity testing, with emphasis on the Ames test, which is the assay most commonly used to evaluate environmental samples. We also describe some possible ways for its use in environmental studies, pointing out key issues. Two case studies provide an illustration of the use of genetic toxicity assays in environmental studies and some of the described issues that arise.

2 Genetic Toxicity Testing: A Brief History

The early efforts in mutagenicity testing were primarily concerned with the potential heritable effects of mutations to the germ cells of exposed individuals, and such testing was proposed for pesticides and other chemicals to which humans would be exposed [12–14]. There was also concern for the carcinogenicity of mutagens but this concern did not become dominant until Ames and his colleagues developed a rapid and sensitive *Salmonella* bacterial test for mutagens, and he and others showed that positive results in the test had a high correlation with carcinogenicity. In those studies, a mutagenic response in the *Salmonella* test procedure was greater than 90% predictive of carcinogenicity [15–19], although subsequent studies showed the predictivity to be 75–80% when a wider range of chemicals was tested

[20–23]. At the same time, it was seen that approximately 50% of the carcinogens tested were not mutagenic in this test [21–23].

Following reports of high correlations between bacterial mutagenicity and mammalian carcinogenicity, a number of international validation exercises with a range of in vitro and in vivo tests using mutation and other genotoxicity endpoints in yeast, fungi, mammalian cells, and rodents were also examined for their ability to identify mutagens for the purpose of identifying carcinogens so that they could be removed from human exposure or the environment (see, e.g., [24, 25]). Based partly on these studies, the scientific and regulatory communities settled on a limited number of in vitro and in vivo tests for measuring gene mutations, chromosome damage, and DNA damage, and incorporated them into regulations for chemicals, including pesticides, drugs, industrial chemicals, and food additives, to which humans would be exposed. It had been determined that chromosome damage and gene mutations are caused by different pathways (with some overlap), and that different substances may trigger different pathways. Therefore, the in vitro test battery includes tests for both pathways. Typically, such an in vitro test battery for assessing chemicals consists of a gene mutation test in bacteria (Ames test), a gene mutation test in mammalian cells (e.g., mouse lymphoma test), and a chromosome damage test in mammalian cells (e.g., a micronucleus test in vitro). Because some in vitro tests (i.e., bacterial and yeast tests) required little, if any, sophisticated equipment, and were rapid and robust, they were seen as ideal for monitoring air, soil, effluents, and water samples in an attempt to identify and control all sources of human and environmental exposure to mutagens. A number of in vitro microbial and mammalian tests are available which do not measure mutation per se, but DNA damage, which is expressed as cell lethality, e.g., *umu* test; SOS Chromo test [26, 27] or DNA strand breaks, e.g., comet assay [28]. Additionally, tests based on gene activation using reporter systems in human cells, e.g., the GreenScreen HC [29], BlueScreen [30], and ToxTracker [31] have been developed and proposed for use. These are not addressed here.

Although all these tests can also be used for screening and testing environmental samples for the presence of mutagenic and genotoxic chemicals, the most commonly used test for this purpose is the Ames test [6]. The reasons for this are that it is the best studied and most robust, quick, and inexpensive assay [32], with a positive predictivity for carcinogenicity of 74% (i.e., approximately 26% of the positive results are false positives [20, 22]). In contrast, mammalian cell mutation assays can be difficult and impractical to perform with environmental samples, mostly because of the high costs involved, considering that in environmental analysis usually more than one sample is tested. In addition, the cell maintenance and growth conditions require more sophisticated equipment and tighter experimental control than the microbial tests, and it takes longer to obtain the test results. Unless there is a very specific situation where a mammalian gene mutation or chromosome damage test would be justified, the Ames test is often considered sufficient to provide an indication on the presence of substances causing gene mutations. The fact that tests for chromosome damage in mammalian cells are not often performed for environmental samples [6] implies that mutagens with only this mechanism of action are not detected.

In vivo animal testing is rarely, if ever, performed with environmental samples because of the high costs and longer time to obtain a response. There are strategies to detect mutagenicity in the environment using field or caged animals such as fish, mussels, rodents, and other sentinel species, but this strategy, although potentially very powerful, is quite expensive, laborious, time-consuming, and requires highly trained personnel.

The focus of this chapter is therefore the Ames test using *Salmonella* and *Escherichia coli* tester strains. It must be kept in mind that these tests detect only gene mutations, and cannot identify substances that produce only chromosomal breaks or rearrangements, or aneuploidy. It is therefore advisable, when feasible, to perform also an assay for chromosomal damage on an environmental sample in situations where the Ames test is negative. The information provided in this chapter, although focusing primarily on the Ames test, highlights how genotoxicity/mutagenicity tests can be used for testing environmental samples.

3 Ames Test in the Analysis of Environmental Samples

The main advantages of using the Ames, or any other mutagenicity test, in environmental studies are the detection of mutagens which may not be identified by targeted chemical analysis, and the possibility to evaluate a complex mixture as a whole without knowing what, if any, classes of chemicals may be present. Because the test can be performed with strains containing different mutation targets and metabolic capacities [33, 34], coupled with different sample preparation [35] and/or fractionation procedures [33, 36], it is also possible to provide information on the specific classes of compounds that are responsible for the mutagenicity of a mixture.

There are several examples of how the Ames test has been used, e.g., in the identification of new environmental mutagens, e.g., MX [37], PBTA [38], 3-nitrobenzantrone [39], evaluation of the efficiency of water and wastewater treatment, helping in the identification and selection of contaminated sites to be prioritized for a comprehensive assessment, evaluation of the effectiveness of remediation processes, and identification of trends and hotspots in monitoring programs, among many others. However, one very important limitation of the assay is that it is not possible to use the data obtained in the Ames test in a traditional toxicological risk assessment to categorize samples from a specific site as a human carcinogen or germ cell mutagen. It is not the case that all environmental samples that come from clean or pristine sites are negative in the Ames test or any other mutagenicity test. There are naturally occurring mutagens such as fungal and plant metabolites, and combustion products from wood fire in natural forests. Chlorinated drinking water coming from a clean surface water containing humic and fulvic acid, if appropriately concentrated, shows mutagenic activity

[40]. There are approaches that help to differentiate the mutagenicity coming from clean and contaminated waters. One example is the use of selective extraction procedures and strains with sensitivities to specific compounds [35]. However, this is not a trivial task and depends on the type of contaminants present in the source water. Therefore, when any researcher or agency decides to analyze environmental samples for genotoxicity/mutagenicity, these considerations must be taken into account so that the assay is used only in a hazard assessment approach.

Traditionally, genetic toxicity data have been used qualitatively, that is, the substance was either mutagenic (which meant that it should be considered a potential carcinogen or germ cell mutagen) or non-mutagenic, which decreased the concern for carcinogenicity and germ cell mutagenicity. There is no evidence to support the relationship between in vitro and in vivo genotoxic potency, and it has been shown that the Ames test is not predictive of potency in other in vitro test systems (e.g., [41]) or for rodent carcinogenicity [42, 43]. However, the magnitudes of the responses can be compared within the same test. For environmental studies, the level of mutagenicity obtained in the Ames test, for example, can be a relevant indication of the level of contamination, and provides a reference for subsequent monitoring studies and remediation measures.

4 Ames Test Protocols for Testing Environmental Samples

4.1 Types of Ames Tests

General protocols for performing the Ames test with *Salmonella* or *E. coli* strains are described in Mortelmans and Zeiger [44] and Mortelmans and Riccio [45]. The *Salmonella* tester strains most widely used in environmental studies are TA98 and TA100, with and without exogenous metabolic activation supplied by S9 mix (i.e., a rodent liver homogenate containing enzyme cofactors). Other strains, e.g., *Salmonella* YG1041 [46] among others [9] have been used as diagnostic strains because they are specific for detecting certain classes of compounds such as aromatic amines and nitro compounds.

There are several protocols of the Ames test that can be used for testing environmental matrices – agar plate assays and liquid fluctuation assays. Agar plate assays include the plate incorporation and the preincubation protocols as described by Mortelmans and Zeiger [44]. Typically those protocols require sample volumes of 100–200 μL per plate. When only small volumes or quantities of sample are available, the microsuspension assay [47, 48] is an alternative because it requires maximum volumes of 2 μL per plate. When the liquid sample needs to be tested in natura or at low concentrations because of the presence of volatiles or low solubility, an in situ concentration method can be applied. In this case, volumes from 1–2 mL can be added to the plate because of the use of a more concentrated top agar [49, 50].

More recently, versions of the Ames fluctuation assay have been developed. One protocol, called MPF (microplate format), has emerged as an alternative to the standard Ames test [51–53]. The main advantages of this version are that it needs less sample and reagents and it enables high-throughput analyses. A round robin test with the Ames fluctuation test has shown acceptable inter-laboratory reproducibility [54] culminating in the publication of ISO Standard 11350:2012 [55]. Variations of this assay, such as the Ames II and MPF Aqua, are commercially available (http://www.drugdevelopment-technology.com/contractors/diagnostics/xenometrix/press2.html).

4.2 Sample Preparation

For pure chemicals, there are established rules for preparation of solutions to be evaluated in the Ames test [56], but much more effort needs to be devoted to prepare environmental samples for testing (see, e.g., [6, 57–61]). In general, both single and composite samples can be analyzed. Composite samples contain a mixture of subsamples taken, e.g., at various time intervals or at various locations. Alternatively, time-averaged samples can be obtained using passive samplers such as blue rayon or POCIS (e.g., [6, 35, 62, 63]). Some samplers can be selective for certain types of substances such as hydrophobic compounds.

It is not always possible to test environmental samples in their natural form in the Ames test or other in vitro cell culture tests because the sample needs to be sterile or contaminating microorganisms interfere with the growth of the tester cells and prevent the identification of mutated cells. Also, because the *Salmonella* and *E. coli* tester strains need histidine and tryptophan, respectively, for their growth, the test sample has to be free of these amino acids and small peptides containing them. It is therefore necessary to remove the microorganisms and interfering substances before the sample can be tested. Because these amino acids have a $\log K_{\mathrm{ow}} < -1$, they do not end up in organic solvent extracts, but this is an issue to be aware of when other sample preparation methods are used.

Liquid samples can be tested unconcentrated after filtration to remove microorganisms, or after being concentrated or chemically extracted. Extractions followed by concentration can be performed in a variety of ways depending on the contaminating substances of concern. Regardless of what type of protocol is chosen, no strategy is able to extract all the substances of interest with 100% efficiency. Additionally, the exact percentage recovery for any of the substances in the final extract is unknown, so that the chemical composition of an extract is considered to be, at best, a reasonable representation of the chemical composition of the environmental sample. The extracts need to be redissolved in a solvent compatible with the bacterial strains used in the Ames test. The solvent most commonly used is dimethylsulfoxide (DMSO). It must be noted that the type and amount of organic solvent used must be chosen with care, as it can affect assay results (e.g., [64]).

4.3 Concentration Factor

The sample preparation method is designed to concentrate the environmental sample, which may be necessary to achieve sufficient sensitivity. Given that, for example, fish and humans filter or consume much larger amounts of water daily than the <2 mL tested in a genotoxicity assay, it is clear that extracts containing the equivalents of higher volumes of water are necessary. The question then automatically arises as to how much the original environmental samples should be concentrated. This is not so easily answered, but very important for how the results of the test are interpreted and can ultimately be used. Different approaches have been used although no consensus has been reached on the best approach.

The first approach is based on the maximum level tested according to OECD Test Guideline 471. When chemical substances are analyzed in the Ames test, a maximum of 5 mg/plate should be tested as top dose unless limited by toxicity or solubility (OECD TG 471). This could be used as a guide for the maximum of water extract to test in this genotoxicity assay. Based on the level of contaminants expected in a sample, the maximum amount of extract dosed to a plate should ensure that these contaminants are dosed up to 5 mg/L. However, this methodology does not consider the amounts of water consumed. Thus, in the case of a very clean drinking water, with very low levels of contaminants, this method would require dosing of a 100-L equivalent of this drinking water to a plate, which would not be practical.

In a second approach, used in the Netherlands for many years for drinking water-related samples, a concentration range around the equivalent volume of 2 L of the original sample per plate was applied in the Ames test. This was based on the assumed daily consumption of 2 L of drinking water per person. This approach assumes that any detectable mutagenicity in the daily portion of drinking water is not desirable, and thus testing at this dose equivalence would have to show no response. However, this approach lacks substantiation that the concentration tested is indeed sufficiently sensitive to detect harmful levels of genotoxic contaminants.

A newer approach is based on the threshold of toxicological concern (TTC). The TTC is the highest dose at which a substance is expected to cause any adverse effect (see, e.g., [65, 66]), originally developed to enable the waiving of testing of food additives present at very low levels in foods. Although the principle is meant to cover all substances, it has been found more practical to determine TTCs for specific groups of substances. The TTC for genotoxic substances is 0.15 μg/person/day (with exclusion of the Cohort of Concern, i.e., highly carcinogenic substances such as aflatoxins, *N*-nitroso-compounds, azoxy-compounds, benzidines, and polyhalogenated dibenzo-*p*-dioxins and dibenzofurans). Translating this to a drinking water level, assuming that 10% of a daily dose comes from drinking water, this becomes 10 ng/L [67]. Thus, the most potent genotoxic substance (excluding the named Cohort of Concern, above) would only be expected to cause a risk above 10 ng/L. This means that a drinking water extract should be dosed in a genotoxicity test at such a concentration, so that if a genotoxic substance is present in the original water at 11 ng/L (i.e., just above the TTC), it should give a positive response, and if present at 9 ng/L, it should give a negative response. For environmental health assessments, a TTC of 0.1 μg/L has been proposed for all contaminants in aqueous systems [68], which may be used in a similar fashion as the TTC for human health to determine concentration factors.

This approach is not ideal, as the TTC is a conservative threshold, and the exclusion of the Cohort of Concern is a limitation. Thus improvement is necessary, but this approach provides practical guidance with the limited knowledge available. Further work on this issue is necessary, but it may be clear that the consumed amounts, the differences in environmental levels of contaminants, and the different potencies of these contaminants in the genotoxicity tests all need to be considered.

4.4 Data Evaluation

There are no formal, universally accepted, procedures for analyzing Ames test mutagenicity responses (e.g., [56]). Because the measured results (colony or fluctuation test well counts) do not follow a normal distribution, the well-known statistical tools for normal distributions, such as the *t*-test or ANOVA, cannot be used. The most widely used method for the Ames plate test is the twofold rule, or modified twofold rule [69]. The twofold rule can be used with most of the bacterial tester strains. With this rule, a dose-related positive response is defined as a dose–response that reaches at least twofold over the background rate. The twofold rule is not appropriate for some of the bacterial strains which exhibit a low background mutant frequency, e.g., <10 mutants/plate; in this case, a modified twofold rule requires the response to be at least threefold over the background. Studies have shown that the twofold rule may be too conservative for bacterial strains with high background rates [69–71], so some laboratories do not look at the fold increase, but require, instead, that there be a reproducible, dose-related increase above the background frequency. In both cases, the variability of the responses of the replicate plates must be taken into consideration. Formal statistical procedures have been developed to evaluate the results (e.g., [54, 69, 72, 73]). Regardless of the rule used, for a test result to be considered negative the positive control must show a positive response.

It is possible to do a regression model from the dose–response relationships to calculate the slope of the response, which is a measure of the mutagenic potency of the sample. There are several models [69] and computer programs that can be used to facilitate the calculations. Specific statistical models have also been developed for the MPF test [73]. Sample potency can be expressed both by the number of revertants per sample unit (e.g., L, m^3, or gram equivalent) as the minimum concentration of the sample able to cause a positive response. Good discussions on these and other types of potency calculations can be found in Escher et al. [74], Gollapudi et al. [75], and Johnson et al. [76]. The potency of a sample can be used to compare different positive samples or sensitivities of different protocols/strains

(Umbuzeiro et al. 2009). To facilitate communication with risk managers, some authors have proposed classifications of the mutagenic activity. They used a relative hazard assessment approach, and created classifications based on the maximum and minimum potencies observed in the literature for different types of environmental matrices. Houk [5] proposed a classification for liquid and solid wastes, and Umbuzeiro et al. [8] for surface water samples. These classifications can be very helpful because they allow the public to understand how hazardous a sample is in comparison with samples elsewhere.

5 Interpretation of Genotoxicity Assay Data

A positive result obtained for an environmental sample indicates there are mutagenic substances present (hazard information) at a level at which risks cannot be excluded. Whether there is a true risk for environmental or human health, however, cannot be deduced from these in vitro test results. In vitro tests, such as the Ames test, do not include some very important processes in organisms of higher trophic levels, such as toxicokinetics, e.g., the uptake, distribution over the organs and tissues, and excretion, of the mutagens. Metabolism is included to some extent in the Ames test by the addition of rat liver S9, but this preparation lacks several other metabolic (activating and detoxifying) enzymes and does not accurately mimic in vivo rodent or human, metabolism.

To determine the potential health risk of a positive sample, either animal testing with the sample itself would be required, or the identification of the responsible substances should be determined and these tested in animals, if data are not already available. An example of the first approach is the testing of a chlorinated drinking water concentrate for effects on development by the U.S. EPA [77]. An example of the second approach is the identification of a range of nitrogenous substances as the suspect chemicals responsible for the mutagenicity increase seen after UV-oxidation of water [78], and the identification of the substances partially responsible for the mutagenic effects [9, 79–81]. These are very resource-intensive procedures, so it may be more cost-effective to investigate ways to remove the genotoxic substance(s) from the environmental area sampled, or to prevent their occurrence.

Another important consideration is that environmental studies often look for a site that is mutagenic. Therefore the focus is always on the behavior of samples of a specific site. Because the levels of mutagenicity may change with the seasons (e.g., because of autumn leaf litter) and can be caused by accidental spills or effluents of industrial processes, it is best to rely on more than one analysis at each site. Monitoring studies are designed to provide such information. In contrast, when there is interest in the effects of a known spill or accident that involved genotoxic compounds, one sample would be sufficient to characterize the event, but to verify the extension and persistence of the contamination more samples are needed. Using a genotoxicity test in a monitoring program allows the identification of hotspots and finding the contamination sources. This information provides support for measures to reduce the mutagenic sources, even without knowing which compounds were causing the effects. CETESB, the environmental agency of Sao Paulo state in Brazil, has a monitoring program of surface waters in place in which several hundreds of samples are analyzed using the Ames test. When hotspots of mutagenicity are found the agency conducts the source identification and subsequently takes enforcement/prevention actions with the aim of reducing or eliminating the observed mutagenicity (see, e.g., www.cetesb.sp.gov.br). A case study describing this approach is presented later in this chapter.

Recently, Altenburger et al. [82] discussed how to adapt tools to deal with mixtures of pollutants in water resource management. In this report, the Ames test along with others is included as an effect-directed analysis detector at the cellular response level. They proposed that the results of these in vitro tests could be used when anchored to specific adverse outcome pathways. The aim of the study was to provide a basis to link exposure assessments and biological and ecological effects of complex mixtures using different analytical tools.

6 Conclusion

In vitro genetic toxicity tests can be valuable assets for identifying contamination of water and other environmental media by substances which have the potential to be human or environmental mutagens and carcinogens. These tests are relatively rapid and inexpensive, easy to perform, and usable with different types of samples. In addition to identifying contaminated samples, they can be used to trace contamination back to its source, monitor changes in contaminant (i.e., genotoxic) activity over time, and follow the progress of attempts to reduce or eliminate the active substances. Sample preparation is an important aspect in these analyses. Although these in vitro methods can be used only for hazard assessment, the results of these in vitro tests can be used as a basis for designing and implementing tests and procedures for risk assessments and mitigation.

7 Case Studies

7.1 *Case 1: Evaluation of Drinking Water Treatment Methods*

A number of different treatment methods and combinations of methods can be chosen for the purification of ground or surface water to produce safe drinking water. Besides physical purification methods (e.g., sand filtration) or absorption methods (e.g., carbon filtration), chemical methods such as chlorination, ozonation,

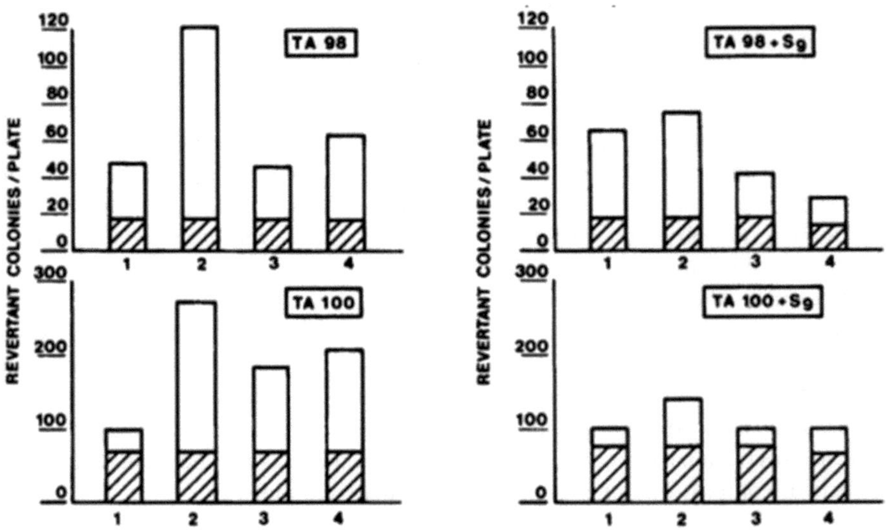

Fig. 1 Ames plate test results of extracts of water samples before and after different treatment steps. Sample 1 = Meuse river water, sample 2 = the same water after breakpoint chlorination, sample 3 = after subsequent carbon filtration, and sample 4 = after final transport chlorination (i.e., finished drinking water). The *striped bars* denote the number of spontaneous mutants. An equivalent of 3 L of water was dosed to each plate. (Adapted from [85])

or UV-oxidation can be used. In the latter methods, microorganisms are killed by the added chemical (disinfection). However, the natural organic matter (NOM) and pollutants present in the water can also react with the treatment chemicals to form a universe of new compounds, some of which may be genotoxic.

Rook [83] was the first to identify the formation of such a disinfection by-product (DBP) of chlorination (trihalomethane) and new chlorination-derived DBPs are still being identified [84]. Investigating the presence of DBPs by chemical analysis, identifying their precise chemical structures, and then determining the toxicity profile of each DBP is clearly a resource-intensive exercise. Instead, the production of harmful DBPs by a treatment method can be screened by a toxicity test, such as for genotoxicity. This also allows a simple comparison between locations, seasons, treatment doses, etc.

Figure 1 shows an example of how an Ames plate test was performed with extracts of water before and after different treatment steps, including chlorination. It shows that the river water (Meuse) was already slightly mutagenic, but that chlorination increased the mutagenicity, most clearly in *Salmonella* strain TA98 in the absence of metabolic activation (S9). This mutagenic activity was subsequently removed by carbon filtration.

As another example, UV-oxidation was introduced as a new drinking water treatment technology which not only disinfects with UV radiation, but also destroys chemical pollutants by hydroxide radicals formed by the UV irradiation of added hydrogen peroxide (e.g., [86]). Important advantages of this technology are that it is

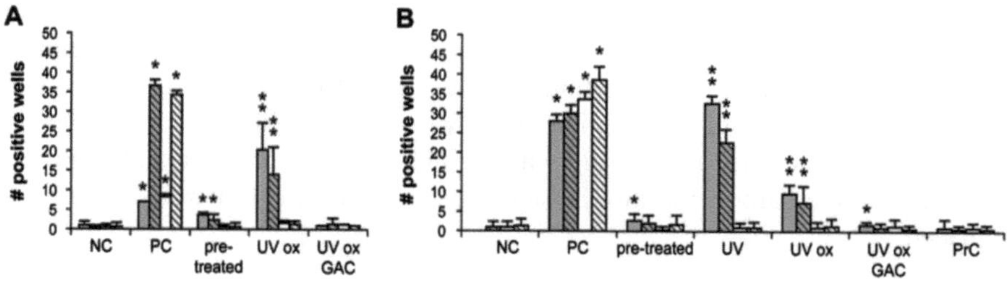

Fig. 2 Results of Ames II tests with water extracts of a bench-scale pilot treatment facility using Meuse river water (**a**) and water extracts from a pilot plant treatment facility using Ohio river water (**b**). Results are given for strains TA98 (*grey bars*) and TAMix (a combination of *Salmonella* strains TA7001 – TA7006; *white bars*), with (*striped bars*) and without (*non-striped bars*) S9. Samples tested were a negative control (NC), positive controls (PC), a procedure control (PrC), and extracts of pre-treated water and water after UV treatment alone (UV), UV/H_2O_2 (UV ox), and after subsequent granular activated carbon filtration (UV ox GAC). *Bars* denote average values, *error bars* denote standard deviations ($n = 3$). *Asterisks* denote responses showing genotoxicity, i.e., deviating from the NC with 99% certainty. *Double asterisks* denote responses showing an increase in genotoxicity by the treatment, i.e., deviating significantly (99%) from the response of the pre-treated water. (Adapted from [72])

more effective at killing *Cryptosporidium* which have produced lethal outbreaks in the past, and this method can also efficiently destroy polar pollutants not removed by carbon filtration. The first plant using this treatment method in the Netherlands was put into operation in 2004 following research showing that none of the known DBPs were formed by this method [87]; this conclusion had been reached using only chemical analysis. However, when Ames fluctuation tests were performed on water extracts before and after UV-oxidation, mutagenic DBPs did appear to be formed (Fig. 2). The carbon filtration step at the end of the process removed all this newly formed mutagenic activity, so that the final drinking water was safe.

These examples show how genotoxicity tests can be used to analyze the effect of the treatment method on the presence of genotoxic substances in the water. They also show what difficulties can arise when a positive response is found in a sample from final drinking water to which consumers are exposed. The questions that then arise are: how does one determine whether this mutagenic activity poses a health risk to these consumers, and how does this health risk of the DBPs compare to the reduced health risk from the removal of pathogens by the treatment?

To calculate the risk of the consumption of a water that was positive in the Ames test, it is necessary to identify the responsible compounds, to determine their concentrations, and to assess their carcinogenic potencies as a function of a dose–response. Then the probability of cancer from the contaminated water can be translated into disability-affected life years (DALYs). This allows a comparison to the DALYs gained from the reduced risk of illness from the pathogens which were present before the disinfection procedure [88]. This is very time-consuming and expensive work; the fact that not all mutagenic DBPs of chlorination have yet been identified [84, 89] since the discovery of the first in 1974 [83] is illustrative.

Alternatively, resources can be spent on finding a way to prevent the formation of the mutagenic DBPs or to remove them in a subsequent treatment step. Figures 1 and 2 show that active carbon filtration is very effective in removing mutagenic DBPs present after both chlorination and UV-oxidation. Mutagenicity assays are the ideal instrument for monitoring whether this goal has been reached, and also when the active carbon filtration system is saturated with DBPs and should be replaced.

7.2 Case 2: Detecting and Identifying Water Contaminants Threatening Drinking Water Quality

In the comprehensive surface water quality monitoring program conducted by CETESB, the Environmental Agency of São Paulo State, Brazil, a hotspot of mutagenic activity was found. This agency, in a pioneer initiative, introduced the *Salmonella* microsome assay in its monitoring program in 1998 [8]. The Cristais River, used as a drinking water source for 60,000 people, was repeatedly found to be mutagenic in contrast to the other rivers analyzed, and was selected for further investigation. The pollution sources were investigated by analyzing sites along the river using different strains of *Salmonella* and selective water extraction procedures [34, 35]. Other ancillary genetic toxicology studies were also carried out (Table 1). Effluent from a dye processing plant located ~6 km upstream from the Drinking Water Treatment Plant (DWTP) intake was identified as the source of mutagenic activity [9, 34]. The river sediment and the sludge of the DWTP were also mutagenic. In 2005, the compounds that were at least partially responsible for the mutagenic activity were identified [9]. These were the dye components of a Black Dye Commercial Product (BDCP), frequently used by the industry. The components identified were CI Disperse Blue 373, CI Disperse Violet 93, and CI Disperse Orange 37. These nitroaminoazobenzene dyes were detected in the raw and treated industrial effluent river waters and sludge of the DWTP using thin layer chromatography (TLC). Using liquid chromatography coupled with a diode array detector, the dyes were identified and quantified in all samples analyzed, including the drinking water [90]. The individual BDCP dye components were tested in the *Salmonella* assay and were found to be mutagenic. Additionally, several mutagenic and carcinogenic aromatic amines, were also found in the effluent and water samples, and the human carcinogen benzidine was found at a concentration of 47 μg/L in the azo dye processing plant effluent. The mutagenic activity of the drinking water extracts indicated additional mutagenicity related to the presence of direct-acting compounds other than halogenated acids, originating from chlorination of humic and fulvic acids or the dyes that were found only in ng/L levels. These direct-acting compounds are most likely non-chlorinated phenylbenzotriazoles (Cl-PBTA) and/or PBTA-type of compounds generated during the chlorination of the identified azo dyes. Other biological systems, besides the *Salmonella* assay,

Table 1 Summary of the results obtained in treated effluent, river, and drinking water from the Cristais river

Samples	Compounds detected using Thin Layer Chromatography (TLC)	Compounds detected using CG/MS/MS related to dyeing activities and disinfection by-products	Compounds detected using LC; LC/DA for dye quantification and LC/ED for aromatic amine quantification	Bioassays and main responses obtained
Treated dye processing plant effluent	CI Disperse Blue 373; CI Disperse Violet 93; CI Disperse Orange 37; unknown fluorescent compounds; other unknown dyes	5-Choro-*o*-anisidine; 2,4,5-trimethylaniline; 2,6-dichloro-1,4-phenylenendiamine; 2,6-dichloro-4-nitroaniline (parental amine of the CI Disperse Orange 37)	CI Disperse Blue 373 (67 μg/L); CI Disperse Orange 37 (126 μg/L); benzidine (10.8–47 μg/L) 3,3-dichlorobenzidine; o-toluidine; 3,3-dimethylbenzidine (<1 μg/L)	*Salmonella* assay: Organic extracts, positive in TA98 and YG1041, positive response with and without S9 but increased with S9 Positive MN results for: *T. pallida*, germinative cells, *in natura*; *Allium cepa*, chromosome aberrations, *in natura*; aberrant crypt foci in Wistar rat colon, *in natura*
River water at the DWTP	CI Disperse Blue 373; CI Disperse Violet 93; CI Disperse Orange 37; unknown fluorescent compounds; other unknown dyes	2,6-Dichloro-4-nitroaniline and *N*-cyanoethyl aniline (both parent amines of CI Disperse Orange 37); *p*-nitroaniline; 2,4,5-trimethylaniline; 2-methyl-mercaptoaniline	CI Disperse Blue 373 (64.9 ng/L); CI Disperse Violet 93 (11.8 ng/L); CI Disperse Orange 37 (397 ng/L); benzidine; 3,3-dichlorobenzidine; *o*-toluidine; 3,3-dimethyl-benzidine (<1 μg/L)	*Salmonella* assay: organic extracts, positive in TA98 and YG1041; decrease in mutagenicity with S9 and increase without S9 compared to the effluent response; positive results for *T. pallida*, MN, germinative cells, *in natura*

(continued)

Table 1 (continued)

Samples	Compounds detected using Thin Layer Chromatography (TLC)	Compounds detected using CG/MS/MS related to dyeing activities and disinfection by-products	Compounds detected using LC; LC/DA for dye quantification and LC/ED for aromatic amine quantification	Bioassays and main responses obtained
Drinking water produced by the DWTP	No contaminating compounds visually detected	2,6-Dichloro-4-nitroaniline (parent amine of CI Disperse Orange 37); 2,4,5-trimethylaniline; 2-methyl-mercaptoaniline; chloral hydrate[a]; 1,5-hexadiene; 1,1,2,5,6,6-hexacloro-2,2-dichloroacetyl chloride[a]	CI Disperse Blue 373 (3.05 ng/L); CI Disperse Violet 93 (1.65 ng/L); CI Disperse Orange 37 ng/L; benzidine, 3,3-dichlorbenzidine, o-toluidine, 3,3′-dimethylbenzidine <1 μg/L; PBTA and non-Cl-PBTA-like compounds	*Salmonella* assay: Results similar to the river water at the DWTP intake, but less mutagenic

[a]Disinfection by product derived from chlorination and humic substances

which were used in the evaluation of the industrial effluent and river water, were the plant assays, *Tradescantia pallida* and *Allium cepa*, for micronuclei (MN) and chromosome aberrations, respectively, and the Wistar rat aberrant crypt foci assay [91–93]. Positive results were obtained, which indicated a concern for human health because drinking water was prepared from this source water. A summary of the water results is presented in Table 1. As requested by the Environmental Agency of São Paulo State, a pipe was constructed to release the industrial effluent to an entry point downstream of the DWTP, so that the contamination no longer impacted the quality of the drinking water. From these results it was concluded that the combination of chemical analysis, selective water extraction procedures, and toxicological assays is a powerful tool for the evaluation of the water quality. The combination of chemical analysis and bioassays was decisive in the elucidation of the sources of the mutagenic activity present in this aquatic system.

References

1. DeMarini DM (2012) Declaring the existence of human germ-cell mutagens. Environ Mol Mutagen 53:166–172
2. Devaux A, Fiat L, Gillet C, Bony S (2011) Reproduction impairment following paternal genotoxin exposure in brown trout (*Salmo trutta*) and Arctic charr (*Salvelinus alpinus*). Aquat Toxicol 101:405–411
3. Chen G, White PA (2004) The mutagenic hazards of aquatic sediments: a review. Mutat Res 567:151–225
4. Claxton LD, Matthews PP, Warren SH (2004) The genotoxicity of ambient outdoor air, a review: *Salmonella* mutagenicity. Mutat Res Rev Mutat Res 567:347–399
5. Houk VS (1992) The genotoxicity of industrial wastes and effluents. Mutat Res 277:91–138
6. Ohe T, Watanabe T, Wakabayashi K (2004) Mutagens in surface waters: a review. Mutat Res 567:109–149
7. White PA, Claxton LD (2004) Mutagens in contaminated soil: a review. Mutat Res 567:227–345
8. Umbuzeiro GA, Roubicek DA, Sanchez PS, Sato MI (2001) The *Salmonella* mutagenicity assay in a surface water quality-monitoring program based on a 20-year survey. Mutat Res 491:119–126
9. Umbuzeiro GA, Freeman HS, Warren SH, Oliveira DP, Terao Y, Watanabe T, Claxton LD (2005) The contribution of azo dyes in the mutagenic activity of the Cristais River. Chemosphere 60:55–64
10. Brack W (2003) Effect-directed analysis: a promising tool for the identification of organic toxicants in complex mixtures? Anal Bioanal Chem 377:397–407
11. Umbuzeiro GA, Machala M, Weiss J (2011) Diagnostic tools for effect-directed analysis of mutagens, ahr agonists, and endocrine disruptors. In: Brack W (ed) Effect-directed analysis of complex environmental contamination, vol 15. Springer, Heidelberg, pp 69–82
12. DHEW (US Department of Health, Education, and Welfare) (1969) Report of the secretary's commission on pesticides and their relationship to environmental health, parts 1 and 2. Washington, DC, US GPO, 677pp
13. Drake JW, Abrahamson S, Crow JF, Hollaender A, Lederberg S, Legator MS, Neel JV, Shaw MW, Sutton HE, Von Borstel RC, Zimmering S (1975) Environmental mutagenic hazards. Science 187:503–514

14. Flamm WG, Valcovic LR, D'Aguanno W, Fishbein L, Green S, Malling HV, Mayer V, Prival M, Wolff G, Zeiger E (1977) Approaches to determining the mutagenic properties of chemicals: risk to future generations. J Environ Pathol Toxicol 1:301–352
15. Ames BN, Durston WE, Yamasaki E, Lee FD (1973) Carcinogens are mutagens: a simple test system combining liver homogenates for activation and bacteria for detection. Proc Natl Acad Sci U S A 70:2281–2285
16. Ames BN, Lee FD, Durston WE (1973) An improved bacterial test system for the detection and classification of mutagens and carcinogens. Proc Natl Acad Sci U S A 70:782–786
17. McCann J, Yamasaki E, Ames BN (1975) Detection of carcinogens in the *Salmonella*/microsome test: assay of 300 chemicals. Proc Natl Acad Sci U S A 72:5135–5139
18. Purchase IF, Longstaff E, Ashby J, Styles JA, Anderson D, Lefevre PA, Westwood FR (1978) An evaluation of 6 short-term tests for detecting organic chemical carcinogens. Br J Cancer 37:873–903
19. Sugimura T, Sato S, Nagao M, Yahagi T, Matsushima T, Seino Y, Takeuchi M, Kawachi T (1976) Overlapping of carcinogens and mutagens. In: Magee PN, Takayama S, Sugimura T, Matsushima T (eds) Fundamentals of cancer prevention. University Park Press, Baltimore, pp 191–215
20. Kirkland D, Aardema M, Henderson L, Muller L (2005) Evaluation of the ability of a battery of three in vitro genotoxicity tests to discriminate rodent carcinogens and non-carcinogens. I. Sensitivity, specificity and relative predictivity. Mutat Res 584:1–256
21. Tennant RW, Margolin BH, Shelby MD, Zeiger E, Haseman JK, Spalding J, Caspary W, Resnick M, Stasiewicz S, Anderson B, Minor R (1987) Prediction of chemical carcinogenicity in rodents from in vitro genetic toxicity assays. Science 236:933–941
22. Zeiger E (1998) Identification of rodent carcinogens and noncarcinogens using genetic toxicity tests: premises, promises and performance. Regul Toxicol Pharmacol 28:85–95
23. Zeiger E, Haseman JK, Shelby MD, Margolin BH, Tennant RW (1990) Evaluation of four in vitro genetic toxicity tests for predicting rodent carcinogenicity: confirmation of earlier results with 41 additional chemicals. Environ Mol Mutagen 16(Suppl 18):1–14
24. Ashby J, de Serres FJ, Draper M, Ishidate M Jr, Margolin BH, Matter BE, Shelby MD (eds) (1985) Evaluation of short-term tests for carcinogens: report of the international programme of chemical Safety's collaborative study on in vitro assays. Elsevier, New York, 752pp
25. De Serres FJ, Ashby J (eds) (1981) Evaluation of short-term tests for carcinogens: report of the international collaborative program. Elsevier, New York, 827pp
26. Fish F, Lampert I, Halachmi A, Riesenfeld G, Herzberg M (1987) The SOS chromotest kit: a rapid method for the detection of genotoxicity. Toxic Assess 2:135–147
27. Oda Y, Nakamura S, Oki I, Kato T, Shinagawa H (1985) Evaluation of the new system (umu-test) for the detection of environmental mutagens and carcinogens. Mutat Res 147:219–229
28. Tice RR, Agurell E, Anderson D, Burlinson B, Hartmann A, Kobayashi H, Miyamae Y, Rojas E, Ryu JC, Sasaki YF (2000) Single cell gel/comet assay: guidelines for in vitro and in vivo genetic toxicology testing. Environ Mol Mutagen 35:206–221
29. Hastwell PW, Chai LL, Roberts KJ, Webster TW, Harvey JS, Rees RW, Walmsley RM (2006) High-specificity and high-sensitivity genotoxicity assessment in a human cell line: validation of the GreenScreen HC GADD45a-GFP genotoxicity assay. Mutat Res 607:160–175
30. Simpson K, Bevan N, Hastwell P, Eidam P, Shah P, Gogo E, Rees S, Brown A (2013) The BlueScreen-384 assay as an indicator of genotoxic hazard potential in early-stage drug discovery. J Biomol Screen 18:441–452
31. Hendriks G, Atallah M, Morolli B, Calléja F, Ras-Verloop N, Huijskens I, Raamsman M, Van De Water B, Vrieling H (2012) The ToxTracker assay: novel GFP reporter systems that provide mechanistic insight into the genotoxic properties of chemicals. Toxicol Sci 125:285–298

32. Claxton LD, de Umbuzeiro GA, DeMarini DM (2010) The *Salmonella* mutagenicity assay: the stethoscope of genetic toxicology for the 21st century. Environ Health Perspect 118:1515–1522
33. DeMarini DM, Brooks LR, Warren SH, Kobayashi T, Gilmour MI, Singh P (2004) Bioassay-directed fractionation and *Salmonella* mutagenicity of automobile and forklift diesel exhaust particles. Environ Health Perspect 112:814–819
34. Umbuzeiro GA, Roubicek DA, Rech CM, Sato MIZ, Claxton LD (2004) Investigating the sources of the mutagenic activity found in a river using the *Salmonella* assay and different water extraction procedures. Chemosphere 54:1589–1597
35. Kummrow F, Rech CM, Coimbrão CA, Roubicek DA, Umbuzeiro GA (2003) Comparison of the mutagenic activity of XAD4 and blue rayon extracts of surface water and related drinking water samples. Mutat Res 541:103–113
36. Mutlu E, Warren SH, Matthews PP, King C, Linak WP, Kooter IM, Schmid JE, Ross JA, Gilmour MI, Demarini DM (2013) Bioassay-directed fractionation and sub-fractionation for mutagenicity and chemical analysis of diesel exhaust particles. Environ Mol Mutagen 54:719–736
37. Meier JR (1988) Genotoxic activity of organic chemicals in drinking water. Mutat Res 196:211–245
38. Nukaya H, Yamashita J, Tsuji K, Terao Y, Ohe T, Sawanishi H, Katsuhara T, Kiyokawa K, Tezuca M, Oguri A, Sugimura T, Wakabayashi K (1997) Isolation and chemical-structural determination of a novel aromatic amine mutagen in water from the Nishitakase river in Kyoto. Chem Res Toxicol 10:1061–1066
39. Enya T, Suzuki H, Watanabe T, Hirayama T, Hisamatsu Y (1997) 3-nitrobenzanthrone, a powerful bacterial mutagen and suspected human carcinogen found in diesel exhaust and airborne particulates. Environ Sci Technol 31:2772–2776
40. Nissinen TK, Miettinen IT, Martikainen PJ, Vartiainen T (2002) Disinfection by-products in Finnish drinking waters. Chemosphere 48:9–20
41. DeMarini DM, Gudi R, Szkudlinska A, Rao M, Recio L, Kehl M, Kirby PE, Polzin G, Richter PA (2008) Genotoxicity of 10 cigarette smoke condensates in four test systems: comparisons between assays and condensates. Mutat Res 650:15–29
42. Fetterman BA, Kim BS, Margolin BH, Schildcrout JS, Smith MG, Wagner SM, Zeiger E (1997) Predicting rodent carcinogenicity from mutagenic potency measured in the Ames *Salmonella* assay. Environ Mol Mutagen 29:312–322
43. Schildcrout JS, Margolin BH, Zeiger E (1999) Predicting rodent carcinogenicity using potency measures of the in vitro sister chromatid exchange and chromosome aberration assays. Environ Mol Mutagen 33:59–64
44. Mortelmans K, Zeiger E (2000) The Ames *Salmonella*/microsome mutagenicity assay. Mutat Res 455:29–60
45. Mortelmans K, Riccio ES (2000) The bacterial tryptophan reverse mutation assay with *Escherichia coli* WP2. Mutat Res 455:61–69
46. Hagiwara Y, Watanabe M, Oda Y, Sofuni T, Nohmi T (1993) Specificity and sensitivity of *Salmonella* typhimurium YG1041 and YG1042 strains possessing elevated levels of both nitroreductase and acetyltransferase activity. Mutat Res 291:171–180
47. DeMarini DM, Dallas MM, Lewtas J (1989) Cytotoxicity and effect on mutagenicity of buffers in a microsuspension assay. Teratog Carcinog Mutagen 9:287–295
48. Kado NY, Langley D, Eisenstadt E (1983) A simple modification of the *Salmonella* liquid incubation assay. Mutat Res 121:25–32
49. Coriel Institute for Medical Research (1986) Ames *Salmonella* mutagenicity assay protocol. Department of Microbiology, CIMR, Camden, NJ, p13
50. ISO (2005) Water quality – determination of the genotoxicity of water and waste water – *Salmonella*/microsome test (Ames test). ISO Standard 16240:2005. Available at http://www.iso.org/iso/iso_catalogue/catalogue_tc/catalogue_detail.htm?csnumber=32160. Accessed 4 Oct 2015

51. Gee P, Maron DM, Ames BN (1994) Detection and classification of mutagens: a set of base-specific *Salmonella* tester strains. Proc Natl Acad Sci U S A 91:11606–11610
52. Gee P, Sommers CH, Melick AS, Gidrol XM, Todd MD, Burris RB, Nelson ME, Klemm RC, Zeiger E (1998) Comparison of responses of base-specific *Salmonella* tester strains with the traditional strains for identifying mutagens: the results of a validation study. Mutat Res 412:115–130
53. Kamber M, Flückiger-Isler S, Engelhardt G, Jaeckh R, Zeiger E (2009) Comparison of the Ames II and traditional Ames test responses with respect to mutagenicity, strain specificities, need for metabolism and correlation with rodent carcinogenicity. Mutagenesis 24:359–366
54. Reifferscheid G, Maes HM, Allner B, Badurova J, Belkin S, Bluhm K, Brauer F, Bressling J, Domeneghetti S, Elad T, Flückiger-Isler S, Grummt HJ, Gürtler R, Hecht A, Heringa MB, Hollert H, Huber S, Kramer M, Magdeburg A, Ratte HT, Sauerborn-Klobucar R, Sokolowski A, Soldan P, Smital T, Stalter D, Venier P, Ziemann C, Zipperle J, Buchinger S (2012) International round-robin study on the Ames fluctuation test. Environ Mol Mutagen 53:185–197
55. ISO (2012) Water quality – determination of the genotoxicity of water and waste water – *Salmonella*/microsome fluctuation test (Ames fluctuation test) ISO 11350:2012. Available at http://www.iso.org/iso/iso_catalogue/catalogue_tc/catalogue_detail.htm?csnumber=50393. Accessed 4 Oct 2015
56. OECD (Organisation for Economic Co-operation and Development) (1997) Guideline for testing of chemicals, test guideline 471: bacterial reverse mutation test, OECD, Paris, France
57. Demeestere K, Dewulf J, De Witte B, Van Langenhove H (2007) Sample preparation for the analysis of volatile organic compounds in air and water matrices. J Chromatogr A 1153:130–144
58. Kolkman A, Schriks M, Brand W, Bauerlein PS, Van Der Kooi MM, Van Doorn RH, Emke E, Reus AA, Van Der Linden SC, De Voogt P, Heringa MB (2013) Sample preparation for combined chemical analysis and in vitro bioassay application in water quality assessment. Environ Toxicol Pharmacol 36:1291–1303
59. Kosikowska M, Biziuk M (2010) Review of the determination of pesticide residues in ambient air. Trends Analyt Chem 29:1064–1072
60. Potts PJ, Robinson P (2003) Sample preparation of geological samples, soils and sediments. In: Comprehensive analytical chemistry, vol 41. Elsevier, pp 723–763
61. Tadeo JL, Pérez RA, Albero B, García-Valcárcel AI, Sánchez-Brunete C (2012) Review of sample preparation atechniques for the analysis of pesticide residues in soil. J AOAC Int 95:1258–1271
62. Alvarez DA, Petty JD, Huckins JN, Jones-Lepp TL, Getting DT, Goddard JP, Manahan SE (2004) Development of a passive, in situ, integrative sampler for hydrophilic organic contaminants in aquatic environments. Environ Toxicol Chem 23:1640–1648
63. Kummrow F, Rech CM, Coimbrão CA, Umbuzeiro GA (2006) Blue rayon-anchored technique/*Salmonella* microsome microsuspension assay as a tool to monitor for genotoxic polycyclic compounds in Santos estuary. Mutat Res 609:60–67
64. Anderson D, McGregor DB (1980) The effect of solvents upon the yield of revertants in the *Salmonella*/activation mutagenicity assay. Carcinogenesis 1:363–366
65. EFSA-WHO (2015) Conclusions and recommendations of the EFSA/WHO expert working group on TTC -draft for public consultation
66. Kroes R, Renwick AG, Cheeseman M, Kleiner J, Mangelsdorf I, Piersma A, Schilter B, Schlatter J, Van Schothorst F, Vos JG, Wurtzen G (2004) Structure-based thresholds of toxicological concern (TTC): guidance for application to substances present at low levels in the diet. Food Chem Toxicol 42:65–83
67. Mons MN, Heringa MB, Van Genderen J, Puijker LM, Brand W, Van Leeuwen CJ, Stoks P, Van Der Hoek JP, Van Der Kooij D (2013) Use of the threshold of toxicological concern (TTC) approach for deriving target values for drinking water contaminants. Water Res 47:1666–1678

68. De Wolf W, Siebel-Sauer A, Lecloux A, Koch V, Holt M, Feijtel T, Comber M, Boeije G (2005) Mode of action and aquatic exposure thresholds of no concern. Environ Toxicol Chem 24:479–485
69. Kim BS, Margolin BH (1999) Statistical methods for the Ames *Salmonella* assay: a review. Mutat Res 436:113–122
70. Cariello NF, Piegorsch WW (1996) The Ames test: the two-fold rule revisited. Mutat Res 369:23–31
71. Margolin BH (1985) Statistical studies in genetic toxicology: a perspective from the U.S. National toxicology program. Environ Health Perspect 63:187–194
72. Heringa MB, Harmsen DJH, Beerendonk EF, Reus AA, Krul CM, Metz DH, IJpelaar GF (2011) Formation and removal of genotoxic activity during UV/H_2O_2-GAC treatment of drinking water. Water Res 45:366–374
73. Piegorsch WW, Simmons SJ, Margolin BH, Zeiger E, Gidrol XM, Gee P (2000) Statistical modeling and analyses of a base-specific *Salmonella* mutagenicity assay. Mutat Res 467:11–19
74. Escher BI, Allinson M, Altenburger R, Bain PA, Balaguer P, Busch W et al (2014) Benchmarking organic micropollutants in wastewater, recycled water and drinking water with in vitro bioassays. Environ Sci Technol 48:1940–1956
75. Gollapudi BB, Johnson GE, Hernandez LG, Pottenger LH, Dearfield KL, Jeffrey AM, Julien E, Kim JH, Lovell DP, MacGregor JT, Moore MM, van Benthem J, White PA, Zeiger E, Thybaud V (2013) Quantitative approaches for assessing dose–response relationships in genetic toxicology studies. Environ Mol Mutagen 54:8–18
76. Johnson GE, Soeteman-Hernández LG, Gollapudi BB, Bodger OG, Dearfield KL, Heflich RH, Hixon JG, Lovell DP, MacGregor JT, Pottenger LH, Thompson CM, Abraham L, Thybaud V, Tanir JY, Zeiger E, Van Benthem J, White PA (2014) Derivation of point of departure (PoD) estimates in genetic toxicology studies and their potential applications in risk assessment. Environ Mol Mutagen 55:609–623
77. Narotsky MG, Pressman JG, Miltner RJ, Speth TF, Teuschler LK, Rice GE, Richardson SD, Best DS, McDonald A, Hunter ES, Simmons JE (2012) Developmental toxicity evaluations of whole mixtures of disinfection by-products using concentrated drinking water in rats: gestational and lactational effects of sulfate and sodium. Birth Defects Res B Dev Reprod Toxicol 95:202–212
78. Kolkman A, Martijn BJ, Vughs D, Baken KA, Van Wezel AP (2015) Tracing nitrogenous disinfection byproducts after medium pressure UV water treatment by stable isotope labeling and high resolution mass spectrometry. Environ Sci Technol 49:4458–4465
79. Brack W, Schirmer K, Erdinger L, Hollert H (2005) Effect-directed analysis of mutagens and ethoxyresorufin-O-deethylase inducers in aquatic sediments. Environ Toxicol Chem 24:2445–2458
80. Lübcke-Von Varel U, Bataineh M, Lohrmann S, Löffler I, Schulze T, Flückiger-Isler S, Neca J, Machala M, Brack W (2012) Identification and quantitative confirmation of dinitropyrenes and 3-nitrobenzanthrone as major mutagens in contaminated sediments. Environ Int 44:31–39
81. Reifferscheid G, Buchinger S, Cao Z, Claus E (2011) Identification of mutagens in freshwater sediments by the Ames-fluctuation assay using nitroreductase and acetyltransferase overproducing test strains. Environ Mol Mutagen 52:397–408
82. Altenburger R, Ait-Aissa S, Antczak P, Backhaus T, Barceló D, Seiler TB, Brion F, Busch W, Chipman K, De Alda ML, De Umbuzeiro AG, Escher BI, Falciani F, Faust M, Focks A, Hilscherova K, Hollender J, Hollert H, Jäger F, Jahnke A, Kortenkamp A, Krauss M, Lemkine GF, Munthe J, Neumann S, Schymanski EL, Scrimshaw M, Segner H, Slobodnik J, Smedes F, Kughathas S, Teodorovic I, Tindall AJ, Tollefsen KE, Walz KH, Williams TD, Van Den Brink PJ, Van Gils J, Vrana B, Zhang X, Brack W (2015) Future water quality monitoring — adapting tools to deal with mixtures of pollutants in water resource management. Sci Total Environ 512–513:540–551

83. Rook JJ (1974) Formation of haloforms during chlorination of natural waters. Water Treat Examin 23(2):234–243
84. Richardson SD, Plewa MJ, Wagner ED, Schoeny R, DeMarini DM (2007) Occurrence, genotoxicity, and carcinogenicity of regulated and emerging disinfection by-products in drinking water: a review and roadmap for research. Mutat Res 636:178–242
85. Kool HJ, Van Kreijl CF, De Greef E, Van Kranen HJ (1982) Presence, introduction and removal of mutagenic activity during the preparation of drinking water in the Netherlands. Environ Health Perspect 46:207–214
86. Kruithof JC, Kamp PC, Belosevic M (2000) UV/H_2O_2-treatment: the ultimate solution for pesticide control and disinfection. Water Sci Technol Water Supply 2:113–122
87. Martijn AJ, Kruithof JC (2012) UV and UV/H_2O_2 treatment: the silver bullet for by-product and genotoxicity formation in water production. Ozone Sci Eng 34:92–100
88. World Health Organisation (WHO) ((2011)) Guidelines for drinking-water quality, 4th edn. http://whqlibdoc.who.int/publications/2011/9789241548151_eng.pdf
89. Richardson SD, Ternes TA (2011) Water analysis: emerging contaminants and current issues. Anal Chem 83:4616–4648
90. Oliveira DP, Carneiro PA, Rech CM, Zanoni MVV, Claxton LD, Umbuzeiro GA (2006) Mutagenic compounds generated from the chlorination of disperse azo-dyes and their presence in drinking water. Environ Sci Technol 40:6682–6689
91. Caritá R, Marin-Morales MA (2008) Induction of chromosome aberrations in the Allium cepa test system caused by the exposure of seeds to industrial effluents contaminated with azo dyes. Chemosphere 72:722–725
92. De Lima ROA, Bazo AP, Salvadori DMF, Rech CM, Oliveira DP, Umbuzeiro GA (2007) Mutagenic and carcinogenic potential of a textile azo dye processing plant effluent that impacts a drinking water source. Mutat Res 626:53–60
93. Umbuzeiro GA, Coimbrão CA, Kummrow F, Lobo DJA, Saldiva PHN (2007) Mutagenic activity assessment of Cristais river, São Paulo, Brazil, using the Blue Rayon/*Salmonella* microsome and the *Tradescantia pallida* micronuclei assays. J Braz Soc Ecotoxicol 2:163–171

Adv Biochem Eng Biotechnol (2017) 157: 81–96
DOI: 10.1007/10_2015_5013

Published online: 10 August 2016

In Vitro–In Vivo Carcinogenicity

Pablo Steinberg

Abstract The evaluation of the carcinogenic potential of chemicals constitutes an essential step in assessing the risk that the chemicals pose to human health. The "gold standard" method to evaluate the carcinogenic potential of chemicals is the carcinogenicity test in laboratory animals. However, because carcinogenicity studies in vivo are extremely time-consuming, expensive, make use of a high number of animals, and cannot be used to screen a high number of compounds at the same time, various different in vitro cell transformation assays have been developed. In this report, procedures to test the carcinogenicity in vivo and in vitro are described, whereby in the latter case three extensively evaluated test systems (the BALB/c 3T3 cell transformation assay, the Bhas 42 cell transformation assay, and the Syrian hamster embryo assay) are presented. Their performance shows that they are a useful complement to in vitro genotoxicity test batteries, can be used to identify non-genotoxic carcinogens, and as screening assays may significantly limit the number of chemicals to undergo an in vivo carcinogenicity testing, thereby strongly reducing the number of laboratory animals to be used. In the future, the development of human cell line-based transformation assays may contribute to increase further their relevance and the willingness to incorporate them into existing in vitro toxicity test batteries.

Keywords BALB/c 3T3 cell transformation assay, Bhas 42 cell transformation assay, Carcinogenicity testing in vivo, Genotoxic carcinogens, Non-genotoxic carcinogens, OECD Guidelines for the Testing of Chemicals, Syrian hamster embryo assay, Tumor initiator, Tumor promoter

P. Steinberg (✉)
Institute for Food Toxicology and Analytical Chemistry, University of Veterinary Medicine Hannover, Bischofsholer Damm 15, 30173 Hannover, Germany
e-mail: pablo.steinberg@tiho-hannover.de

Contents

1 Introduction

Whenever a new drug is developed, its human safety has to be assessed. In this context the evaluation of its carcinogenic potential constitutes an essential step in the risk assessment process. Up to the present time, the "gold standard" method to evaluate the carcinogenic potential of chemicals is the carcinogenicity test in laboratory animals. However, this procedure is extremely time-consuming, expensive, makes use of a high number of animals, and cannot be used to screen a high number of compounds at the same time. Because of these limitations, in the last few years great efforts have been made to develop test systems to evaluate the carcinogenic potential of chemicals in vitro. In this report, procedures to test the carcinogenicity in vivo and in vitro are described, whereby in the latter case three extensively evaluated test systems (the BALB/c 3T3 cell transformation assay, the Bhas 42 cell transformation assay, and the Syrian hamster embryo assay) are presented. Furthermore, the added value of the three above-mentioned assays in the context of the risk assessment of chemicals, particularly when wanting to assess their carcinogenic potential, is discussed.

2 Carcinogenicity Testing In Vivo

To test the potential carcinogenicity of chemicals, two internationally accepted test guidelines published by the Organization for Economic Co-Operation and Development (OECD), the OECD Test Guideline 451 [1] and the OECD Test Guideline 453 [2], are available. Moreover, upon request from the European Commission, the European Food Safety Authority (EFSA) prepared a scientific report to support the future establishment of protocols for chronic toxicity and/or carcinogenicity studies in rodents with whole food/feed [3]. The carcinogenicity testing in vivo is based on the daily administration of the test substance in graduated doses to several groups of test animals for the majority of their life span, and the animals are closely observed for signs of toxicity and for the development of neoplastic lesions.

OECD Test Guideline 451 primarily covers the assessment and evaluation of the carcinogenicity of chemicals in rodents [1]. The preferred rodent species is the rat, although other rodent species, e.g., the mouse, may be used. The use of non-rodent species may be considered when available data suggest that they are more relevant for the prediction of health effects in humans. Rats and mice have been preferred as experimental models because of their relatively short life span, their widespread use in pharmacological and toxicological studies, their susceptibility to tumor induction, and the availability of sufficiently characterized strains. The test compound should be tested in both female and male animals. A sufficient number of animals should be used so that a thorough biological and statistical evaluation is possible. Therefore, each dose group and concurrent control group should consist of at least 50 animals of each sex. At least three dose levels and a concurrent control should be used. The highest dose level should be chosen to identify the principal target organs and toxic effects, simultaneously avoiding suffering, severe toxicity, morbidity, or death. The control group should be an untreated group or a vehicle-control group in the case where a vehicle is used to administer the test substance. The test substance is normally administered orally (in the diet, in the drinking water, or by gavage), whereby the route and method of administration is dependent on the purpose of the study, the physical/chemical properties of the test substance, its bioavailability, and the predominant route of exposure to humans. The duration of the study is normally 24 months for rodents, this period covering a major part of the normal life span of the animals to be used. For certain strains of mice (e.g., the AKR/J, C3H/J, or C57BL/6J strains), a duration of 18 months may be more appropriate.

All animals are weighed at the start of treatment, at least once a week for the first 13 weeks and at least once monthly thereafter. Measurements of food consumption and food efficiency should be made at least weekly for the first 13 weeks and at least once monthly thereafter. Water consumption should be measured at least weekly for the first 13 weeks and at least once monthly thereafter when the substance is administered in drinking water. If considered appropriate, blood sampling for hematological and clinical chemistry determinations and urine analysis may be conducted as part of an interim kill and at study termination on a minimum of 10 animals per sex per group. All animals in the study should be subjected to a full, detailed gross necropsy, which includes careful examination of the external surface of the body, all orifices, and the cranial, thoracic, and abdominal cavities and their contents. Organ weights are not normally part of a carcinogenesis study because geriatric changes and, at later stages, the development of tumors confound the usefulness of organ weight data. If needed (e.g., to perform a weight of evidence evaluation), they should be collected no later than 1 year after initiation of the study. After fixation and staining with eosin/hematoxylin, a microscopic examination of the following tissues/organs was performed: (1) all tissues (listed in Table 1) from the high dose and control groups; (2) all tissues of animals dying or killed during the study; (3) all tissues showing macroscopic abnormalities including tumors; (4) if treatment-related histopathological changes are observed in the high dose group, those same tissues are to be examined from all animals in all

Table 1 List of tissues to be examined histopathologically in an in vivo carcinogenicity study

All gross lesions	Harderian gland	Peripheral nerve (sciatic) preferably in close proximity to the muscle	Sternum with bone marrow
Adrenal glands	Heart	Prostate	Stomach
Aorta	Kidneys (left and right)	Rectum	Testes
Brain (representative regions including cerebrum, cerebellum, medulla/pons and pituitary)	Lacrimal gland	Salivary glands	Thymus
Cecum	Large intestine	Section of bone marrow and/or a fresh bone marrow aspirate	Thyroid
Cervix	Liver	Seminal vesicles	Tongue
Coagulating gland	Lymph nodes (submandibular and mesenteric)	Skeletal muscle	Trachea and lungs inflated with fixative and then immersed in formalin
Epididymides	Esophagus	Skin with mammary gland area	Urinary bladder
Eyes	Ovaries	Small intestine (including the Gut-Associated Lymphoid Tissue, GALT)	Uterus
Femur (femoro-tibial joint)	Pancreas	Spinal cord (cervical, mid-thoracic and lumbar regions)	Vagina
Gonads (testes, left and right; ovaries, left and right)	Parathyroid	Spleen	Additional tissues may need to be investigated based on clinical or any other findings

other dose groups; (5) in the case of paired organs, e.g., kidney, adrenal, both organs should be examined.

The combined chronic toxicity/carcinogenicity test described in the OECD Test Guideline 453 [2] consists of two parallel phases, a chronic phase, normally of 1 year duration, and a carcinogenicity phase, normally of 2 years duration. For the chronic phase, the test substance is administered daily in graduated doses to several groups of test animals, one dose level per group. The 1-year duration is chosen to be sufficiently long to allow any effects of cumulative toxicity to become manifest without the confounding effects of geriatric changes. In the course of the carcinogenicity phase, all examinations previously described in the context of the OECD Test Guideline 451 are performed.

3 Carcinogenicity Testing In Vitro

If one takes into account the limitations of the carcinogenicity testing in vivo (time-consuming, costly, and the use of large number of animals), in vitro assays, being quicker and cheaper, could be of great help in detecting animal (rodent) carcinogens and, if the assays were highly specific and sensitive, could be integrated into a future carcinogenicity screening strategy. In this context, the availability of so-called cell transformation assays, in which primary cell cultures (e.g., Syrian hamster embryo cells), cell lines (e.g., BALB/c 3T3 cells), or genetically modified cell lines (e.g., Bhas 42 cells) are incubated with the test compounds and then the treated cells are analyzed regarding the expression of certain traits associated with tumor development in vivo (see below), has become increasingly important.

The malignant transformation of cells held in culture has been described as a progressive multistep process, in which a series of in vitro events, very reminiscent of those associated with carcinogenicity in vivo, leads to the transition of the cells from a normal to a transformed state [4–8]. Various different cellular alterations associated with the in vitro transformation process have been described, including: (1) the acquisition of an infinite lifespan (immortalization); (2) changes in morphology (e.g., fusiform morphology); (3) changes in the growth pattern (e.g., criss-cross and multilayered growth of the cultured cells); (4) aneuploidy and genetic instability; (5) anchorage-independent growth (e.g., colony formation in soft agar); (6) the ability to form tumors in vivo (e.g., after subcutaneous injection in nude mice) [8].

In 2007, the OECD published a Detailed Review Paper on "Cell Transformation Assays for Detection of Chemical Carcinogens' [9, 10], in which it recommended, based on the performance of the assays, the development of OECD test guidelines for the BALB/c 3T3 cell transformation and the Syrian hamster embryo assay. Moreover, a cell transformation assay using v-Ha-*ras*-transfected BALB/c 3T3 cells, which is able to detect the tumor initiating and tumor promoting activity of chemicals depending on the protocol used (see below), has been established in recent years, and a corresponding OECD test guideline draft is also available. In the following sections the BALB/c 3T3 cell transformation assay, the Bhas 42 cell transformation assay, and the Syrian hamster embryo assay are described, whereby it should be noted that each assay is concisely presented. Experimental details can be taken from the recommended protocols cited in the corresponding sections.

3.1 BALB/c 3T3 Cell Transformation Assay

The BALB/c 3T3 cell transformation assay [11] makes use of the subclone A31-1-1 [12] of the mouse fibroblast cell line BALB/c 3T3 [13, 14]. The test consists of two steps, a preliminary dose-range finding experiment and the BALB/c 3T3 cell transformation assay itself. In the preliminary dose-range finding experiment, one dose between the "No Observable Effect Level" (NOEL) and the concentration leading to 50% cytotoxicity (i.e., the IC_{50} value) and one dose between the IC_{50} and the concentration leading to a 90% cytotoxicity (the IC_{90} value) should be tested. Cytotoxicity can be measured by performing the crystal violet assay as described by Sasaki et al. [11]. In the BALB/c 3T3 cell transformation assay, one concentration below the NOEL, two concentrations between the NOEL and the IC_{50}, and two or three concentrations between and the IC_{50} and the IC_{90} are used.

A scheme of the recommended BALB/c 3T3 cell transformation assay protocol [11] is shown in Fig. 1. Briefly, three controls, namely a medium control (in which no test compound or solvent is present), a solvent control (medium containing the solvent), and a positive control (4 μg/mL 3-methylcholanthrene dissolved in DMSO), are run in each experiment. On day 0, 2×10^4 BALB/c 3T3 cells are seeded in 10 mL of minimum essential medium supplemented with 10% fetal bovine serum, 10,000 units penicillin/mL, and 10 mg streptomyin/mL (M10F medium) in a 100-mm Petri dish, and ten dishes per dose and control are used. After 24 h, the cell culture medium in the different dishes is replaced by M10F medium including the different test compound concentrations, and the cells are incubated with the test compound for 72 h. Thereafter, the cell culture supernatant is replaced by fresh compound-free M10F medium. On day 7, the M10F medium is replaced by Dulbecco's modified Eagle's medium/F12 supplemented with 2% fetal bovine serum, 10,000 units penicillin/mL, 10 mg streptomycin/mL, and 2 mg insulin/mL (DF2I2F medium), and thereafter the medium is changed twice a week until day 24 or 25. On day 31 or 32 (i.e., 1 week after the last medium change), cells are fixed in methanol, stained with a Giemsa solution, and air-dried after removing the staining solution.

With the help of a stereomicroscope, so-called type III foci with a diameter greater than 2 mm are counted (Fig. 2). Type III foci are characterized by the following morphological features: (1) strong basophilic staining of spindle-shaped cells that are morphologically clearly different from those building the background

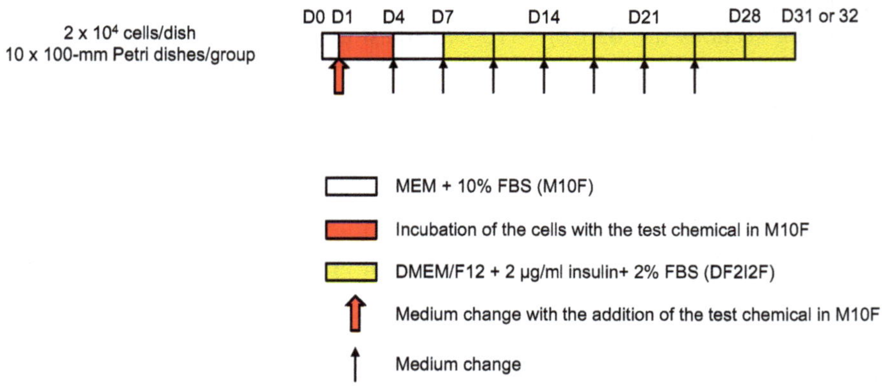

Fig. 1 Scheme of the various steps in the BALB/c 3T3 cell transformation assay (from [12], with permission of Elsevier Ltd)

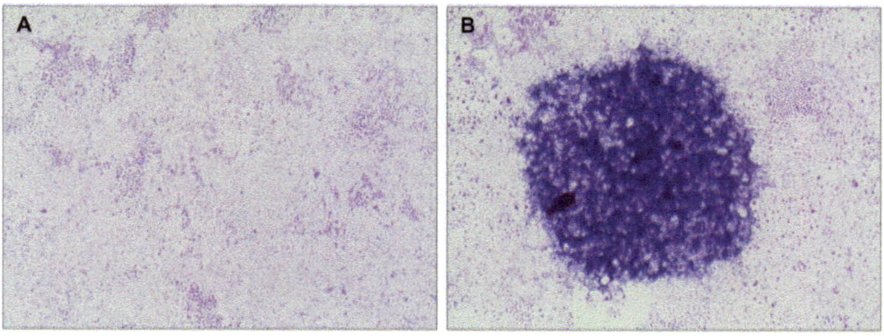

Fig. 2 (**a**) Monolayer of non-transformed BALB/c 3T3 cells. (**b**) Type III focus consisting of malignantly transformed BALB/c 3T3 cells

cell monolayer; (2) dense multilayering of cells (the "piling up"); (3) random orientation of the cells at the edge of the foci (the "criss-cross" pattern of cell growth) [11, 15]. The correct scoring of the type III foci has proven to be a critical step in the BALB/c 3T3 cell transformation assay and requires adequate training. It should be pointed out that a photo catalogue which helps in assessing the morphology of the foci and scoring the type III foci has been published [16]. Moreover, a statistical analysis based on the negative binomial distribution combined with William's-type protected tests has been recommended to evaluate the BALB/c 3T3 cell transformation assay data [17].

The performance characteristics of the BALB/c 3T3 cell transformation assay including concordance, sensitivity, specificity, positive, and negative predictivity, as well as the proportion of false negatives and false positives summarized in the OECD Detailed Review Paper on "Cell Transformation Assays for Detection of Chemical Carcinogens" [9] are shown in Table 2. Very recently the BALB/c 3T3 cell transformation assay has been used to test the carcinogenic potential not only of chemicals but also of wood combustion particulate matter [19], soil organic matter [20], and metal nanoparticles [21].

3.2 *Bhas 42 Cell Transformation Assay*

Tumor promotion is viewed as the phase in the multistep process of tumor development in which initiated cells (i.e., cells in which DNA alterations have been induced by chemical, physical, or biological agents) are stimulated to proliferate. The Bhas 42 cell transformation assay was originally developed by Ohmori et al. [22] to evaluate the potential tumor promoting activity of chemicals. Thereafter, the assay was complemented in such a way that it can be used to detect tumor initiators as well as tumor promoters (see below) [23]. The test system uses the so-called Bhas 42 cell clone, which was isolated after the transfection of BALB/c 3T3 cells with an activated *ras* oncogene, thereby strongly enhancing the transformation frequency of the BALB/c 3T3 cells [24].

Table 2 Performance characteristics of the BALB/c 3T3 cell transformation assay [9], the Syrian hamster embryo assay at pH 6.7 and pH ≥ 7.0 [9], and the Bhas 42 cell transformation assay [18]

	BALB/c 3T3 cell transformation assay	Syrian hamster embryo assay at pH 6.7	Syrian hamster embryo assay at pH ≥ 7.0	Bhas 42 cell transformation assay
Number of chemicals tested	149	88	204	98
Prevalence of carcinogens[a]	68%	61%	85%	53%
Concordance[b]	68%	74%	85%	78%
Sensitivity[c]	75%	66%	92%	73%
Specificity[d]	53%	85%	66%	84%
Positive predictivity[e]	77%	88%	88%	86%
Negative predictivity[f]	50%	62%	75%	69%
False positives	47%	15%	34%	27
False negatives (%)	25%	33%	8%	16
Number of chemicals not included[g]	28	2	12	0

[a]Proportion of carcinogens in the database
[b]Percentage agreement with the in vivo carcinogenicity study
[c]Percentage carcinogens that are positive (number of positives/number of tested carcinogens)
[d]Percentage non-carcinogens that are negative (number of negatives/number of tested non-carcinogens)
[e]Percentage positives that are carcinogens (number of true positives/number of positive cell transformation assays)
[f]Percentage negatives that are non-carcinogens (number of true negatives/number of negative cell transformation assays)
[g]Not included because the results of the assay were not clearly positive or negative, or because the chemicals were only tested for tumor promotion

The Bhas 42 cell transformation assay can be subdivided into an "initiation assay" and a "promotion assay" to examine the tumor initiating and the tumor promoting activity of chemicals, respectively [18, 23]. To set the concentrations of the compound to be tested in the initiation and promotion assays, cell growth assays using the crystal violet staining method are first performed [18]. In the initiation assay, one concentration below the NOEL, two concentrations between the NOEL and the IC_{50}, and two concentrations between the IC_{50} and the IC_{90} values are at least evaluated. In the promotion assay and when wanting to test chemicals that markedly enhance cell growth, one concentration below the NOEL, three concentrations in the range leading to cell growth stimulation, and one concentration in the range leading to a weak growth inhibition are tested, whereas in the case of compounds that do not markedly enhance cell growth, concentrations ranging

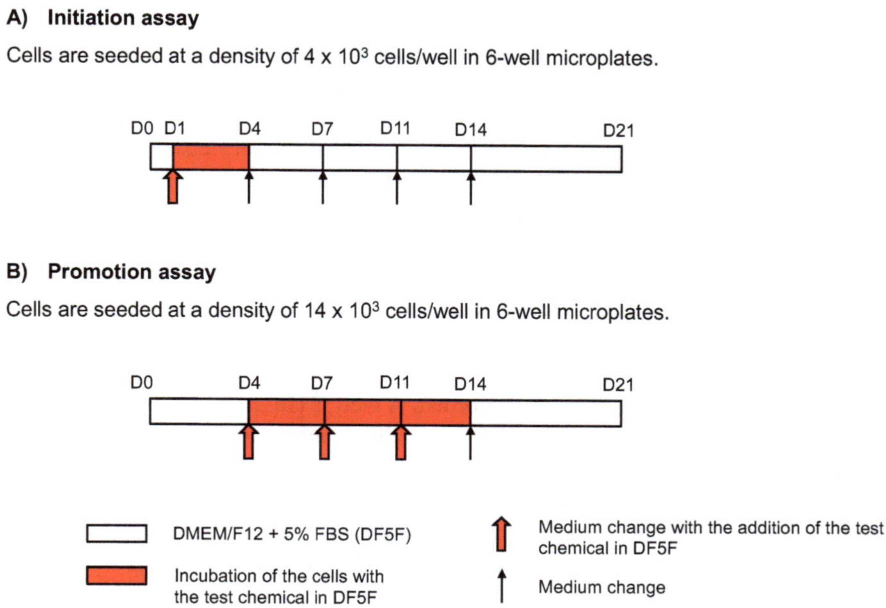

Fig. 3 Scheme of the Bhas 42 cell transformation assay. (**a**) Initiation assay. (**b**) Promotion assay (from [18], with permission of Elsevier Ltd)

from two to three levels below the NOEL to a concentration leading to a cell survival rate below 50% are evaluated [18].

The Bhas 42 cell-based initiation and promotion assays are schematically shown in Fig. 3. In the initiation assay, 2,000 cells/mL (i.e., 4,000 cells/well) are seeded in 6-well microplates (6 wells per treatment group) and 24 h later the cell culture supernatant is substituted by cell culture medium containing the different concentrations of the test chemical. After a 3-day incubation period the cell culture medium is removed and fresh chemical-free medium is added. The cells are cultured until day 21, medium changes being made on days 7, 11, and 14, then fixed in methanol and stained with Giemsa's solution. In each assay a positive control (1 μg 3-methylcholanthrene/mL) and a negative control (0.1% DMSO) are also tested. Thereafter, the type III foci are scored under a stereomicroscope, as is the case in the BALB/c 3T3 cell transformation assay. The test compound-induced transformation frequency is statistically analyzed by multiple comparisons using the one-sided Dunnett test ($p < 0.05$, one-sided).

In the promotion assay, 7,000 cells/mL (i.e., 14,000 cells/well) are seeded in 6-well microplates (6 wells per treatment group) and cultured for 96 h without changing the cell culture medium. On days 4, 7, and 11 the cell culture medium is replaced by cell culture medium containing the test chemical. On day 14 the cell culture supernatant is substituted by the compound-free cell culture medium, and on day 21 the cells are fixed in methanol and stained with Giemsa's solution. In each assay a positive control (0.05 μg 12-*O*-tetradecanoylphorbol-13-acetate/mL) and a negative control (0.1% DMSO) are also tested. The type III foci are scored under a stereomicroscope, and the test compound-induced transformation frequency is statistically analyzed by multiple comparisons using the one-sided Dunnett test.

In the meantime, a protocol to perform the Bhas 42 cell transformation assay in 96-well culture plates has been established [25]. The protocol steps are basically the same as those when the assay is performed in 6-well microplates with the exception of the number of cells seeded per well and the volume of cell culture medium per well. In the initiation assay, 200 cells/0.05 mL/well are seeded, whereas in the promotion assay 400 cells/0.1 mL are added to each well. Moreover, in the initiation assay 0.05 mL of medium containing a test chemical at double its final concentration are added to 0.05 mL of cell culture medium in each well. In the 96-well assay format, the number of wells in which a transformed focus is observed vs the number of analyzed wells is recorded. A well is classified as "transformed" if it contains at least one transformed focus, independently of the presence of other transformed and non-transformed foci in the same well. The transformation frequency is expressed as the number of wells with transformed foci/total number of analyzed wells. The test compound-induced transformation frequency is statistically analyzed by the chi-square test with Bonferroni adjustment ($p < 0.05$, one sided).

The concordance, sensitivity, specificity, positive and negative predictivity, and proportion of false negatives and false positives in the Bhas 42 cell transformation assay have been determined by Sakai et al. [18] and are shown in Table 2. In an international validation study [26] the assay was shown to be transferable and reproducible between laboratories, and its within-laboratory reproducibility was also confirmed.

3.3 *Syrian Hamster Embryo Assay*

The Syrian hamster embryo (SHE) assay [27] makes use of normal diploid primary cells isolated from Syrian hamster embryos at day 13 of gestation. These cells are metabolically competent in the sense that they retain the ability to metabolize a wide range of chemicals [9, 10, 28]. From an animal welfare point of view it must be pointed out that Syrian hamsters need to be sacrificed to perform the assay. However, it should be taken into account that one Syrian hamster provides cells to perform 50–100 cell transformation assays.

In the SHE assay, morphologically transformed cell colonies are recorded as the earliest identifiable in vitro endpoint related to carcinogenesis. The transformed cell colonies are characterized by a decreased cytoplasm-to-nucleus ratio, a stronger basophilic staining than their normal counterparts, a spindle-shaped morphology, loss of intercellular communication, criss-crossing, and stacking. The "classical assay" is performed at a pH of 7.1–7.3, although LeBoeuf et al. [29] run it at a pH of 6.7, which provides optimal growth conditions for the SHE cells. At a pH of 6.7 the SHE cells remain longer in a less differentiated state, adopt a more spindle-shaped morphology, and make the criss-cross growth pattern of the transformed cell colonies more visible.

The SHE assay consists of two steps, a preliminary dose-range-finding experiment and the SHE assay itself. In the preliminary dose-range-finding experiment, at least ten concentrations, covering a wide toxicity range, should be tested, whereby at least five and preferably ten dishes per concentration should be used. Cytotoxicity is determined by measuring the decrease in plating efficiency and/or colony density and size of the treated SHE cells when compared to the solvent-treated cell cultures. In the SHE assay, one concentration not affecting the plating efficiency, a high concentration maximally leading to 50% cytotoxicity and three or four concentrations in between should at least be tested.

To obtain the primary SHE cell cultures [27], embryos are removed from the uterine horns of pregnant Syrian hamsters. After excising the head, limbs, and viscera, the remaining embryonic tissues are minced and dissociated, e.g., with a dispase (2 units/mL) solution in buffered saline. The dissociated cells are plated and after reaching 60–80% confluence the cells are trypsinized (e.g., 0.25% [w/v] trypsin in buffered saline), washed and cryopreserved. When wanting to perform the assay, cells are thawed and feeder cells need to be prepared. The feeder cells, which are of the same origin as the target cells, are X-rayed at 50 grays, so that they are viable but are unable to replicate, and are used to support the growth of the target cell colonies. Twenty-four hours after seeding the feeder cells, target cells are plated on them and another 24 h later (on the 3rd day) the exposure of the cells to the test compounds is started and continues up to the 10th day. Then, the medium is removed, and the cells are washed with phosphate-buffered saline, fixed in absolute ethanol or methanol, and stained with a 10% aqueous Giemsa solution.

The number of normal and morphologically transformed cell colonies is determined by using a stereomicroscope. In a next step, the morphological transformation frequency (= number of transformed colonies/total number of colonies × 100) and cytotoxicity, evaluated on the basis of the plating efficiency (= total number of colonies per dish/total number of target cells seeded per dish × 100), the relative plating efficiency (= plating efficiency of the treated group/plating efficiency of the solvent group × 100), and colony size and density (= number of cells per colony), are calculated. Photo catalogues with pictures of normal and morphologically transformed cell colonies in the SHE assay at pH 6.7 [30] and pH 7.0 [31] are available. Three controls, namely a solvent control (medium containing the solvent), a positive control (1–5 μg/mL benzo[*a*]pyrene dissolved in DMSO), and a feeder cell control (at least five dishes with feeder cells only to confirm that they do not proliferate and form colonies), are run in each experiment. To determine whether the increase in the number of morphologically transformed SHE cell colonies after treating the cells with the test compound is statistically significant when compared to the number of colonies in the solvent group, the one-sided Fisher's exact test is performed.

The visual scoring of morphologically transformed cell colonies in the SHE assay has been considered to be subjective [32], and an alternative approach to objectively classify SHE cell colonies involving the application of attenuated total reflection Fourier-transform infrared (ATF-FTIR) spectroscopy coupled to a multivariate analysis has recently been proposed [32–34].

The performance characteristics of the SHE assay including concordance, sensitivity, specificity, positive and negative predictivity, and the proportion of false negatives and false positives summarized in the OECD Detailed Review Paper on "Cell Transformation Assays for Detection of Chemical Carcinogens" [9] are shown in Table 2. The transferability and the inter- and intra-laboratory reproducibility of the SHE assay at pH 6.7 [35] and pH 7.0 [36] have been reported. In a very recent study, Benigni et al. [37] concluded that the SHE test is able to identify non-genotoxic carcinogens with an efficiency of 80–90%, and in this context may play an important role in alternative testing strategies combining in vitro test systems with different endpoints [37–40].

4 Discussion

Carcinogens are divided into genotoxic and non-genotoxic carcinogens. In the case of genotoxic carcinogens, their genotoxicity might be based on a number of different effects, ranging from the induction of gene mutations to the induction of chromosomal aberrations. Nowadays, various short-term genotoxicity assays with different endpoints are available. The induction of gene mutations can be determined by performing in vitro assays such as the bacterial reverse mutation test [41], the mouse lymphoma assay [42], and the hypoxanthine guanine phosphoribosyl transferase test [42], among others, and chromosomal damage can be detected by an in vitro chromosome aberration test [43] or an in vitro micronucleus test [44]. The in vitro genotoxicity assays show a high sensitivity, but a low specificity (i.e., they identify many compounds as genotoxic, which are neither genotoxic nor carcinogenic in vivo) [45, 46]. When wanting to evaluate the carcinogenic potential of chemicals, cell transformation assays may help in delimiting the number of false positive genotoxic compounds detected in vitro that should be further tested in vivo. In fact, Benigni and Bossa [38] have shown that a test battery, consisting of the bacterial reverse mutation test, the analysis of structural alerts for DNA-reactive carcinogens, and the SHE assay performed at pH 7.0, allows the identification of up to 95% of rodent carcinogens, at the same time strongly increasing specificity. Moreover, the same test battery has successfully been used to correctly identify compounds classified by the International Agency for Research on Cancer as recognized (Class 1), probable (Class 2A), or possible (Class 2B) human carcinogens [40].

A high number of chemicals, so-called non-genotoxic carcinogens, are carcinogenic via diverse mechanisms such as the stimulation of cell proliferation, inhibition of apoptosis, and modulation of gene expression, among others, which do not involve DNA damage. As previously mentioned, the SHE test as well as the Bhas 42 cell transformation assay have proven to be able to detect tumor promoters with a very high efficiency [18, 37] and should therefore be taken into account when developing an in vitro test battery to evaluate the carcinogenic potential of chemicals [37].

In a review by Vanparys et al. [47] possible applications of cell transformation assays in the pharmaceutical, chemical, food, and cosmetic industries were described. In the pharmaceutical industry, cell transformation assays could be used, e.g., as a tool to screen for carcinogens during the early phases of drug development or to follow up positive in vitro genotoxicity assay results, although in the chemical industry, cell transformation assays may be performed, e.g., to screen for carcinogens during the early phases of chemical development, to evaluate the carcinogenic potential of genotoxic compounds, or to determine whether a non-genotoxic chemical is a potential non-genotoxic carcinogen [47]. Furthermore, Vanparys et al. [47] pointed out that cell transformation assays were considered by EFSA to be useful to assess the significance of positive results in in vitro genotoxicity assays, to analyze compounds with structural alerts for carcinogenicity, and to demonstrate differences or similarities across chemical categories of food ingredients in the frame of food and feed safety evaluation [48]. In the case of cosmetic ingredients, cell transformation assays can be used to complement information obtained in in vitro genotoxicity tests [49], and the SHE assay has been included in a test strategy to follow up cosmetic ingredients tested positive in in vitro genotoxicity assays [50].

The lower sensitivity and specificity of the BALB/c 3T3 cell transformation assay as compared to the SHE assay and the relatively high rate of false positives in the BALB/c 3T3 cell transformation assays are explained by a number of shortcomings of the elder protocols applied [9]. Future studies will show whether the use of the recommended BALB/c 3T3 cell transformation assay [11] described in the present report, in which the shortcomings have been eliminated, results in an improved performance of the assay. Benigni et al. [39] postulated that the higher sensitivity of the SHE assay, when compared to other cell transformation assays, might be explained by the fact that the cells are metabolically competent, the cell cultures comprise a variety of different cell types that can be transformed via a number of different toxicity pathways, and the assay detects more basic and unspecific transformation mechanisms.

It should be mentioned that, in the scientific literature, further cell transformation assays, such as the one based on the use of C3H/10T1/2 cells, have been described. In the case of the C3H/10 T1/2 cell transformation assay, it was found useful in elucidating molecular mechanisms of cell transformation, but because of the limited data regarding its reproducibility it was not possible to recommend the preparation of an OECD Test Guideline at that time [9]. It is foreseeable that, in the future, cell transformations assays based on the use of human cell lines should be developed and validated. By doing so, the relevance and predictability of cell transformation assays may increase even more.

In conclusion, cell transformation assays are a useful complement to in vitro genotoxicity test batteries. They can be used to identify non-genotoxic carcinogens and, as screening assays, may significantly limit the number of chemicals having to undergo in vivo carcinogenicity testing, thereby strongly reducing the number of laboratory animals used. Lastly, the development of human cell line-based transformation assays may contribute to increase further their relevance and the willingness to incorporate them into existing in vitro toxicity test batteries.

References

1. OECD (2009) Guideline for the testing of chemicals, test guideline 451: carcinogenicity studies. OECD, Paris. doi:10.1787/9789264071186-en. ISBN 9789264071186
2. OECD (2009) Guideline for the testing of chemicals, test guideline 453: combined chronic toxicity/carcinogenicity studies. OECD, Paris. doi:10.1787/9789264071223-en). ISBN 9789264071223
3. EFSA (2013) Scientific report of EFSA – considerations on the applicability of OECD TG 453 to whole food/feed testing. EFSA J 11:3347
4. Kuroki T, Sato T (1968) Transformation and neoplastic development in vitro of hamster embryonic cells in cultures by 4-nitroquinoline-1-oxide and its derivatives. J Natl Cancer Inst 41:53–71
5. Kakunaga T, Kamahora J (1969) Analytical studies on the process of malignant transformation of hamster embryonic cells in cultures with 4-nitroquinoline-1-oxide. Symp Cell Chem 20:135–148
6. Mondal S, Brankow DW, Heidelberger C (1976) Two-stage chemical oncogenesis in cultures of C3H/10T1/2. Cancer Res 36:2254–2260
7. Barrett JC, Ts'o POP (1978) Evidence for the progressive nature of neoplastic transformation in vitro. Proc Natl Acad Sci 75:3761–3765
8. Schechtman LM (2012) Rodent cell transformation assays – a brief historical perspective. Mutat Res 744:3–7
9. OECD (2007) Detailed review paper on cell transformation assays for detection of chemical carcinogens. OECD environment, health and safety publications. Series on testing and assessment no. 31. www.oecd.org/officialdocuments/publicdisplay documentpdf/?doclanguage = en&cote = env/jm/mono(2007)18
10. Vasseur P, Lasne C (2012) OECD detailed review paper (DRP) number 31 on "cell transformation assays for detection of chemical carcinogens": main results and conclusions. Mutat Res 744:8–11
11. Sasaki K, Bohnenberger S, Hayashi K, Kunkelmann T, Muramatsu D, Phrakonkham P, Poth A, Sakai A, Salovaara S, Tanaka N, Thomas BC, Umeda M (2012) Recommended protocol for the BALB/c 3T3 cell transformation assay. Mutat Res 744:30–35
12. Kakunaga T, Crow JD (1980) Cell variants showing differential susceptibility to ultraviolet light-induced transformation. Science 209:505–507
13. Aaronson SA, Todaro GJ (1968) Development of 3T3-like lines from Balb/c mouse embryo cultures: transformation susceptibility to SV40. J Cell Physiol 72:141–148
14. Aaronson SA, Todaro GJ (1968) Basis for the acquisition of malignant potential by mouse cells cultivated in vitro. Science 162:1021–1026
15. IARC/NCI/EPA Working Group (1985) Cellular and molecular mechanisms of cell transformation and standardization of transformation assays of established cell lines for prediction of carcinogenic chemicals: overview and recommended protocols. Cancer Res 45:2395–2399
16. Sasaki K, Bohnenberger S, Hayashi K, Kunkelmann T, Muramatsu D, Poth A, Sakai A, Salovaara S, Tanaka N, Thomas BC, Umeda M (2012) Photo catalogue for the classification of foci in the BALB/c 3T3 cell transformation assay. Mutat Res 744:42–53
17. Hoffmann S, Hothorn LA, Edler L, Kleensang A, Suzuki M, Phrakonkham P, Gerhard D (2012) Statistical analysis of BALB/c 3T3 cell transformation assay data. Mutat Res 744:36–41

18. Sakai A, Sasaki K, Muramatsu D, Arai S, Endou N, Kuroda S, Hayashi K, Lim YM, Yamazaki S, Umeda M, Tanaka N (2010) A Bhas 42 cell transformation assay on 98 chemicals: the characteristics and performance for the prediction of chemical carcinogenicity. Mutat Res 702:100–122, erratum in Mutat Res 703:209–226 (2010)
19. Gauggel-Lewandowski S, Heussner AH, Steinberg P, Pieterse B, van der Burg B, Dietrich DR (2013) Bioavailability and potential carcinogenicity of polycyclic aromatic hydrocarbons from wood combustion particulate matter in vitro. Chem Biol Interact 206:411–422
20. Baderna D, Colombo A, Romeo M, Cambria F, Teoldi F, Lodi M, Diomede L, Benfenati E (2014) Soil quality in the Lomellina area using in vitro models and ecotoxicological assays. Environ Res 133:220–231
21. Sighinolfi GL, Artoni E, Gatti AM, Corsi L (2015) Carcinogenic potential of metal nanoparticles in BALB/3T3 cell transformation assay. Environ Toxicol (in press). doi:10.1002/tox.22063
22. Ohmori K, Sasaki K, Asada S, Tanaka N, Umeda M (2004) An assay method for the prediction of tumor promoting potential of chemicals by the use of Bhas 42 cells. Mutat Res 557:191–202
23. Asada S, Sasaki K, Tanaka N, Takeda K, Hayashi M, Umeda M (2005) Detection of initiating as well as promoting activity of chemicals by a novel cell transformation assay using v-Ha-*ras*-transfected BALB/c 3T3 cells (Bhas 42 cells). Mutat Res 588:7–21
24. Sasaki K, Mizusawa H, Ishidate M (1988) Isolation and characterization of *ras*-transfected BALB/3T3 clone showing morphological transformation by 12-O-tetradecanoyl-phorbol-13-acetate. Jpn J Cancer Res 79:921–930
25. Arai S, Sakai A, Hayashi K, Sasaki K, Muramatsu D, Endou N, Umeda M, Tanaka N (2013) A high-throughput cell transformation assay applicable to automation for detecting potential chemical carcinogens using Bhas 42 cells. AATEX 18:1–18
26. Sakai A, Sasaki K, Hayashi K, Muramatsu D, Arai S, Endou N, Kuroda S, Pot A, Bohnenberger S, Kunkelmann T, Asakura M, Hirose H, Ishii N, Mizuhashi F, Kasamoto S, Nagai M, Pant K, Bruce SW, Sly JE, Yamazaki S, Umeda M, Tanaka N (2011) An international validation study of a Bhas 42 cell transformation assay for the prediction of chemical carcinogenicity. Mutat Res 725:57–77
27. Maire M-A, Pant K, Phrakonkham P, Poth A, Schwind K-R, Rast C, Bruce SW, Sly JE, Bohnenberger S, Kunkelmann D, Schulz M, Vasseur P (2012) Recommended protocol for the Syrian hamster embryo (SHE) cell transformation assay. Mutat Res 744:76–81
28. Pienta RJ, Poiley JA, Lebherz WB III (1977) Morphological transformation of early passage golden Syrian hamster embryo cells derived from cryopreserved primary cultures as a reliable in vitro bioassay for identifying diverse carcinogens. Int J Cancer 19:642–655
29. LeBoeuf RA, Kerckaert GA, Aardema MJ, Gibson DP (1990) Multistage neoplastic transformation of Syrian hamster embryo cell cultures at pH 6.70. Cancer Res 50:3722–3729
30. Bohnenberger S, Bruce SW, Kunkelmann T, Pant K, Perschbacher S, Schwind KR, Sly J, Poth A (2012) Photo catalogue for the classification of cell colonies in the Syrian hamster embryo (SHE) cell transformation assay at pH 6.7. Mutat Res 744:82–96
31. Maire M-A, Rast C, Vasseur P (2012) Photo catalogue for the classification of cell colonies in the Syrian hamster embryo (SHE) cell transformation assay at pH 7.0. Mutat Res 744:97–110
32. Ahmadzai AA, Trevisan J, Fullwood NJ, Carmichael PL, Scott AD, Martin FL (2012) The Syrian hamster embryo (SHE) assay (pH 6.7): mechanisms of cell transformation and application of vibrational spectroscopy to objectively score endpoint alterations. Mutagenesis 27:257–266
33. Ahmadzai AA, Trevisan J, Pang W, Patel II, Fullwood NJ, Bruce SW, Pant K, Carmichael PL, Scott AD, Martin FL (2012) Classification of test agent-specific effects in the Syrian hamster embryo assay (pH 6.7) using infrared spectroscopy with computational analysis. Mutagenesis 27:375–382
34. Ahmadzai AA, Trevisan J, Pang W, Riding MJ, Strong RJ, Llabjani V, Pant K, Carmichael PL, Scott AD, Martin FL (2015) Classification of agents using Syrian hamster embryo (SHE) cell transformation assay (CTA) with ATR-FTIR spectroscopy and multivariate analysis. Mutagenesis (in press)

35. Pant K, Bruce SW, Sly JE, Kunkelmann T, Kunz-Bohnenberger S, Poth A, Engelhardt G, Schulz M, Schwind KR (2012) Prevalidation study of the Syrian hamster embryo (SHE) cell transformation assay at pH 6.7 for assessment of carcinogenic potential of chemicals. Mutat Res 744:54–63
36. Maire MA, Pant K, Poth A, Schwind KR, Rast C, Bruce SW, Sly JE, Kunz-Bohnenberger S, Kunkelmann T, Engelhardt G, Schulz M, Vasseur P (2012) Prevalidation study of the Syrian hamster embryo (SHE) cell transformation assay at pH 7.0 for assessment of carcinogenic potential of chemicals. Mutat Res 744:64–75
37. Benigni R, Bossa C, Tcheremenskaia O, Battistelli CL, Giuliani A (2015) The Syrian hamster embryo cells transformation assay identifies efficiently nongenotoxic carcinogens, and can contribute to alternative, integrated testing strategies. Mutat Res 779:35–38
38. Benigni R, Bossa C (2011) Alternative strategies for carcinogenicity assessment: an efficient and simplified approach based on in vitro mutagenicity and cell transformation assays. Mutagenesis 26:455–460
39. Benigni R, Bossa C, Tcheremenskaia O (2013) In vitro cell transformation assays for an integrated, alternative assessment of carcinogenicity: a data-based analysis. Mutagenesis 28:107–116
40. Benigni R, Bossa C, Battistelli CL, Tcheremenskaia O (2013) IARC classes 1 and 2 carcinogens are successfully identified by an alternative strategy that detects DNA-reactivity and cell transformation ability of chemicals. Mutat Res 758:56–61
41. OECD (1997) Guideline for the testing of chemicals, test guideline 471: bacterial reverse mutation test. OECD, Paris. doi:10.1787/9789264071247-en. ISBN 9789264071247
42. OECD (1997) Guideline for the testing of chemicals, test guideline 476: in vitro mammalian cell gene mutation test. OECD, Paris. doi:10.1787/9789264071322-en. ISBN 9789264071322
43. OECD (1997) Guideline for the testing of chemicals, test guideline 473: in vitro mammalian chromosome aberration test. OECD, Paris. doi:10.1787/9789264071261-en, 2074-5788
44. OECD (2010) Guideline for the testing of chemicals, test guideline 487: in vitro mammalian cell micronucleus test. OECD Publishing, Paris. doi:10.1787/9789264091016-en, 2074-5788
45. Kirkland D, Aardema M, Henderson L, Müller L (2005) Evaluation of the ability of a battery of three in vitro genotoxicity tests to discriminate rodent carcinogens and non-carcinogens I. Sensitivity, specificity and relative predictivity. Mutat Res 584:1–256
46. Kirkland D, Aardema M, Müller L, Hayashi M (2006) Evaluation of the ability of a battery of three in vitro genotoxicity tests to discriminate rodent carcinogens and non-carcinogens II. Further analysis of mammalian cell results, relative predictivity and tumour profiles. Mutat Res 608:29–42
47. Vanparys P, Corvi R, Aardema MJ, Gribaldo M, Hayashi M, Hoffmann S, Schechtman L (2012) Application of in vitro cell transformation assays in regulatory toxicology for pharmaceuticals, chemicals, food products and cosmetics. Mutat Res 744:111–116
48. EFSA (2011) Scientific opinion on genotoxicity testing strategies applicable in food and feed safety assessment. EFSA J 9:2379
49. Adler S, Basketter D, Creton S, Pelkonen O, van Benthem J, Zuang V, Andersen KE, Angers-Loustau A, Aptula A, Bal-Price A, Benfenati E, Bernauer U, Bessems J, Bois FY, Boobis A, Brandon E, Bremer S, Broschard T, Casati S, Coecke S, Corvi R, Cronin M, Daston G, Dekant W, Felter S, Grignard E, Gundert-Remy U, Heinonen T, Kimber I, Kleinjans J, Komulainen H, Kreiling R, Kreysa J, Leite SB, Loizou G, Maxwell G, Mazzatorta P, Munn S, Pfuhler S, Phrakonkham P, Piersma A, Poth A, Prieto P, Repetto G, Rogiers V, Schoeters G, Schwarz M, Serafimova R, Tähti H, Testai E, van Delft J, van Loveren H, Vinken M, Worth A, Zaldivar JM (2011) Alternative (non-animal) methods for cosmetic testing: current status and future prospects. Arch Toxicol 85:367–385
50. Pfuhler S, Kirst A, Aardema M, Banduhn N, Goebel C, Araki D, Costabel-Farkas M, Dufour E, Fautz R, Harvey J, Hewitt NJ, Hibatallah J, Carmichael P, Macfarlane M, Reisinger K, Rowland J, Schellauf F, Schepky A, Scheel J (2010) A tiered approach to the use of alternatives to animal testing for the safety assessment of cosmetics: genotoxicity. A COLIPA analysis. Regul Toxicol Pharmacol 57:315–324

Adv Biochem Eng Biotechnol (2017) 157: 97–134
DOI: 10.1007/10_2016_4

Published online: 13 September 2016

Advanced Approaches to Model Xenobiotic Metabolism in Bacterial Genotoxicology In Vitro

Yoshimitsu Oda

Abstract During the past 30 years there has been considerable progress in the development of bacterial test systems for use in genotoxicity testing by the stable introduction of expression vectors (cDNAs) coding for xenobiotic-metabolizing enzymes into bacterial cells. The development not only provides insights into the mechanisms of bioactivation of xenobiotic compounds but also evaluates the roles of enzymes involved in metabolic activation or inactivation in chemical carcinogenesis. This review describes recent advances in bacterial genotoxicity assays and their future prospects, with a focus on the development and application of genetically engineering bacterial cells to incorporate some of the enzymatic activities involved in the bio-activation process of xenobiotics. Various genes have been introduced into bacterial *umu* tester strains encoding enzymes for genotoxic bioactivation, including bacterial nitroreductase and *O*-acetyltransferase, human cytochrome P450 monooxygenases, rat glutathione *S*-transferases, and human *N*-acetyltransferases and sulfotransferases. Their application has provided new tools for genotoxicity assays and for studying the role of biotransformation in chemical carcinogenesis in humans.

Keywords Bacterial recombinant tester strain, Genotoxicity, Metabolic activation, Mutagenicity, SOS response, *umu* test, Xenobiotic metabolism

Contents

Y. Oda (✉)
Institute of Life and Environmental Sciences, Osaka Shin-Ai College, 6-2-28 Tsurumi, Tsurumi-ku 538-0053, Japan
e-mail: oda.1948@hotmail.com

1 Introduction

Most genotoxic carcinogens in the environment are chemically inert; they require metabolic activation by drug metabolizing enzymes to electrophilic species, which covalently bind to proteins, RNA, and DNA. The majority of bacterial cell systems used in genotoxicity tests are inadequate for xenobiotic metabolism because they do not possess appropriate metabolization competence. This led to the need to introduce mammalian-derived metabolic systems to mimic mammalian metabolism, which enhanced the sensitivity of these assays to detect genotoxic carcinogens. By 1938, Fieser and colleagues [1] had postulated that metabolism of carcinogens plays a major role in the etiology of cancer. A few years later an important discovery in the understanding of chemical carcinogenesis came from the investigation of Miller and Miller [2] who found that many carcinogens are not intrinsically carcinogenic, but require metabolic activation to be carcinogenic.

In early bacterial genotoxicity assays, the cells used did not possess endogenous xenobiotic metabolism capability. It was therefore necessary to incorporate an exogenous metabolic activation system to mimic mammalian xenobiotic metabolism.

In the metabolic activation systems using *Salmonella typhimurium* (*S. typhimurium*) as test species, Malling first showed that a mouse liver homogenate could activate *N*-nitrosodimethylamine to a genotoxic metabolite [3]. Since then, incorporation of the capability for metabolism led to further success of the bacterial Ames test [4, 5] for mutagenicity and testing for carcinogenic potential. This new approach then allowed the wide use of the results for screening and for the identification and quantification of potential genotoxins for regulation. Subsequently, many investigators have used various types of S9 fraction ($9{,}000 \times g$ supernatant preparation from rodent liver homogenate) and microsomal preparations for metabolic activation. Although these exogenous metabolic activation systems significantly contributed to the detection of genotoxic chemicals, these systems had some limitations. Because the test compounds are metabolized by S9 in the medium but not inside the cells, only a portion of the highly reactive and short-lived intermediates penetrate through cell membranes to reach the target macromolecules. In recent studies of genotoxicity assay systems, excellent results have been brought about by approaches based on improved understanding of the molecular basis of mutagenesis and advances in bacterial gene technology. Bacteria are especially suitable for the development of metabolic competent tester systems because of the introduction of mammalian genes by an appropriate transduction method. This review describes recent advances in bacterial genotoxicity assays and their future prospects, with a focus on the development and application of genetic engineering in bacterial cells to incorporate some of the enzymatic activities involved in the activation process of xenobiotics.

2 Bacterial Genotoxicity Assays

From the early 1970s, the *Salmonella*/microsomal mutagenicity assay, called the Ames test, was used routinely to detect a wide range of chemical compounds that can produce genetic damage leading to gene mutations. This test observes the reverse mutation from histidine auxotrophy to prototrophy in several *Salmonella* mutant strains, each carrying different mutations in various genes of the histidine operon. These mutations act as hot spots for mutagens that cause DNA damage via different mechanisms. The carcinogenicity of chemicals can be evaluated by bioactivating the compound with mammalian liver extracts before exposing it to the bacterial cells [4]. The most advance in the Ames test was the development of a unique tester strain, bearing the so-called plasmid pKM101 (a derivative of a naturally-occurring bacterial drug-resistance factor). This plasmid encodes *mucAB* gene (bearing a *umuDC* homologous gene), which is involved in error-prone DNA repair and, in some cases, is absolutely necessary for mutagenesis. The presence of pKM101 enhances the mutagenicities of many mutagens [5]. So far, many studies have been performed to estimate the sensitivity and correlation of the Ames test with animal carcinogenicity studies. In the early years of mutagenicity testing with the Ames test it was estimated that the correlation between *Salmonella* mutagenicity and carcinogenicity was from 90% [5] to 77% [6]. However, several validations showed that the test fails to detect a few classes of carcinogens such as polychlorinated pesticides [7–9]. As another reversion test, one of the most extensively used is the WP2 test based on the *Escherichia coli* (*E. coli*) WP2 strain [10]. This test is based on the reversion of tryptophan auxotrophy in *E. coli*. A comparison between the WP2 test and the Ames test showed that both tests had comparable sensitivities to detect carcinogens [11].

The bacterial SOS response is a global response to DNA damage in which the cell cycle is arrested and DNA repair and mutagenesis are induced [12]. Two important genes involved in the SOS response are the *lexA* gene and the *recA* gene. LexA encodes a repressor for all the SOS genes and *recA* gene encodes a protein to promote the inactivation of the LexA repressor, thereby inducing the responses. Since 1980, several representative assays have been developed to monitor the SOS response through the activation of the RecA protein, the cleavage of LexA or λ cI repressor, and expression of any of the SOS genes. The Biochemical Induction Assay [13] using *E. coli* BR513 is based on a tester strain containing a derivative of the λ phage. The *lacZ* gene is under the control of the P_L promotor of the λ phage. Cleavage of the λ CI repressor results in the production of β-galactosidase. This test is a simple and rapid (less than 5 h) assay and has been used to screen for antitumor agents.

Another SOS assay, the SOS chromotest, uses a *uvrA, rfa* strain of *E. coli* PQ37. In the bacterial genome, the test induces synthesis of the reporter enzyme β-galactosidase (LacZ) from a translational fusion between *sfiA* gene (the SOS cell division inhibitor gene) and *lacZ* gene in response to genotoxins (i.e., on triggering the SOS induction) [14]. In addition, the constitutive synthesis of alkaline phosphatase enables the measurement of the toxicity of tested chemicals. The test takes only 6 h. A paper has reviewed the testing of 751chemicals [15]. For 452 chemicals, the results obtained in the SOS chromotest were compared to those obtained in the Ames test. A total of 373 (82%) of these chemicals gave similar responses in both tests. For 65 confirmed carcinogens, the sensitivity (the capacity to identify carcinogens) was 62% with SOS chromotest and 77% with the Ames test. For 44 suspected carcinogens, the sensitivity was 66% with the SOS chromotest and 68% with the Ames test.

We first proposed the *umu* test, which is based on tester strain *S. typhimurium* TA1535/pSK1002 harboring a multicopy plasmid pSK1002 with an *umuC::lacZ* gene fusion [16]. The cell toxicity is evaluated by monitoring the optical density of the culture at 600 nm. The strain used has a deletion of the normal *lac* region, so that β-galactosidase activity is dependent on *umu* gene expression (Fig. 1). The strain also contains *rfa* and *uvrB* mutations that enhance cell permeability and eliminate the DNA excision-repair activity. The product of *umuC* gene is involved in the process of mutagenesis in *E. coli* [12, 17, 18], suggesting that this product named

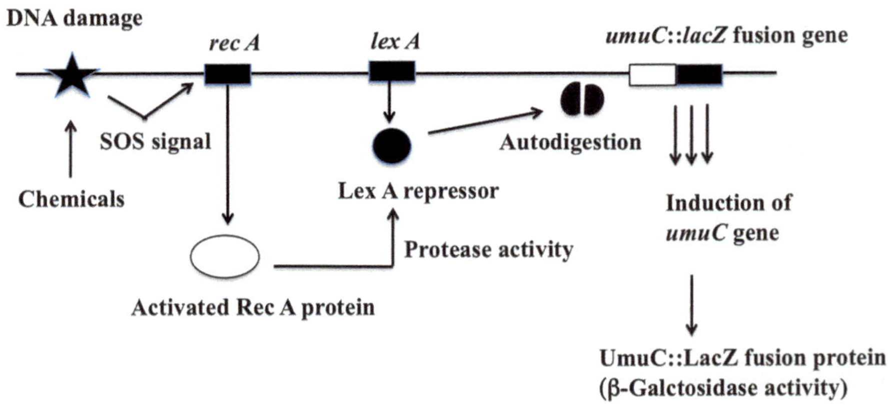

Fig. 1 Principle of the *umu* test

DNA polymerase V catalyzes error-prone translesion synthesis [19, 20]. We have published extensive validation studies of the *umu* test [16]. Using 38 chemicals with different structures and mode of action, including 31 known animal carcinogens, evaluation of the system was performed. The threshold sensitivity of the *umu* test was approximately equal to that of the Ames test for genotoxic chemicals in both tests. It was found that the *umu* test using the single tester strain could detect many types of DNA-damaging agents. Further, we evaluated the abilities of 151 chemicals to induce *umu* gene expression in *S. typhimurium* TA1535/pSK1002 [21]. Some of the chemicals tested, such as dimethyl sulfoxide, *m*-dioxan, 5-fluorouracil, and paraquat, which have been reported to be non-mutagenic in the Ames test, were found to be positive in the *umu* test. McDaniels and colleagues [22] determined the sensitivity and reproducibility of each of the Ames test, *umu* test, and SOS chromotest methods. The Ames test with strains TA98 and TA100 was ranked as the most sensitive method more often than the others, but the results indicated that the *umu* test was statistically equivalent to the Ames test. The *umu*-microplate test was the most suitable for screening large numbers of environmental samples. Reifferscheid and Heil [23] showed a comprehensive update of all *umu* test results published by 1996. The available data of 486 chemicals tested with the *umu* test were compared with the Ames test (274 compounds) as well as rodent carcinogenicity data (179 compounds). The concordance between the *umu* test and Ames test results was about 90%. The *umu* test detected 86% of the Ames test positive mutagens. The agreement between carcinogenesis and *umu* response was 65%. This result agrees with earlier comparative studies, which use results of the rodent carcinogenicity test and the Ames test showing a concordance of about 62% [24] and 58% [25]. In addition, Yasunaga et al. [26] have investigated the DNA-damaging effects of 83 National Toxicology Program (NTP) chemicals, including noncarcinogens and carcinogens, using the *umu* test. The overall concordance between genotoxicity in the *umu* test and carcinogenicity was 67%, which was similar to the concordance between Ames test results and carcinogenicity (63%) using the same 83 NTP chemicals. In some recent studies, new electrochemical genotoxicity assays, which enable the analysis of turbid samples, have been developed [27–29]. They are based on the *umu* test using a rotating disk electrode in a microtiter droplet. The results indicated that the signal detection in the genotoxicity assay based on hydrodynamic voltammetry was less influenced by the presence of colored components and sediment particles in the samples when compared to the usual colorimetric signal detection.

Brinkmann and Eisentraeger [30] showed that the automated *umu* test is highly applicable for the assessment of non-volatile samples with strong or moderate genotoxic effects using a RoboSeqR 4204 SE pipetting station. We have recently developed a new *umu* test kit called *Umulac AT*® using *S. typhimurium* NM2009 strain. This kit has been successfully used to screen for the presence of genotoxic substances in a broad range of materials and environments such as new drugs, foods, cosmetic products, and the working environment.

In 1997, the SOS *lux* test was developed using *Photobacterium leiognathi luxCDABE* as a reporter. This test responded sensitively to various genotoxins

Table 1 Bacterial genotoxicity assays

Test method	Bacterium	SOS gene	Index	References
Ames test	*S. typhimurium*	*mucAB*	his	[5]
WP2 test	*E. coli*	*mucAB*	trp	[10]
Induct test	*E. coli*	Prophage	λ phage	[13]
SOS chromotest	*E. coli*	*sfiA::lacZ*	β-gal	[14]
umu test	*S. typhimurium*	*umuC::lacZ*	β-gal	[16]
SOS *lux* test	*E. coli*	*luxCPABFE*	lux	[31]
VITOTOX® test	*S. typhimurium*	*recN2-4::lux*	lux	[32]
Geno-Tox test	*S. typhimurium*	*cda::lux*	lux	[34]

S. typhimurium Salmonella typhimurium, *E. coli Escherichia coli*, *his* histidine, *trp* tryptophane, *β-gal* β-galactosidase, *lux* luciferase

such as MMC, MNNG, dimethylsulfate, H_2O_2, and CH_2O and with UV and γ radiation [31]. Similarly, Verschaeve et al. [32] developed the Vitotox test using *S. typhimurium* TA104pr 1 with a *recN::luxCDABE* construct fusion. This assay is based on luminometric detection and correlated well with either the Ames test or the SOS chromotest. The *cda* GenoTox assay was recently developed by Norman and colleagues [33, 34] who used *S. typhimurium* TGO1 harboring the fusion plasmid pANO1::*cda*. This assay showed a high sensitivity to *N*-methyl-*N′*-nitro-*N*-nitrosoguanidine for a low background activity of the *cda* promoter [34]. Table 1 summarizes the bacterial genotoxicity assays developed to date using tester strains of *E. coli* or *S. typhimurium.*

3 Applications of the *umu* Test for the Detection of Genotoxins in Environmental Samples

Because environmental genotoxins are usually present in very low concentrations, the development of small-scale, rapid, and sensitive assay systems is required for detection of these environmental toxins. We developed a high-throughput *umu*-microplate test system using the standard *umu* test strain to detect the genotoxicity of complex mixtures such as water, soil, and wastewater [35–37]. The test was also standardized according to the German Institute of Standardization (DIN 38415-3: 1995), Japan (1993), Malaysia (MS ISO 13829: 2008), and the International Standardization Organization (ISO 13829: 2000) in the Technical Committee 147 Water Quality. Table 2 summarizes some reported detection of genotoxicants in many environmental samples exhibited by the *umu* test described in this review. The results are presented including sample source, country, type of sample, genotoxic responses, and reference. The sample sources are described as drinking, tap, and source waters, industrial, hospital, municipal, native, and paper mills wastewaters, petroleum hydrocarbon-contaminated, surface soils, complex mixtures, volatile chemicals, air pollutants, etc. Because a detailed discussion of all the test results

Table 2 Some representative examples of the application of the *umu* test system in environmental samples

Sample source	Country	Type of sample	Genotoxic responses	References
Bleached kraft mill effluents	Canada	Extracts	Direct acting genotoxicity	[38]
Hospital, municipal and various industrial wastewaters	Switzerland and Germany	Native wastewaters	13% positive results	[36]
Native water samples	Germany	Native wastewaters	Positive results in large numbers of samples	[39]
Raw tannery wastewaters	Sweden, Spain, and Germany	Domestic wastewaters and industrial effluents	High genotoxicity	[40]
Water samples from the selected rivers and the primary and secondary effluents of some sewage treatment plant	Germany	Water, primary and secondary effluents	Only samples of primary effluents caused genotoxicity	[41]
Native hospital wastewaters from five German clinics	Germany	Extracts	Detection of genotoxic fluoroquinolone antibiotics in hospital wastewaters	[42, 43]
Wastewater samples from paper mills	Germany	Wastewaters	No genotoxicity	[44]
Water and soil samples	Germany	Water and soil	Industrial sewage water was positive without metabolic activation	[30]
Native samples of wastewaters, surface waters, and portable waters	Germany	Wastewaters, surface and portable waters	33% of wastewaters, 25% of surface waters, and 12.5% of portable waters were genotoxic	[45]
Petroleum hydrocarbon-contaminated soil	Germany	Soil	The *umu* test was more sensitive than the SOS chromotest in soil	[46]
Industrial sludge containing a complex mixture of nitroaromatic compounds	Germany	Industrial sludge	The material from the non-aerated system increased genotoxicity in the acetone-soluble fraction after treatment	[47]

(continued)

Table 2 (continued)

Sample source	Country	Type of sample	Genotoxic responses	References
Organic micropollutants from water recycling	Germany	Drinking water	The toxic equivalent concentrations concept was applied to the *umu* test	[48]
Tap water samples	China	Tap water	The boiled water displayed strong genotoxic potential compared to its original tap water	[49]
Oilfield produced waters	China	Wastewater	Direct and indirect genotoxic substances were observed	[50]
Wastewater samples from industrial effluents	China	Wastewater	The polycyclic aromatic hydrocarbons were minor contributors to the genotoxicity in the effluents	[51]
Sixteen centralized source waters	China	Source waters	Eight samples showed both indirect and direct genotoxic effects. Another four samples showed indirect effects. *o*-Phenyl was identified as a genotoxin	[52]
A chlorinated secondary effluent of municipal wastewaters	China	Municipal wastewaters	The presence of bromide decreased the genotoxicity in the secondary effluent during chlorination	[53–55]
Municipal secondary effluent	China	Wastewaters	Reverse osmosis, filtration, and ozonation removed the genotoxic effect of secondary effluent, whereas chlorination elevated the genotoxicity	[56]
Chlorination disinfection of cephalosporins	China	Chlorinated products	The genotoxicity of cephalosporin was enhanced after chlorination	[57]
Some pure chemicals and water samples	China	Pure chemicals and water	BugBuster Master Mix increased the detection sensitivities of the selected	[58]

(continued)

Table 2 (continued)

Sample source	Country	Type of sample	Genotoxic responses	References
			genotoxins and environmental water samples	
Surface soil samples	China	Soil	The amounts of soil weight required for the extracts to lead to positive results	[59, 60]
Soils irrigated with wastewater	China	Soil	Significant increases in genotoxic effects in soils irrigated with wastewater	[61]
Surface sediment samples	China	Soil extracts	Soil extracts were positive results	[62]
Complex mixtures, volatile chemicals and air pollutants	USA	Complex mixture	The extracts of all complexes showed positive results	[63, 64]
Antineoplastic drugs and workbench wipe samples	Japan	Work environmental contamination	Of 19 drugs, 8 drugs induced genotoxicity	[65]
Treated effluent samples	Australia	Wastewaters	A genotoxic response was observed in half of the samples without metabolic activation and 75% of samples with metabolic activation	[66]

shown in Table 2 is beyond the scope of this book, the description on the following pages is restricted to these studies that reported remarkable genotoxicity.

Plaza et al. [46] evaluated the genotoxicity of petroleum hydrocarbon-contaminated soil following bioremediation treatment using SOS chromotest and the *umu* test with and without metabolic activation (S9). The *umu* test was more sensitive than the SOS chromotest in the analysis of this soil. The results suggested that the combined test systems used in this study were useful tools for the genotoxic examination of remediated petroleum hydrocarbon-contaminated soil.

Macova et al. [48] evaluated the applicability of a bioanalytical test battery for monitoring micropollutants across all seven barriers of an indirect potable reuse scheme. The toxic equivalent concentrations (TEQ) concept was applied for the first time to the *umu* test, indicative of genotoxicity using 4-nitroquinoline as the reference compound for direct genotoxicity and benzo[*a*]pyrene for genotoxicity after metabolic activation. The results indicated that bioanalytical results expressed as TEQ are useful to assess removal efficiency of micropollutants throughout all treatment steps of water recycling.

Fang et al. [51] reported the genotoxicity of various effluents from textile and dyeing plants, electronic and electroplate factories, pulp and paper mills, fine chemical factories, and municipal wastewater treatment plants in the Pearl River Delta region by using *umu* test combined with chemical analysis. The genotoxic effects expressed as benzo[*a*]pyrene equivalent concentrations varied between below detection limit and 88.2 μg/L, with a mean of 8.76 μg/L in all effluents. The polycyclic aromatic hydrocarbons were minor contributors to the genotoxicity in the effluents, and some unidentified compounds in the effluents were responsible for the measured genotoxicity. The authors concluded that, in terms of genotoxicity, discharge of these effluents could pose high risks to aquatic organisms in the receiving environments.

Hu et al. [53–55] examined the genotoxicity in a chlorinated secondary effluent of a municipal wastewater treatment plant in China. In the study, the effect of bromide on genotoxicity during chlorination was evaluated by the *umu* test. The presence of bromide notably decreased the genotoxicity in the secondary effluent during chlorination, especially under conditions of high ammonia concentrations. By fractionating dissolved organic matter in the secondary effluent into different fractions, the fractions containing hydrophilic substances and hydrophobic acids contributed to the decrease in genotoxicity induced by bromide. Cao and colleagues [56] investigated the effectiveness of several technologies, that is, combination of coagulation and sand filtration, ultraviolet irradiation, chlorination, ozonation, ultrafiltration, and reverse osmosis filtration (RO) on the removal of genotoxicity against activity from the municipal secondary effluent. The secondary effluent exhibited a genotoxic effect on *S. typhimurium* strain TA1535/pSK1002. RO and ozonation demonstrated remarkable removals of the genotoxic effect, whereas chlorination could elevate genotoxicity.

Soil screening could be a process of identifying and defining areas, contaminants, and condition at the sites warranting further attention for developing ecological risk assessments. Xiao et al. [59, 60] sampled a total of 41 surface soil samples from Tianjin, China and the soil organic extracts were evaluated using a battery of in vitro cell bioassays. The results have shown that the amounts of soil weight required for the extracts to lead to a positive result (induction ratios 2.0) in the *umu* test were between 3.9 and 31.3 mg (dry weight) per well. In addition, the genotoxic effects in Tianjin area exhibited a strong positive correlation with each other. It has been concluded that the toxicity assessment of surface soil using a battery of in vitro cell bioassays could provide meaningful information for ecological risk assessment. The study also evaluated the genotoxicity of field soils in the Tianjin area, one of the most industrialized and contaminated areas in northeast China. The genotoxicity of organic extracts of soils was assayed using the *umu* test. The results obtained demonstrated that the genotoxicity expressed as induction ratios ranged from 1.00 to 4.60.

Allinson et al. [66] reported that samples of treated effluent collected at the points of discharge to the environment from 39 wastewater treatment plants (WWTPs) located across Victoria in Australia had their genotoxicity assessed with a high-throughput luminescent *umu* test method using *S. typhimurium*

TL210 strain, with and without addition of a commercially available metabolic activation system. A genotoxic response was observed in half the samples tested without the metabolic activation system. On addition of the metabolic activation system, 75% of samples elicited a genotoxic response, the majority of responses being stronger than without metabolic activation. However, the type of WWTP had no effect on genotoxicity. This study showed that the high speed luminescent *umu* test worked well as an initial investigative tool.

4 Bioactivation of Genotoxic Chemicals

When a xenobiotic chemical enters the bloodstream, the body attempts to remove it using different metabolic processes. All chemicals entering the bloodstream from the gastrointestinal tract first pass through the liver, the site of most xenobiotic metabolism. The liver expresses a wide variety of enzymes that chemically alter blood-borne toxins and chemicals. Active transport or carrier proteins import drugs and other chemicals into and export metabolites out of liver cells. In contrast, numerous genotoxic chemical classes including polycyclic aromatic hydrocarbons, aromatic amines, aflatoxin, and nitrosamines are bioactivated by mainly cytochrome P450 monooxygenase (P450) to genotoxic metabolites capable of covalent binding to DNA and proteins. The enzymes responsible for bioactivation also participate in drug metabolism and biotransformation.

The P450 enzymes involved in phase I drug metabolism first modify these chemicals with functional groups by oxidation, reduction, and hydrolysis. Furthermore, the phase I intermediates are metabolized by glutathione *S*-transferases, acetyltransferases, and sulfotransferases involved in phase II drug metabolism. Ultimately, this process generates larger and more polar metabolites than the original xenobiotic compound, making them easier to excrete.

5 Application of Mammalian Liver Microsome Enzymes in the *umu* Test

The following examples show that promutagens and procarcinogens can be detected by the *umu* test after their metabolic conversion to reactive electrophiles by a liver 9,000 g supernatant fraction (S9) [16]. To compare the abilities of a wide variety of environmental chemicals to induce *umu* gene expression, however, more detailed experiments using kinetic analysis and the possible role of activating enzymes such as multiple forms of cytochrome P450 (P450) were required. In the late 1980s, Shimada and colleagues began studies on the roles of P450 in the activation of carcinogens with the use of rat and human liver microsomes and the *umu* test [67–72].

Shimada and Nakamura [67] first developed a simple and sensitive procedure for the determination of P450-mediated activation of chemical procarcinogens to DNA-damaging products using *S. typhimurium* TA1535/pSK1002 strain. They reported that rat liver microsomes or a reconstituted monooxygenase system containing three forms of purified P450 catalyzed the activation of compounds including heterocyclic amines, aromatic amines, polycyclic aromatic hydrocarbons, and aflatoxin B_1 (AFB_1) to genotoxic metabolites. Data were also presented showing that a high spin form of P450 isolated from 3-methylcholanthrene-treated rats (later identified as P450 1A1) has a profound role for most of the chemicals examined. Shimada et al. [68, 69] examined rat liver and human liver cytochrome P450-mediated activation of AFB_1 to genotoxic products, which subsequently cause induction of *umu* gene expression in *S. typhimurium* TA1535/pSK1002. The results indicated that the constitutive forms of P450 have very important roles for the genotoxicity and mutagenic activation of AFB_1. Further evidence from Guengerich's laboratory using a bioactivation assay with *S. typhimurium* TA1535/pSK1002 indicated that the human P-450 IIIA4 (P450 NF, later termed P450 3A4) family can activate many important procarcinogens such as AFB_1 and AFG_1, sterigmatocystin, 6-aminochrysene, 7,8-dihydroxy-7,8-dihydrobenzo[*a*]pyrene (both + and − diastereomers), 3,4-dihydroxy-3,4-dihydrobenz[*a*]anthracene, 3,4-dihydroxy-3,4-dihydro-7,12-dimethylbenz[*a*]anthracene, 9,10-dihydroxy-9,10-dihydrobenzo[*b*]fluoranthene, and tris(2,3-dibromopropyl)phosphate [70, 71, 73]. The ability of cigarette smoke condensate to induce a genotoxic response in the *umu* test was examined in liver microsomal and reconstituted monooxygenase systems containing rat and human P450 enzymes. The results showed that cigarette smoke condensate contains both inhibitors of P450 enzymes and procarcinogens capable of being activated by P450 enzymes and that P450 1A2 appears to be the most important catalyst for the activation reaction of a copper phthalocyanine cellulose extract of cigarette condensate [74]. Yamazaki et al. examined the role of individual rat and human P450 enzymes in the bioactivation of the potent hepatocarcinogen 3-methoxy-4-aminoazobenzene (3-MeO-AAB) in an *S. typhimurium* TA1535/pSK1002 system. The results showed that multiple P450 enzymes in rat and human liver microsomes are involved in the bioactivation of 3-MeO-AAB, regardless of its selective induction of the rat P4501A2 gene [75].

Yamazaki and colleagues [76] investigated the genotoxicity of four samples of diesel exhaust particle (DEP) extracts (DEPE) and nine nitroarenes found in DEPE using the *umu* test after activation catalyzed by human cytochrome P450 family 1 enzymes co-expressed with NADPH-cytochrome P450 reductase in *E. coli* membranes. Apparent genotoxic activities of DEPE were very low compared with standard nitroarenes in the presence of P450s, possibly because unknown component(s) of DEPE had inhibitory effects on the bioactivation of 1-NP and 1,8-DNP catalyzed by human P450 1B1. These results suggested that environmental chemicals existing in airborne DEP, in addition to 1-NP, 1,6-DNP, 1,8-DNP, 2-NF, and 3-NF, can be activated by human P450 1B1. They further examined the genotoxic potential of benzophenone and its metabolically related compounds, benzhydrol and *p*-benzoylphenol, using human P450 enzymes. Benzophenone and its two metabolites showed a suppression of bacterial growth without any P450 system, but no induction of *umu* gene expression was observed in *S. typhimurium* TA1535/pSK1002. Human liver microsomes induced bacterial cytotoxicity of these compounds without any *umu* gene expression. On the other hand, with the addition of *E. coli* membranes expressing recombinant human P450 2A6 and NADPH-cytochrome P450 reductase (NPR), benzophenone showed *umu* gene expression. Activation of benzhydrol and *p*-benzoylphenol by the P450/NPR system was similar to that of benzophenone. These results suggested that benzophenone and its metabolically related benzhydrol and *p*-benzoylphenol can be bioactivated by P450 2A6 and P450 family 1 enzymes [77].

By adding purified and reconstituted mammalian enzymes, it was possible to screen for their abilities to detoxify direct-acting mutagens. In this case, metabolic deactivation of furylfuramide by human and rat liver microsomal P450 enzymes was examined. Both human and rat liver microsomes catalyzed the metabolism of furylfuramide to inactive forms which are incapable of inducing *umu* gene expression. These results suggested that P450 1A1 and P450 1A2 in rats, and P450 1A2 in humans, are the major enzymes involved in the deactivation of furylfuramide in liver microsomes and also that furylfuramide can be degraded very rapidly through the aerobic metabolism by liver microsomes [78]. Additionally, human and rat liver microsomes deactivated 1,3-, 1,6, and 1,8-dinitropyrene, which are major components present in diesel exhaust. With human liver microsomes the activity of 1,3-dinitropyrene was most strongly inhibited, whereas with rat liver microsomes the genotoxicity of all three dinitropyrenes was inhibited to a similar extent [79]. These results suggested that P450 enzymes are involved in the detoxification of different dinitropyrene congeners. Ueng et al. [80] examined the protective effects of baicalein and wogonin (the main active flavonoids of *Scutellariae radix*, which is one of the main constituents of a Kampo medicine) against benzo[*a*]pyrene- and aflatoxin B_1-induced genotoxicity using the *umu* test. The results showed that baicalein and wogonin decreased the genotoxicity of benzo[*a*]pyrene and aflatoxin B_1 as monitored by *umu* expression in *S. typhimurium* TA1535/pSK1002.

6 Genetically-Engineered *umu* Tester Strains Over-Expressing Bacterial Enzymes

E. coli and S. typhimurium used for genotoxicity assays have little capacity for bioactivation of chemicals. Therefore the assays require the use of exogenous mammalian enzyme systems such as S9 mix. However, in the case of certain classes of chemicals, bacterial enzymes are deeply involved in the genotoxic activation.

Nitroarenes were detected in extracts of automobile emissions, fly ash particles, cigarette smoke condensates, and the urban atmosphere [81, 82]. Some of these

Table 3 Establishment of bacterial nitroreductase- / or *O*-acetyltransferase-overexpressing strains and the enzymes-deficient strains

S. typhimurium	Character	Detection	References
NM1011	Nitroreductase-overexpressing strain	Nitroarenes	[88–90]
NM2009	*O*-Acetyltransferase-overexpressing strain	Aromatic amines	[87, 89, 90]
NM3009	Nitroreductase and *O*-acetyltransferase-overexpressing strain	Nitroarenes and aromatic amines	[37, 88]
NM1000	Nitroreductase-deficient strain		[86–88]
NM2000	*O*-Acetyltransferase-deficient strain		[88–90]

chemicals are potent mutagens and environmental carcinogens. They do not require metabolic activation by liver microsomal enzymes but are activated to mutagens by reduction to arylhydroxylamine intermediates by bacterial nitroreductase (NR). Their arylhydroxylamine derivatives are themselves DNA-reactive, but further activated by *O*-acetyltransferase (*O*-AT) to convert hydroxylamine metabolites into more reactive species: *N*-acetoxyesters, N–O bond heterolysis of *N*-acetoxyesters presumably generates nitrenium ions [83], electrophilic intermediates which react with DNA. The O-acetylation of arylhydroxylamines is catalyzed by acetyl CoA: aromatic amine *O*-AT. However, little was known about the genotoxic effects of these chemicals in systems based on SOS responses in bacteria. Watanabe et al. [84] developed *S. typhimurium* strains (YG1021 and YG1026) with an elevated level of nitroreductase activity by cloning the corresponding gene into plasmid pBR322. The resulting strains were extremely sensitive to several carcinogenic nitroarenes [85].

To construct further highly sensitive bacterial strains for nitroarenes, we subcloned the NR gene or both NR and *O*-AT genes into plasmid pACYC184, and subsequently introduced the plasmids carrying these genes into TA1535 having a fusion gene *umu::lacZ* (pSK1002). The new strains NM1011, NM2009, and NM3009 overexpressing NR and/or /*O*-AT have been developed to increase the sensitivity against specific genotoxins such as nitroarenes and aromatic amines [86–89] (Table 3). Among six strains tested, NM3009 showed the highest sensitivity toward such chemicals as 1-nitronaphthalene, 2-nitrofluorene, 3,7-dinitrofluoranthene, 3-nitrofluoranthene, 5-nitroacenaphthene, 2-nitronaphthalene, 1-nitropyrene, 1,6-dinitropyrene, 3,9-dinitrofluoranthene, 4,4′-dinitrobiphenyl, 1,8-dinitropyrene, *m*-dinitrobenzene, 2,4-dinitrotoluene, and 1,3-dinitropyrene. The combined use of the enzyme-overexpressing and enzyme-deficient strains provides excellent tools in evaluating the possible role of enzymes in metabolic activation and inactivation of the various mutagenic nitroarene compounds and identifying of the mutagenic principles in complex mixtures. The following studies have been applied to environmental samples using these tester strains.

Ozturk and Durusoy [91] compared the SOS chromotest and the *umu* test system with NM2009 and NM3009 strains to evaluate three weakly genotoxic monocyclic nitroarenes (*m*-nitrocinnamic acid, *m*-dinitrobenzene, 2,4-dinitrotoluene) and the well-known genotoxic nitro compound 4-NQO. The results found that the *umu* test strains NM2009 and NM3009 may be somewhat more useful in genotoxicity tests with nitroarenes than the SOS chromotest strain *E. coli* PQ37. They further applied these *umu* test systems for screening genotoxicity of surface water in Turkey and examined the SOS-inducing activity of the organic extracts in the Meric delta, which is highly polluted by industrial and agricultural wastes. The results showed that the Meric delta was highly polluted, especially in spring and summer. However, it was also confirmed that the pollutants were not derived from nitroarenes but from other organics [92]. We detected genotoxic activity in atmospheric particles in urban areas using a relatively small sample load with *S. typhimurium* NM2009, NM3009, and parental strains. The results indicated that the test system could detect slight increases in induced genotoxicity in atmospheric particles and that genotoxicity was detected mainly in the fine fraction but also partially in the coarse fraction. The pattern of the response suggested that the genotoxic activity of the particulate extract was caused primarily by nitrated polycyclic aromatic hydrocarbons. These results indicated that the present microplate test may be a useful way of carrying out rapid screening for genotoxicity in small-volume environmental samples [37, 93]. Ma et al. [94] used these strains to characterize potential genotoxins in river and adjacent ground waters in the Jialu River basin, China. The major genotoxic activities of river water and adjacent ground water occurred in the same two fractions when assayed using the strain TA1535/pSK1002, although the genotoxicity in the river was stronger than the ground water. LC-MS/MS analysis identified that flumequine was one of the causal agents producing the genotoxicity. The specific response to the strain NM3009 compared with the strain TA1535/pSK1002 demonstrated the presence of nitroarenes in the river sample, although the exact chemicals could not be identified by analyzing the nitroarenes commonly detected in the environment.

In the following, some representative examples using *S. typhimurium* NM2009 are shown. Ono et al. [95, 96] carried out the experimental investigation of effluents from municipal wastewater treatment plants and nightsoil treatment plants using *S. typhimurium* NM strains in the *umu* test to evaluate the strategies and regulations for wastewater reuse. The toxicity of aromatic amines could be detected in the matter contained in raw nightsoil. They suggested that human feces contain some genotoxic substances and the genotoxicity could not be reduced through biological treatment with nitrification and denitrification nor removed by the ultrafilter separation process. However, genotoxicity could be reduced to a negative level by ozonation. Ohe and Nukaya [97] separated a diethyl ether extract recovered from Yodo river water in Japan by the XAD-2 resin column method into neutral, acid, and basic fractions, and the neutral fraction accounted for 52% of the genotoxicity of the extract in the absence of the metabolic activation system for strain NM2009. The genotoxicity of the benzene fraction accounted for 80% of the neutral fraction. They showed that mutagenic nitroarenes might be contained in the benzene fraction of the neutral one and that the concentration of 1-nitropyrene in municipal river water was 1 ng/L, accounting for only 1% of the total genotoxicity. Ohe [98] quantified four mutagenic/carcinogenic heterocyclic amines (HCAs) (2-amino-

3,8-dimethylimidazo[4,5-*f*]quinoxaline, 3-amino-1,4-dimethyl-5*H*-pyrido[4,3-*b*]indole, 3-amino-1-methyl-5*H*-pyrido[4,3-*b*]indole, and 2-amino-1-methyl-6-phenylimidazo[4,5-*b*]pyridine) in organic extracts obtained by blue rayon hanging method from the Yodo River water, Japan. The total amounts of the four HCAs accounted for a mean of 24% of the genotoxicity of blue rayon extracts evaluated by the *umu* test using a strain NM2009.

In 1992, we demonstrated that tester strain *S. typhimurium* NM2009 is highly sensitive toward carcinogenic aromatic amines, aminoazo compounds, and heterocyclic aromatic amines, when compared with the parent strain TA1535/pSK1002 and the *O*-AT-defective strain NM2000 [89, 90, 99]. Unexpectedly, expression of bacterial *O*-acetyltransferase also made the *umu* test system more sensitive to oxidation products of nitrosamines [100, 101], but another group has reported that they were unable to confirm the phenomenon in another system [102].

Yamazaki et al. [103–105] also examined the roles of rat and human liver P450 enzymes and the mechanism of activation of 6-aminochrysene (6-AC) and its 1,2-dihydro-1,2-dihydroxy-6-aminochrysene (6-AC-diol) by rat and human P450 enzymes in the *S. typhimurium* strains TA1535/pSK1002 and NM2009. 6-AC may be *N*-hydroxylated principally by P450 2B enzymes in rats and P450 3A4 and 2B6 in humans and activation to its ultimate metabolites may proceed through esterification of the *N*-hydroxy metabolites by an *N*-acetyltransferase. The authors suggested that the 6-AC-diol is metabolized to its ultimate diolepoxide product by P450 1A enzymes in rat and human liver microsomes and P450 3A4 (human) and P450 3A2 (rats) may also contribute to some extent in the activation of it. They further carried out comparisons of the activation of procarcinogens such as heterocyclic arylamines and aminoazodyes, polycyclic aromatic hydrocarbons, and mycotoxins in reconstituted monooxygenase systems containing modified P450 3A4 and 3A5 enzymes expressed in *E. coli* using the *umu* test. It was found that P450 3A4 had similar activities to or higher rates than P450 3A5 for these procarcinogens [106].

Shimada et al. [107] examined the catalytic properties of human P450 1B1 for carcinogen activation using recombinant P450 1B1 in yeast microsomes. The results suggested that P450 1B1 is involved in the activation of a variety of procarcinogenic chemicals to DNA-damaging products in the *umu* test using *S. typhimurium* NM2009 [107, 108]. Procarcinogens identified as being activated by P450 1B1 include polycyclic aromatic hydrocarbons (PAH) and arylamines. They also compared recombinant P450 1A1 and P450 1B1 enzymes in *E. coli* (in which the NADPH-P450 reductase was co-expressed) with regard to their abilities to activate 19 procarcinogens to genotoxic metabolites in the strain NM2009. The results indicated that P450 1A1 and P450 1B1 have relatively similar activities for procarcinogen activation, except for the activities when benzo[*c*]phenanthrene-3,4-diol, chrysene-1,2-diol, 7,12-dimethylbenz[*a*]anthracene, MeIQ, MeIQx, 3-methoxy-4-aminoazobenzene, and Trp-P-1 were used as substrates [109]. They further compared activities of metabolic activation of a number of PAHs and PAH dihydrodiols and other procarcinogens by recombinant human P450 enzymes using a genotoxicity assay based on *S. typhimurium* NM2009. The results supported the importance of P450 1A1 and P450 1B1 in the activation of PAHs and PAH dihydrodiols; other P450 enzymes such as P450 1A2, 2C9, and 3A4 have the ability to catalyze PAH compounds at much slower rates [110]. Recently, Shimada et al. [111] examined the metabolic activation of PAHs and aryl- and heterocyclic amines to genotoxic products in *S. typhimurium* NM2009, and found that P450 2A13 and 2A6 (and P450 1B1) were able to activate several of these procarcinogens. The former two enzymes were particularly active in catalyzing 2-aminofluorene and 2-aminoanthracene activation. These results suggested that P450 2A enzymes, and P450 1 enzymes including P450 1B1, are major enzymes involved in activating PAHs, aryl- and heterocyclic amines, and tobacco-related nitrosamines. Hatanaka et al. [112] studied the metabolic activation pathway of 1-nitropyrene by human P450 enzymes in different *Salmonella umu* tester strains to define the abilities of individual human P450 family 1 enzymes. The results indicated that 1-nitropyrene can be activated by human 1B1 to a genotoxic agent by nitroreduction/O-acetylation at low substrate concentration and probably by epoxidation at high concentration.

Shimada et al. [113] examined the effects of several organoselenium compounds (1,2-, 1,3-, and 1,4-phenylenebis(methylene)selenocyanate (XSCs)) and inorganic sodium selenite on the activities of xenobiotic oxidation and procarcinogen activation by human liver microsomes and by recombinant human P450 1A1, 1A2, and 1B1 enzymes using *S. typhimurium* NM2009 tester strain. The three XSCs were found to be very potent inhibitors of metabolic activation of Trp-P-1, MeIQ, and 2-aminoanthracene, catalyzed by P450 1A1, 1A2, and 1B1, respectively. These inhibitory actions may, in part, account for the mechanisms responsible for cancer prevention by organoselenium compounds in laboratory aminals. Additionally, Shimada and Guengerich [114] determined whether individual PAHs and other procarcinogens affect the activities of human P450 1A1, 1A2, and 1B1 by measuring 7-ethoxyresorufin O-deethylation (EROD) activity and metabolic activation of PAH dihydrodiols and 2-amino-3,5-dimethylimidazo[4,5-*f*]quinoline (MeIQ) to genotoxic metabolites in an *S. typhimurium* NM2009 system. They found that three selected PAHs (5-methylchrysene, B[*a*]P, and B[*a*]A) inhibited metabolic activation of 5-methylchrysene-1,2-diol, (±)-B[*a*]P-7,8-diol, dibenzo[*a*,*l*]pyrene-11,12-diol, and MeIQ to genotoxic metabolites catalyzed by P450 1A1, 1B1, and 1A2, respectively, in *S. typhimurium* NM2009. Thus, individual PAHs may affect their own metabolisms and those of other carcinogens catalyzed by P450 1A1, 1A2, and 1B1. Wu et al. [115] also reported that human P450 2W1, which is found in tumor tissues from various organs, catalyze activation of PHA-diols to genotoxic metabolites in *S. typhimurium* NM2009, although the rates were much slower than those by P450 1B1.

In conclusion, we suggested that SOS activation and deactivation assays using *umu* tester strains can analyze a variety of genotoxic carcinogens in terms of the catalytic specificity of mammalian P450 enzymes toward their activation as mentioned in Sect. 5 and this section.

The procarcinogens are known to undergo bioactivation by P450-directed oxidation, which then become substrates for the UDP-glucuronosyltransferases

(UGTs). Yueh et al. [116] analyzed 11 human UGTs for their ability to modulate the mutagenic actions of *N*-hydroxy-2-acetylaminofluorene (*N*-hydroxy-2-AAF) and 2-hydroxyamino-1-methyl-6-phenylimidazo[4,5-*b*]pyridine (*N*-hydroxy-PhIP) formed by P450 1A2 with the *umu* test using *S. typhimurium* NM2009. In the presence of uridine 5′-diphosphoglucuronic acid (UDPGA), UGT1A9 inhibited the genotoxicity of *N*-hydroxy-2-AAF when incubated at 25 μM and completely abolished genotoxicity at lower concentrations. In contrast to the glucuronidation of *N*-hydroxy-2-AAF, UGT1A9 was unable to interfere with the genotoxicity of *N*-hydroxy-PhIP. This may be because of the dramatic differences in the formation of UGT1A9-generated glucuronide.

7 Genetically-Engineered *umu* Tester Strains Expressing Human Cytochrome P450 Enzymes

As the drug metabolism by P450s shows considerable species differences in rodents and humans, it is difficult to extrapolate the data from rodents to humans. It is therefore necessary to use human P450s to investigate the activation of various chemicals by P450. Although S9 fraction and human liver microsomes have been used to analyze the genotoxicity of chemicals, the use of these preparations is limited by several factors such as the short-lived metabolic intermediates produced outside the target cell and the low level of metabolizing enzymes in human liver samples. Thus, it is necessary to develop an alternative method(s) to overcome the species differences and to evaluate bioactivation of chemicals in humans. Many attempts have been made to develop more sensitive genotoxicity assays by introducing cDNAs of several drug-metabolizing enzymes, such as P450s, in bacteria and mammalian cells [117–124].

To develop a new tester strain for detecting environmental promutagens and procarcinogens, we introduced two plasmids into *S. typhimurium* TA1535; one containing the cDNAs of human P450 1A2 and NADPH-P450 reductase and the other pOA101, a *umuC″lacZ* fusion gene. The newly developed tester strain *S. typhimurium* OY1001/1A2 was found to activate heterocyclic amines (e.g., IQ, MeIQ, and MeIQx) to reactive metabolites that cause induction of *umu* gene expression in a concentration-dependent manner. We demonstrated that the established strain OY1001/1A2 can be of use for the detection of the genotoxicity of aromatic amines without the addition of metabolic activation enzymes [125]. To enhance further the sensitivity of the strain toward procarcinogenic heterocyclic aromatic amines (HCAs), we developed *S. typhimurium* OY1002/1A2 by introducing pCW″/1A2:hNPR (bicistronic construct co-expressing human P450 1A2 and the reductase) and pOA102 (constructed by subcloning the *Salmonella O*-AT gene in the pOA101-expressing *umuC″lacZ* gene) in *S. typhimurium* TA1535. We also developed the OY1003/1A2 as an *O*-AT-deficient strain. By using strains OY1001/1A2, OY1002/1A2, and OY1003/1A2, we compared the *umu* induction of HCAs

Table 4 Establishment of *umu* tester strains expressing mammalian metabolic enzymes

S. typhimurium	Character	Detection	References
OY1002/1A1	Human P4501A1 and NPR, and *O*-AT	PAH, arylamines	[127–129]
OY1002/1A2	Human P4501A2 and NPR, and *O*-AT	Arylamines	[126–129]
OY1002/1B1	Human P4501B1 and NPR, and *O*-AT	PAH, arylamines	[127–129]
OY1002/2C9	Human P4502C9 and NPR, and *O*-AT		[127, 128]
OY1002/2D6	Human P4502D6 and NPR, and *O*-AT		[127, 128]
OY1002/2E1	Human P4502E1 and NPR, and *O*-AT	Nitrosoamines	[127, 128]
OY1002/3A4	Human P4503A4 and NPR, and *O*-AT	Aflatoxins	[127–129]
NM6001	Human *N*-acetyltransferase 1	Arylamines, nitroarenes	[129–131]
NM6002	Human *N*-acetyltransferase 2	Arylamines, nitroarenes	[129–131]
NM7001	Human sulfotransferase 1A1	Arylamines	[132]
		Benzylic alcohols	
NM7002	Human sulfotransferase 1A2	Arylamines	[132]
		Benzylic alcohols	
NM7003	Human sulfotransferase 1A3	Alkenylbenzenes	[132]
NM5004	Rat glutathione *S*-transferase 5-5	Dihaloalkenes	[133, 134]

NPR NADPH-P450 reductase, *O-AT O*-acetyltransferase

and observed that the OY1002/1A2 strain was more sensitive than the OY1001/1A2 strain toward HCAs. However, their genotoxicity was not detected with OY1003/1A2 strain. These results indicated that strain OY1002/1A2 can be used to detect potential genotoxic aromatic amines requiring bioactivation by P450 1A2 and *O*-acetyltransferase [126]. To clarify the roles of different P450 enzymes in the bioactivation of HCAs and other procarcinogens, we selected seven of the major human P450 enzymes (1A1, 1A2, 1B1, 2C9, 2D6, 2E1, and 3A4). We further established seven strains (OY1002/1A1, OY1002/1A2, OY1002/1B1, OY1002/2C9, OY1002/2D6, OY1002/2E1, and OY1002/3A4) by introducing two plasmids into *S. typhimurium* TA1535, one carrying both P450 and the reductase cDNAs in a bicistronic construct under control of an IPTG-inducible double tac promoter and the other, pOA102, carrying *O*-AT and *umuC″lacZ* fusion genes (Table 4). An outline of the *umu* test systems is shown in Fig. 2.

Among all homo- and heterocyclic aromatic amines tested, 2-aminoanthracene (2-AA), 2-aminofluorene, GluP-1, MeIQx, MeIQ, and IQ exhibited high genotoxicity in the OY1002/1A2 strain, and genotoxicity of IQ and 2-AA was detected in the OY1002/1A1 strain. Aflatoxin B_1 exhibited genotoxicity in the OY1002/1A2, OY1002/1A1, and OY1002/3A4 strains. β-Naphthylamine and benzo[*a*]pyrene did not exhibit genotoxicity in any of the strains. These results suggested that P450 1A2 is the major P450 enzyme involved in bioactivation of

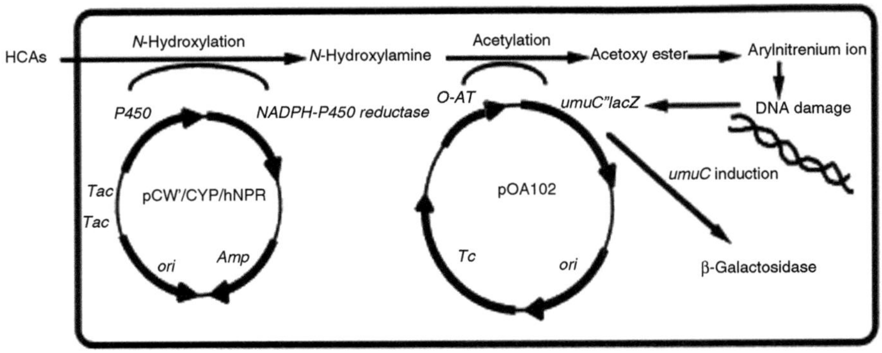

Fig. 2 Schema of the *umu* test system. The *rectangle* shows the bacterial cell and two plasmids are depicted with a *circle*. *Tc* tetracycline resistance gene, *Amp* ampicillin resistance gene, *Tac* tac promoter, *ori* the origin of replication, *O-AT O*-acetyltransferase gene, *umuC*::*lacZ*: *umuC*::*lacZ* fusion gene under control of *umu* operon

HCAs [127]. These strains could provide a sensitive means of assessing the genotoxicity of procarcinogens requiring activation by P450s, and useful tools for studying the role of human P450 enzymes involved in biotransformation of xenobiotic chemicals. We recently suggested that these strains provide the possibility of a high-throughput *umu* test system.

Other mutagenicity studies using genetically engineered bacteria strains expressing human P450s have been conducted by many researchers. Josephy et al. [119] introduced an expression plasmid carrying human P450 1A2 into *S. typhimurium* YG1019 strain to detect the mutagenicity of HCAs and arylamines. The mutagenicity of 2-aminoanthracene (2-AA) and 2-aminofluorene was detectable with this system. Kranendonk et al. [120] reported on the development of an *E. coli* tester strain, BMX100, which can express active human P450 1A2, alone or fused to rat liver NADPH P450 reductase. The strain can detect bioactivation of 2-AA, AFB_1, and IQ. Suzuki et al. [121] introduced an expression plasmid (p1A2OR) carrying human P450 1A2 and the human NADPH-P450 reductase cDNAs and an expression plasmid (pOAT) carrying *S. typhimurium O*-AT, a derivative of pACYC184 vector, to *S. typhimurium* TA1538 strain to yield the TA1538/ARO strain. The TA1538/ARO strain showed a high sensitivity to mutagenic HCAs and the mutagenic activation of MeIQ, IQ, MeIQx, and 2-amino-1-methyl-6-phenylimidazo[4,5-*f*]pyridine (PhIP) was seen from the concentration at around picomole order. Kushida et al. [123, 124] also developed *Salmonella* tester strains YG7108 2E1/OR and YG7108 2A6/OR highly sensitive to promutagenic *N*-nitrosamines by introducing a plasmid carrying human P450 2A6 and NADPH-P450 reductase (OR) cDNAs or human P450 2E1 and OR cDNAs, respectively, into the ada- and ogt-deficient strain YG7108. P450 2E1 expressed in the YG7108 2E1/OR cells showed mutagen-activating capacity, as indicated by induced revertants/min/pmol cytochrome P450, for *N*-nitrosodimethylamine (NDMA), *N*-nitrosodiethylamine (NDEA), *N*-nitrosodipropylamine (NDPA),

N-nitrosodibutylamine (NDBA), *N*-nitrosopyrrolidine (NPYR), and 4-(methylnitrosamino)-1-(3-pyridyl)-1-butanone (NNK), but not *N*-nitrosomethylphenylamine (NMPhA) and *N*-nitrosonornicotine (NNN). P450 2A6 activated NDMA, NDEA, NDPA, NDBA, NMPhA, NPYR, NNN. and NNK. They suggested that human P450 2E1 is mainly responsible for the metabolic activation of *N*-nitrosamines with a relatively short alkyl chain(s), whereas P450 2A6 was predominantly responsible for the metabolic activation of *N*-alkylnitrosamines possessing a relatively bulky alkyl chain(s). Using the same host strains, Cooper and Porter [122] have constructed two mutagenicity tester strains that co-express full-length human cytochrome P450 2E1 and P450 reductase in *S. typhimurium* lacking *ogt* and *ada* methyltransferases (YG7104ER, *ogt*$^-$ and YG7108ER, *ogt*$^-$, *ada*$^-$). These strains were sensitive to nitrosamines with longer alkyl side chains including diethylnitrosamine, dipropylnitrosamine, and dibutylnitrosamine. Mutagenicity decreased with alkyl chain length, consistent with the stringency of the *ada*-encoded enzyme for methyl and ethyl DNA adducts. These strains may prove useful in the evaluation of nitrosamine contamination of food and environmental samples. Taking all these reports, obtained in the last decade, into account, the bacterial tester strains harboring human P450 may be a useful tool to investigate the roles of P450 on the metabolism of drugs and metabolic activation of chemicals in humans.

8 Genetically-Engineered *umu* Tester Strains Expressing Phase II Metabolic Enzymes

In addition to P450s, there are other enzymes involved in xenobiotic metabolism. So far, three types of phase II enzymes (mammalian *N*-acetyltransferases, sulfotransferases, and glutathione *S*-transferases) have been expressed in bacterial strains. These enzymes are known to play important roles in the metabolism of a variety of toxic and carcinogenic compounds and have been functionally expressed in prokaryotic cells.

8.1 Rat Glutathione S*-Transferase 5-5*

Multiple forms of glutathione *S*-transferase (GST) are found in subcellular fractions of all mammalians, and numerous studies have established that cytosolic GST enzymes can be classified into four groups e.g. alpha, pi, mu, and theta on the basis of structural similarity of isolated genes [135, 136]. With regard to xenobiotic metabolism, alpha, pi, mu, and theta are the most important classes. GST enzymes primarily act to detoxify toxic electrophilic chemicals, including mutagenic metabolites, by conjugating them with the nucleophilic sulfhydryl peptide glutathione

(GSH) [137], but in several cases the enzymes are involved in the formation of reactive GSH conjugates that alkylate cellular macromolecules [138]. Among these classes, a theta-class rat GST 5-5 enzyme, which shares 82% sequence homology with human theta-class GSTT1 [139], has been reported to be involved in the activation of dihalomethanes [140]. We introduced a fusion plasmid containing rat GST 5 gene and *umu* operon into *S. typhimurium* TA1535. First, the fragment of *umu* operon was subcloned into a multicopy vector pKK233-2 containing rat GST 5 cDNA and the resulting plasmid was designated as pYO100. After the plasmid was modified by introducing into *S. typhimurium* SJ1002 and again introduced into *S. typhimurium* TA1535, we developed a GST-overexpressing *umu* tester strain NM5004 [133]. The strain exhibited a 52-fold higher GST activity than the parent strain TA1535/pSK1002 toward the model substrate 1,2-epoxy-3-(4′-nitrophenoxy) propane and was found to detect the genotoxicity of ethylene dibromide, 1-bromo-2chloroethane, 1,2-dichloroethane, and methylene dichloride (Table 3). Our results were in good agreement with the results reported by Thier et al. [140] who showed that the dihaloalkanes are mutagenic in Ames tester strain TA1535 which produces rat GST 5-5 protein. Additionally, ten chemicals (1,2-dibromoethane, *N*-(2,3-epoxypropyl)phthalimide, 1,3-dichloroacetone, CH_2I_2, 1,2-epoxy-3-phenoxypropane, 2,3-epoxypropyl *p*-methoxyphenyl ether, 1-bromo-2-chloroethane, 1-bromo-2,3-dichloropropane, CH_2BrCl, and CH_2Br_2) were found to enhance induction of *umu* gene expression in the NM5004 as compared with the TA1535/pSK1002 strain [134] (Table 5). Simula et al. [141] have reported that CH_2Cl_2 could not be activated by human GST alpha and pi classes of enzymes. This finding suggested that that class GST enzymes may have more important roles in the activation of CH_2Cl_2 than other GST enzymes. In the case of 1-nitropyrene and 2-nitrofluorene, however, NM5004 strain showed weaker *umu* gene expression responses than the TA1535/pSK1002 strain. Further, *trans*-1,2-dihydro-1,2-dihydroxy-6-aminochrysene, AFB_1, sterigmatocystin, and the (+)- and (−)-enantiomers of B(*a*)P-7,8-diol could be trapped as inactivated GSH conjugates by rat liver microsemes in the NM5004 strain. These results indicated that the theta class rat GST 5-5 enzyme participates in the activation and inactivation of potential environmental carcinogenic chemicals. CH_2Cl_2 and 1,2-dichloropropane, which are widely used as industrial solvents, have recently been found to develop novel human bile duct cancer in workers at a printing company in Japan. This tester strain might therefore be useful in further studies on the mechanism of genotoxicity of dihaloalkanes and of the role of the GST in human cancer risk.

8.2 Human N-Acetyltransferases

Numerous studies have suggested that carcinogenic nitroarenes and aromatic amines are present in the environment and workplaces [142, 143]. These compounds are strong mutagens in bacteria and are carcinogenic in rodents [81, 82]. Aromatic amines are activated to *N*-hydroxy metabolites by P450 enzymes, and

Table 5 Comparison of genotoxicity activities of a variety of chemicals in *S. typhimurium* NM5004 strain, expressing rat GST 5-5 and the control strain. TA1535/pSK1002 and NM5004 strains [133, 134]

Chemicals	NM5004 [GST(+)]	TA1535/pSK1002 [GST(−)]
1,2-Dibromoethane	+++	−
N-(2,3-Epoxypropyl)phthalimide	+++	+
1,3-Dichloroacetone	++	+
CH_2I_2	++	−
1,2-Epoxy-3-phenoxypropane	+	−
2,3-Epoxypropyl *p*-methoxyphenyl ether	+	−
1-Bromo-2-chloroethane	+	−
1-Bromo-2,3-dichloropropane	+	−
CH_2BrCl	+	−
CH_2Br_2	+	−
1,2-Epoxy-3-(4′-nitrophenoxy)-propane	−	++
2,3-Dibromo-1-chloropropane	±	+
1,4-Dibromo-2,3-epoxybutane	+	++
1,2-Epoxy-3-bromopropane	−	+
1,2-Epoxy-3-chloropropane	−	±
1,2,3,4-Diepoxybutane	+	+
2,3-Dibromopropionaldehyde	+	+
1,4-Dibromo-2,3-dihydroxybutane	±	−
1,4-Dibromobutane	−	−
1,3-Dibromoacetone	+	+
2,3-Dibromo-1-propanol	±	±
1,2-Epoxy-4-bromobutane	±	±
CH_2Cl_2	++	−
1,3-Dibromo-2-propanol	−	−
1-Bromo-2,3-propanediol	−	−
4-Vinylcyclohexene dioxide	−	−
Cyclohexene oxide	−	−
1,2-Epoxybutane	−	−
1-Bromo-2-fluoroethane	−	−

Efficacies of chemicals in *umu* test were ranked as follows: (−) 0–50; (±) 50–100; (+) 100–250; (++) 250–450; (+++) 450 for *umu* expression (units)

subsequently acetylated by *N,O*-acetyltransferase to give acetoxy esters, which form an arylnitrenium ion that binds to DNA. Thus, *N*-acetyltransferases (NATs) play an important role as phase II enzymes in the metabolic activation of aromatic amines. In 1993 we succeeded in the construction of the highly sensitive *umu* tester strain NM2009 overexpressing a *Salmonella O*-AT for the detection of aromatic amines, HCAs, and aminoazo compounds by the bacterial SOS response [88, 89]. We developed *umu* test systems to clarify the roles of two human NATs (NAT1 and NAT2) in the genotoxicity of aromatic amines and nitroarenes [130]. These strains NM6001 and NM6002 were constructed by introducing human NAT1 and NAT2 cDNAs, respectively, into NM6000 (TA1538/1,8-DNP/pSK1002), which is a derivative of TA1538 devoid of endogenous *Salmonella O*-AT and harbors the *umuC″lacZ* fusion gene (Table 4). The human NAT1-expressing strain NM6001 showed much higher sensitivity than the NAT2-expressing NM6002 strain to the cytotoxic and genotoxic effects of 2-nitrofluorene and 2-aminofluorene (2-AF). These results were in good agreement with those reported by Grant et al. [144], who showed that 2-AF exhibited the mutagenic response in an *S. typhimurium* strain expressing human NAT1 in the presence of rat liver S9. Minchin et al. [145] also reported that NAT1 efficiently catalyzed *O*-acetylation of *N*-hydroxy-2-aminofluorene in vitro. In contrast, the NM6002 strain showed higher sensitivity than the NM6001 strain to the cytotoxicity and genotoxicity of 1,8-dinitropyrene, 6-aminochrysene, and MeIQ. Similarly, Watanabe et al. [146] observed that CHL cells expressing human NAT2 were much more sensitive to the clastogenicity of 1,8-dinitropyrene in the in vitro micronucleus test. These results suggested that the human NAT strains can be employed for studies on the mechanisms of genotoxicity of a variety of nitroarenes and aromatic amines, along with the assessment of cancer risk to humans. Interestingly, we found that the bladder carcinogenic arylamines 4-aminobiphenyl, 2-acetylaminofluorene, β-naphthylamine, *o*-toluidine, *o*-anisidine, and benzidine are mainly activated by the NAT1 enzyme to produce DNA damage rather than NAT2 [131].

In the late 2000s, our laboratories reported the roles of human P450s and human NATs enzymes in the metabolic activation of various carcinogenic chemicals as follows. The β-carboline compounds norharman (9*H*-pyrido[3,4-*b*]indole) and harman (1-methyl-9*H*-pyrido[3,4-*b*]indole) are formed in the pyrolysis of the amino acid tryptophan and are shown to be present at much higher levels than heterocyclic amines in tobacco condensates and cooked foods [147, 148]. These chemicals are non-mutagenic to *Salmonella* strains, but show co-mutagenicity with S9 mixture in the presence of aniline and *o*-toluidine. The resulting aminophenylnorharman, aminomethylphenylnorharman, and aminophenylharman are produced by coupling of norharman and aniline, norharman and *o*-toluidine, and harman and aniline in the presence of S9 mixture. We determined the genotoxicity of these coupling chemicals using *umu* tester strains established in our laboratory. These chemicals were found to be mainly bioactivated by P450 1A2 and NAT2 [128]. 3-Nitrobenzanthrone (3-NBA) is a mutagenic and carcinogenic compound identified in diesel exhaust, airborne particulate matter, soil, and water. We constructed the *S. typhimurium* OY1022 strain by selecting resistant colonies of TA1535NR capable of growth in the presence of 1,8-dinitropyrene to reduce the direct sensitivity to 3-NBA and established *S. typhimurium* strains OY1022/1A1, OY1022/1A2, OY1022/1B1, and OY1022/3A4 expressing four recombinant human P450s by introducing two plasmids into the OY1022, one carrying both P450 and NPR cDNAs in a bicistronic construct under control of an IPTG-inducible double tac promoter and the other, pOA102, carrying *O*-AT and *umuC″lacZ* fusion gene. Using these strains, we determined using these strains whether any human P450 enzymes are involved in the genotoxic activation of 3-NBA to genotoxic metabolites. We demonstrated that the activation of 3-NBA can be catalyzed by human P4503A4, 1A1, 1A2, and 1B1 and NPR to genotoxin(s) in the presence of bacterial *O*-AT, probably via nitroreduction [149].

Four 2-phenyl benzotriazole (PBTA)-type compounds (PBTA-4, PBTA-6, PBTA-7, and PBTA-8) were identified as major mutagens in blue cotton/rayon-absorbed substances collected at sites below textile dyeing factories or municipal water treatment plants treating domestic water and effluents from textile dyeing factories in several rivers in Japan [150]. We examined the genotoxicity of four PBTA derivatives using parental strain TA1535/pSK1002 and *O*-AT-overexpressing strain NM2009 and determined the genotoxic activation of these chemicals by recombinant human or rat P450 enzymes in NM2009. We further determined the potential role of human NATs (NAT1 and NAT2) in the bioactivation of them using NM6000 (parent strain), NM6001, and NM6002. We suggested that P4501A1 and NATs are important enzymes participating in the genotoxic activation of PBTA-type compounds [151].

3,6-Dinitrobenzo[*e*]pyrene (DNBeP) is a potent mutagen identified in surface soil in two metropolitan areas of Japan [152]. Using *umu* tester strains of *S. typhimurium* OY1022/1A1, OY1022/1A2, OY1022/1B1, and OY1022/3A4 expressing human P450s, we investigated the role of human P450 enzymes in the bioactivation of DNBeP to genotoxic metabolites. We suggested that nitroreduction by human P4501A2, P4503A4, and P4501A1 and *O*-acetylation by human NAT2 contribute to the activation of DNBeP [129].

8.3 Human Sulfotransferases

Sulfo conjugation has been shown to be a major pathway in the biotransformation of numerous xenobiotics and endobiotics such as drugs, chemical carcinogens, hormones, bile acids, neurotransmitters, peptides, and lipids [153, 154]. Sulfotransferases (SULTs) transfer the sulfo moiety from the cofactor 5′-phosphoadenosine-3′-phosphosulphate to nucleophilic groups of their substrates. In the case of most xenobiotics and small endogenous substrates, sulfonation has generally been considered as a detoxification pathway leading to more water-soluble products, thereby facilitating their excretion via kidney or bile [155, 156]. However, for xenobiotics such as *N*-hydroxy arylamines, *N*-hydroxy heterocyclic amines, and hydroxymethyl polycyclic aromatic hydrocarbons the enzymes activate them to highly reactive sulfate esters which bind covalently to DNA [157, 158]. In humans, SULTs consist of four families (SULT1, 2, 4, and 6) and contain at least 13 proteins [159]. SULT 1A1, 1A2, 1A3, 1C2, 1E1, and 2A1 are the major enzymes to catalyze the conjugation of xenobiotic chemicals, including carcinogens [160].

We developed a newly *umu* assay system to investigate the functions of three different human SULTs (SULT 1A1, 1A2, and 1A3) in the bioactivation of aromatic amines, nitroarene compounds, benzylic and allylic alcohols, and estrogens-like compounds to genotoxins [132]. To express the three different

SULT enzymes in *S. typhimurium*, we subcloned human SULT 1A1, 1A2, and 1A3 cDNA genes into the multicopy plasmid vector pTrc99A^{KM}. The generated plasmids were introduced into the *S. typhimurium* *O*-AT-deficient strain NM6000 (TA1538/1,8-DNP/pSK1002), resulting in the tester strains NM7001, NM7002, and NM7003 (Table 6). These test systems are highly sensitive for SULT-dependent carcinogens without external supply of the cofactor 3′-phosphoadenosine 5′-phosphosulfate and $MgSO_4$. We compared the sensitivities of these three strains with the parental strain NM7000 against 51 chemicals with and without S9 mix. 2-Amino-3-methyl-9*H*-pyrido[2,3-*b*]indole (MeAαC) and 2-amino-6-methyl-dipyrido[1,2-α:3′,2′-*d*]imidazol (Glu-P-1) exhibited strong genotoxicity in the strain NM7001 in the presence of liver S9 mix compared with the strains NM7002, NM7003, and NM7000 (Table 6). Our results were consistent with Glatt et al. [161] who reported that MeAαC showed strongly enhanced mutagenicity in an *S. typhimurium* strain expressing SULT1A1 in the presence of rat liver postmitochondrial fraction. Furthermore, in the case of Glu-P-1, Chu et al. [162] showed that *N*-hydroxy-Glu-P-1 was selectively sulfonated by a human liver thermostable phenol SULT purified from human liver, probably SULT 1A1 or a mixture of SULT 1A1 and 1A2. These results suggested that human SULT 1A1 is involved in the bioactivation of MeAαC and Glu-P-1 to genotoxic metabolites. In contrast, 2-aminoanthracene, 2-acetylaminofluorene, and 2-amino-1-methyl-6-phenylimidazo[4,5-*b*]pyridine exhibited stronger genotoxicity in the strain NM7002 compared with the strains NM7001 and NM7003. The results were in agreement with reports by Glatt and colleagues, suggesting that the *N*-hydroxy-2-acetylaminofluorene is activated most efficiently by SULT 1A2 expressed in *S. typhimurium* [163]. Aromatic amines such as 2-aminoanthracene, 4-aminobiphenyl, APNH, and 3-methoxy-4-aminoazobenzene showed a similar genotoxic potential in strains NM7001 and NM7002, suggesting that these chemicals are bioactivated by SULT 1A1 and SULT 1A2. NM7001, NM7002, and NM7003 strains were found to be of similar sensitivities toward 2-amino-9*H*-pyrido[2,3-*b*]indole and β-naphthylamine. In the cases of 6-aminochrysene, 2-amino-3,5-dimethylimidazo[4,5-*f*]quinoline, 3-amino-1,4-dimethyl-5*H*-pyrido [4,3-*b*]indole, and 3-amino-1-methyl-5*H*-pyrido[4,3-*b*]indole, all strains used showed similar sensitivities.

Of the 15 nitroarenes, 5-nitroacenaphthene, 3-nitrobenzanthrone, and 3,9-dinitrofluoranthene (3-NBA) show the highest genotoxic potential in the strain NM7001. Arlt et al. [164] reported that human SULT 1A1 is involved in the formation of DNA adducts by 3-NBA using a Chinese hamster lung cell line that expresses human SULT 1A1.

The strain NM7002 was highly sensitive to 2-nitrofluorene, 1-nitropyrene, and 2-nitropropane. The genotoxicity of 2-nitropropane in rat has been attributed to the SULT-mediated formation of DNA-reactive arylnitrenium ions from propane 2-nitronate [165]. 4,4′-Dinitrobiphenyl showed the same genotoxic potential in the strains NM7001 and NM7002. However, in the case of other nitroarenes, such as furylfuramide, 3-nitrofluoranthene, nitrofurazone, 1-nitronaphthalene,

Table 6 Comparison of substrate specificity of human SULTs (1A1, 1A2, and 1A3) expressed in *S. typhimurium* TA1538/1,8-DNP/pSK1002 toward a variety of chemicals [132]

Chemicals	S9	SULT isoforms
Aromatic amines		
2-Aminoanthracene	+	1A1 = 1A2
2-Aminofluorene	+	1A2
2-Acetylaminofluorene	+	1A1 < 1A2
4-Aminobiphenyl	+	1A1 = 1A2
6-Aminochrysene	+	SR
Aminophenylnorharman	+	1A1 < 1A2
AαC	+	1A3 = 1A2 < 1A1
Glu-P-1	+	1A1
IQ	+	SR
MeAαC	+	1A2 < 1A1
MeIQ	+	SR
3-MeO-AAB	+	1A1 = 1A2
β-Naphthylamine	+	1A3 < 1A2 = 1A1
PhIP	+	1A1 < 1A2
Trp-P-1	+	SR
Trp-P-2	+	SR
Nitro compounds		
Furylfuramide	–	1A2
5-Nitroacenaphthene	–	1A1
3-Nitrobenzanthrone	–	1A2 < 1A1
2-Nitrofluorene	–	1A2
Nitrofurazone	–	SR
3-Nitrofluoranthene	–	1A1
1-Nitronaphthalene	–	SR
1-Nitropyrene	–	1A2
2-Nitropropane	–	1A2
4-Nitroquinoline 1-oxide	–	SR
2-Nitrotriphenylene	–	SR
4,4′-Dinitrobiphenyl	–	1A1 = 1A2
3,7-Dinitrofluoranthene	–	1A2 < 1A1
3,9-Dinitrofluoranthene	–	1A2 < 1A1
1,6-Dinitropyrene	–	SR
Benzylic and allylic alcohols (and their precursors)		
Estragole	+	1A3
Hycanthone	–	1A1 = 1A3
1′-Hydroxysafrole	–	1A3
1-Hydroxymethylpyrene	–	1A3 < 1A2 < 1A1
Other compounds		
4-Hydroxyestradiol	–	SR
Acrolein	–	SR

(continued)

Table 6 (continued)

Chemicals	S9	SULT isoforms
MNNG	–	SR
MX	–	SR

SR presents same response in all strains
AαC 2-amino-9*H*-pyrido[2,3-*b*]indole, *Glu-P-1* 2-amino-6-methyl-dipyrido[1,2-*a*:3',2'-*d*]imidazole, *IQ* 2-amino-3-methylimidazo[4,5-*f*]quinoline, *MeAαC* 2-amino-3-methyl9*H*-pyrido[2,3-*b*] indole, *MeIQ* 2-amino-3,5-dimethylimidazo[4,5-*f*]quinoline, *3-MeO-AAB* 3-methoxy-4-aminoazobenzene, *PhIP* 2-amino-1-methyl-6-phenylimidazo[4,5-*b*]pyridine, *Trp-P-1* 3-amino-1,4-dimethyl-5*H*-pyrido[4,3-*b*]indole, *Trp-P-2* 3-amino-1-methyl-5*H*-pyrido[4,3-*b*]indole, *MNNG* *N*-methyl-*N'*-nitro-*N*-nitrosoguanidine, *MX* 3-chloro-4-(dichloromethyl)-5-hydroxy-2(5*H*)-furanone

4-nitroquinoline 1-oxide, 2-nitrotriphenylene, 3,7-dinitrofluoranthene, and 1,6-dinitropyrene, the genotoxicity was almost equal in all strains.

Among numerous benzylic alcohols, 1′-hydroxysafrole and 1′-hydroxyestragole were strongly activated in the strain NM7003 that expresses the human SULT 1A3. The result was the first indication that human SULT 1A3 plays an important role in the metabolic activation of benzylic alcohols to genotoxic intermediates. Interestingly, another congener, 1′-hydroxymethyleugenol, was activated by human SULT 1A1, 1A2, and 1C2, but not by human SULT 1A3 expressed in *S. typhimurium* TA100 [166].

Finally, our study also showed that the genotoxic potential of several chemicals is reduced by SULT enzymes. For example, the genotoxicity of Glu-P-1, 2-amino-1-methyl-6-phenylimidazo[4,5-*b*]pyridine, 2-nitrofluorene, 3-nitrofluoranthene, 1-nitropyrene, and 3,7-dinitrofluoranthene was inhibited by SULT 1A3. In the case of acrolein, the genotoxicity was inhibited by SULT 1A1 and SULT 1A3. These findings suggested that SULT 1A1 or SULT 1A3 enzymes were involved in the detoxification of several genotoxic compounds. The *umu* test system with overexpressed human SULT enzymes may be useful for a further investigation of the SULT-function in the metabolic inactivation of carcinogens.

9 Future Perspectives of Bacterial Assay Systems

Bacterial genotoxicity bioassay systems have long been of use for detecting genotoxic agents and alternate models of mammalian systems. We have developed numerous new *umu* tester strains by introducing various genes encoding enzymes of genotoxic bioactivation, including bacterial nitroreductase and *O*-acetyltransferase, phase I enzymes human cytochrome P450s, phase II enzymes rat glutathione *S*-transferase, human *N*-acetyltransferases and sulfotransferases. However, although the modification of *umu* test systems expressing mammalian metabolic enzymes has been mentioned above, there is still room for further improvements. To date, a few genes have been introduced to the *umu* assay systems. Coexpression systems in the same cell for P450s and other enzymes involved in the metabolic activation and inactivation of genotoxins are expected to be developed for the estimation of the genotoxicological roles of the enzymes. They are also appropriate for high-throughput genotoxicity screening. For example, hydroxylation of aromatic amines and heterocyclic amines and their subsequent *O*-acetylation or sulfation may be carried out by phase I enzyme P450 1A2 and phase II enzymes NAT1/2 or SULT 1A1/1A2. These test systems could find application as useful screening tools for genotoxicity in numerous fields and also be suitable for basic studies on the possible roles of respective enzymes. These systems also have many advantages: they are simple, use small sample volumes (a few microliters), are ideally suited for high throughput applications, and are finished within 5 h. They can therefore provide new tools for genotoxicity assays and for studying the role of biotransformation of chemicals in human carcinogenesis.

References

1. Fieser LF (1938) Carcinogenic activity, structure and chemical reactivity of polynuclear hydrocarbons. Am J Cancer 34:37–127
2. Miller EC, Miller JA (1947) The presence and significance of bound amino azodyes in the liver of rats fed *p*-dimethylaminoazobenzene. Cancer Res 71:468–480
3. Malling HV (1971) Dimethylnitrosamine: formation of mutagenic compounds by interaction with mouse liver microsomes. Mutat Res 13:425–429
4. Ames BN, Durston WE, Yamazaki E, Lee FD (1973) Carcinogens are mutagens: a simple test system combining liver homogenates for activation and bacteria for detection. Proc Natl Acad Sci U S A 70:2281–2285
5. McCann J, Springarn NE, Kobori J, Ames BN (1975) Detection of carcinogens as mutagens: bacterial tester strains with R-factor plasmid. Proc Natl Acad Sci U S A 72:979–983
6. Zeiger E (1998) Identification of rodent carcinogens and noncarcinogens using genetic toxicity tests: premises, promises, and performance. Regul Toxicol Pharmacol 28:85–95
7. Rinkus SJ, Legator MS (1979) Chemical characterization of 465 known or suspected carcinogens and their correlation with mutagenic activity in the *Salmonella typhimurium* system. Cancer Res 39:3289–3318
8. Rinkus SJ, Legator MS (1981) *Salmonella* revised: a reply to Ames and McCann. Cancer Res 41:4196–4203
9. Ames BN, McCann J (1981) Validation of the *Salmonella* test: a reply to Rinkus and Legator. Cancer Res 41:4192–4196
10. Venitt S, Crofton-Sleigh C, Forrester R (1984) Bacterial mutation assays using reverse mutation. In: Venitt S, Parry JM (eds) Mutagenicity testing, a practical approach. IRL Press, Oxford, pp 45–98
11. Dyrby T, Ingvardsen P (1983) Sensitivity of different *E. coli* and *Salmonella* strains in mutagenicity testing calculated on the basis of selected literature. Mutat Res 123:47–60
12. Walker GC (1984) Mutagenesis and inducible responses to deoxyribonucleic acid damaged in *Escherichia coli*. Microbiol Rev 48:60–93
13. Elesperu RK, White RJ (1983) Biochemical prophage induction assay: a rapid test for antitumor agents that interact with DNA. Cancer Res 43:2819–2830
14. Quillardet P, Huisman O, D'ari R, Hofnung M (1982) SOS chromotest, a direct assay of induction of an SOS function in *Escherichia coli* K12 to measure genotoxicity. Proc Natl Acad Sci U S A 79:5971–5975
15. Quillardet P, Hofnung M (1993) The SOS chromotest: a review. Mutat Res 297:235–279

16. Oda Y, Nakamura S, Oki I, Kato T, Shinagawa H (1985) Evaluation of the new system (*umu*-test) for the detection of environmental mutagens and carcinogens. Mutat Res 147:219–229
17. Kato T, Shinoura Y (1977) Isolation and characterization of mutants of *Escherichia coli* deficient in induction of mutations by ultraviolet light. Mol Gen Genet 156:121–131
18. Shinagawa H, Kato T, Ise T, Makino K, Nakata A (1983) Cloning and characterization of the *umu* operon responsible for inducible mutagenesis in *Escherichia coli*. Gene 23:167–174
19. Reuven NB, Arad G, Maor-Shoshani A, Livneh Z (1999) The mutagenesis protein UmuC is a DNA polymerase activated by UmuD′, RecA and SSB and is specialized for translesion replication. J Biol Chem 274:31763–31766
20. Tang M, Shen X, Frank EG, O'Donnell M, Woodgate R, Goodman MF (1999) $UmuD'_2C$ is an error-prone DNA polymerase, *Escherichia coli* pol V. Proc Natl Acad Sci U S A 96:8919–8924
21. Nakamura S, Oda Y, Shimada T, Oki I, Sugimoto K (1987) SOS-inducing activity of chemical carcinogens and mutagens in *Salmonella typhimurium* TA1535/pSK1002: examination with 151 chemicals. Mutat Res 192:239–246
22. McDaniels AE, Reyes AL, Wymer LJ, Rankin CC, Stelma GN Jr (1990) Comparison of the *Salmonella* (Ames) test, *umu* tests, and the SOS chromotests for detecting genotoxins. Environ Mol Mutagen 16:204–215
23. Reifferscheid G, Heil J (1996) Validation of the SOS/*umu* test using test results of 486 chemicals and comparison with the Ames test and carcinogenicity data. Mutat Res 369:129–145
24. Tennant RW, Margolin BH, Shelby MD, Zeiger E, Haseman JK, Spalding J, Caspary W, Rensnick M, Stasiewicz S, Anderson B, Minor R (1987) Prediction of chemical carcinogenicity in rodent from *in vitro* genetic toxicity assays. Science 236:933–941
25. Zeiger E (1987) Carcinogenicity of mutagens: predictive capability of the *Salmonella* mutagenesis assay for rodent carcinogenicity. Cancer Res 47:1287–1296
26. Yasunaga K, Kiyonari A, Oikawa T, Abe N, Yoshikawa K (2004) Evaluation of the *Salmonella umu* test with 83 NTP chemicals. Environ Mol Mutagen 44:329–345
27. Matsui N, Kaya T, Nagamine K, Yasukawa T, Shiku H, Matsue T (2006) Electrochemical mutagens screening using microbial chip. Biosens Bioelectron 21:1202–1209
28. Buchinger S, Grill R, Morosow V, Yoav HB, Diamand YS, Biran A, Pedahzur R, Belkin S, Reifferscheid G (2010) Evaluation of chrone-amperometric signal detection for the analysis of genotoxicity by a whole cell biosensor. Anal Chim Acta 659:122–128
29. Kuramitz H, Sazawa K, Nanayama Y, Hata N, Taguchi S, Sugawara K, Fukushima M (2012) Electrochemical genotoxicity assay based on a SOS/*umu* test using hydrodynamic voltammetry in a droplet. Sensors 12:17414–17432
30. Brinkmann C, Eisentraeger A (2008) Completely automated short-term genotoxicity testing for the assessment of chemicals and characterization of contaminated soils and waste waters. Environ Sci Pollut Res 15:211–217
31. Ptitsyn LR, Horneck G, Komava O, Kozubek S, Krasavin EA, Bonev M, Rettberg P (1997) A biosensor for environmental genotoxins screening based on an SOS lux assay in recombinant *Escherichia coli* cells. Appl Environ Microbiol 63:4377–4387
32. Verschaeve L, Van Gompel J, Thilemans L, Regniers Vanparys P, van der Lelie D (1999) VITOTOX bacterial genotoxicity and toxicity test for the rapid screening of chemicals. Environ Mol Mutagen 33:240–248
33. Norman A, Hansen LH, Sorensen SJ (2006) A flow cytometry-optimized assay using an SOS-green fluorescent protein (SOS-GFP) whole-cell bacterial sensors for the detection of genotoxins in complex environments. Mutat Res 63:164–172
34. Østrgaard TG, Hansen LH, Binderup ML, Norman A, Sørensen SJ (2007) The *cda* GenoTox assay: a new and sensitive method for detection of environmental genotoxins, including nitroarenes and aromatic amines. Mutat Res 63:77–84
35. ISO (2000) Water quality-determination of the genotoxicity of water and wastewater using the *umu*-test. ISO 13829. ISO, Geneva

36. Giuliani F, Koller T, Wurgler FE, Widmer RM (1996) Detection of genotoxic activity in native hospital waste water by the *umuC* test. Mutat Res 368:49–57
37. Oda Y, Funasaka K, Kitano M, Nakama A, Yoshikura T (2004) Use of a high-throughput *umu*-microplate test system for rapid detection of genotoxicity produced by mutagenic carcinogens and airborne particulate matter. Environ Mol Mutagen 43:10–19
38. Rao SS, Quinn BA, Burnison BK, Hayes MA, Metcalfe CD (1995) Assessment of the genotoxic potential of pulp mill effluent using bacterial, fish and mammalian assays. Chemosphere 31:3567–3574
39. Reifferscheid G, Heil J, Oda Y, Zahn RK (1991) A microplate version of the SOS/*umu*-test for rapid detection of genotoxins and genotoxic potentials of environmental samples. Mutat Res 253:215–222
40. Castillo M, Alonso MC, Riu J, Reinke M, Kloter G, Dizer H, Fischer B, Hansen PD, Barcelo D (2001) Identification of cytotoxic compound in European wastewaters during a field experiment. Anal Chim Acta 426:265–277
41. Dizer H, Wittekindt E, Fischer B, Hansen PD (2002) The cytotoxic and genotoxic potential of surface water and wastewater effluents as determined by bioluminescence, *umu*-assays and selected biomarkers. Chemosphere 46:225–233
42. Hartmann A, Alder AC, Koller T, Widmer RM (1998) Identification of fluoroquinolone antibiotics as the main source of *umuC* genotoxicity in native hospital waste water. Environ Toxicol Chem 17:377–382
43. Hartmann A, Golet EM, Gartiser S, Alder AC, Koller T, Widmer RM (1999) Primary DNA damage but not mutagenicity correlates with ciprofloxacin concentrations in German hospital waste waters. Arch Environ Contam Toxicol 36:115–119
44. Gartiser S, Hafner C, Hercher C, Konenberger-Schafer K, Paschke A (2010) Whole effluent assessment of industrial wastewater for determination of bat compliance. Part 1: paper manufacturing industry. Environ Sci Pollut Res 17:856–865
45. Zegura B, Heath E, Cernosa A, Filipic M (2009) Combination of *in vitro* bioassays for the determination of cytotoxic and genotoxic potential of wastewater, surface water and drinking water samples. Chemosphere 75:1453–1460
46. Plaza G, Nalecz-Jawecki G, Ulfig K, Brigmon RL (2005) Assessment of genotoxic activity of petroleum hydrocarbon-bioremediated soil. Ecotoxicol Environ Saf 62:415–420
47. Gustavsson L, Engwall M (2006) Genotoxic activity of nitroarene-contaminated industrial sludge following large-scale treatment in aerated and non-aerated sacs. Sci Total Environ 367:694–703
48. Macova M, Toze S, Hodgers L, Mueller JF, Bartkow M, Escher BI (2011) Bioanalytical tools for the evaluation of organic micropollutants during sewage treatment, water recycling drinking water generation. Water Res 45:4238–4247
49. Shen L, Wu JY, Lin GF, Shen JH, Westendorf J, Huehnerfuss H (2003) The mutagenic potentials of tap water samples in Shanghai. Chemosphere 52:1642–1646
50. Li J, Cui Q, Wang Z (2008) Assessing the potential risk of oil-field produced waters using a battery of bioassays/biomarkers. Bull Environ Contam Toxicol 80:492–496
51. Fang YX, Ying GG, Zhao JL, Clen F, Liu S, Zhang LJ, Yang B (2012) Assessment of hormonal activities and genotoxicity of industrial effuents using *in vitro* bioassays combined with chemical analysis. Environ Toxicol Chem 31:1273–1282
52. Yan Y, Jain W, Li N, Ma M, Wang D, Wang Z, Rao K (2014) Assessing of genotoxicity of 16 centralized source-waters in China by means of the SOS/*umu* assay and the micronucleus test: initial identification of the potential genotoxicants by use of a GC/MS method and the QSAR Toolbox 3.0. Mutat Res 763:36–43
53. Wu GY, Li Y, Hu HY, Sun YX, Zhao FY (2010) Reduced effect of bromide on the genotoxicity of secondary effluent of a municipal wastewater treatment plant during chlorination. Environ Sci Technol 44:4924–4929

54. Wu GY, Li Y, Hu HY, Ding YN, Huang H, Zhao FY (2012) Removal of genotoxicity in chlorinated secondary effluent of a domestic wastewater treatment plant during dechlorination. Environ Sci Pollut Res Int 19:1–7
55. Tang F, Hu HY, Wu QY, Sun YX, Shi XL, Huang JJ (2013) Effects of chemical agent injection on genotoxicity of wastewater in a microfiltration-reverse osmosis membrane process for wastewater reuse. J Hazard Mater 15:231–237
56. Cao N, Yang M, Zhang Y, Hu J, Ike M, Hirotsuji J, Matsui H, Inoue D, Sei K (2009) Evaluation of wastewater reclamation technologies based on in vitro and in vivo bioassays. Sci Total Environ 407:1588–1597
57. Li L, Wei D, Du Y (2013) Transformation of cefazolin during chlorination process: products mechanism and genotoxicity assessment. J Hazard Mater 262:48–54
58. Tian Z, Oda Y, Zhang Y, Yang M, Li H (2015) Use of a new enzyme extraction system to improve the sensitivity of SOS/*umu* test and application to environmental samples. Bull Environ Contam Toxicol 94:370–375
59. Xiao R, Wang Z, Wang C, Yu G (2006) Soil screening for identifying ecological risk stressors using a battery of *in vitro* cell assays. Chemosphere 64:71–78
60. Xiao R, Wang Z, Wang C, Yu G, Zhu Y (2006) Genotoxic risk identification of soil contamination at a major industrialized city in northeast China by a combination of *in vitro* and *in vivo* bioassays. Environ Sci Technol 40:6170–6175
61. Yu G, Xiao R, Wang D, Zhou J, Wang Z (2008) Assessing the ecological risk of soil irrigated with wastewater using *in vitro* cell bioassays. J Environ Sci Health A Tox Hazard Subst Environ Eng 43:1618–1627
62. Lei B, Kang J, Wang X, Liu Q, Yu Z, Zeng X, Fu J (2015) The toxicity of sediments from Taihu Lake evaluated by several in vitro bioassays. Environ Sci Pollut Res Int 22:3419–3430
63. Whong WZ, Wen YF, Stewart J, Ong TM (1986) Validation of the SOS/*umu* test with mutagenic complex mixtures. Mutat Res 175:139–144
64. Ong TM, Stewart J, Wen YF, Whong WZ (1987) Application of SOS *umu*-test for the detection of genotoxic volatile chemicals and air pollutants. Environ Mutagen 9:171–176
65. Yoshida J, Kosaka H, Tomioka K, Kumagai S (2006) Genotoxic risks to nurses from contamination of the work environment with antineoplastic drugs in Japan. J Occup Health 48:517–522
66. Allinson M, Kageyama S, Nakajima D, Kamata R, Shiraishi F, Goto S, Salzman SA, Allison G (2012) A pilot survey of 39 Victorian WWTP effluents using a high speed luminescent *umu* test in conjunction with a novel GC-MS-database technique for automatic identification of micropollutants. Water Sci Technol 66:768–774
67. Shimada T, Nakamura S (1987) Cytochrome P-450 mediated activation of procarcinogens and promutagens to DNA-damaging products by measuring expression of *umu* gene in *Salmonella typhimurium* TA1535/pSK1002. Biochem Pharmacol 36:1979–1987
68. Shimada T, Nakamura S, Imaoka S, Funae Y (1987) Genotoxic and mutagenic activation of aflatoxin B_1 by constitutive forms of cytochrome P-450 in rat liver microsomes. Toxicol Appl Pharmacol 91:13–21
69. Shimada T, Okuda Y (1988) Metabolic activation of environmental carcinogens and mutagens by human liver microsomes; role of cytochrome P-450 homologous to a 3-methylcholanthrene-inducible isozyme in rat liver. Biochem Pharmacol 37:459–465
70. Shimada T, Guengerich FP (1989) Evidence for cytochrome P-450_{NF}, the nifedipine oxidase, being the principal enzyme involved in the bioactivation of aflatoxins in human liver. Proc Natl Acad Sci U S A 86:462–465
71. Shimada T, Iwasaki M, Martin MV, Guengerich FP (1989) Human liver microsomal cytochrome P-450 enzymes involved in the bioactivation of procarcinogens detected by *umu* gene response in *Salmonella typhimurium* TA1535/pSK1002. Cancer Res 49:3218–3228
72. Shimada T, Martin MV, Pruess-Schwartz D, Marnett LJ, Guengerich FP (1989) Roles of individual human cytochrome P-450 enzymes in the bioactivation of benzo(a)pyrene,

7,8-dihydroxy-7,8-dihydrobenzo(a)pyrene, and other dihydrodiol derivatives of polycyclic aromatic hydrocarbons. Cancer Res 49:6304–6312

73. Brian WR, Sari MA, Iwasaki M, Shimada T, Kaminsky L, Guengerich FP (1990) Catalytic activities of human liver cytochrome P-450 IIIA4 expressed in *Saccharomyces cerevisiae*. Biochemistry 29:11280–11292
74. Guengerich FP, Shimada T (1991) Activation of amino-α-carboline, 2-amino-1-methyl-6-phenylimidazo[4,5-*b*]pyridine and a copper phthalocyanine cellulose extract of cigarette smoke condensate by cytochrome P450 enzymes in rat and human liver microsomes. Cancer Res 51:5284–5291
75. Yamazaki H, Degawa M, Funae Y, Imaoka S, Inui Y, Guengerich FP, Shimada T (1991) Roles of different cytochrome P450 enzymes in bioactivation of the potent hepatocarcinogen 3-methoxy-4-aminoazobenzene by rat and human liver microsomes. Carcinogenesis 12:133–139
76. Yamazaki H, Hatanaka N, Kizu R, Hayakawa K, Shimada N, Guengerich FP, Nakajima M, Yokoi T (2000) Bioactivation of diesel exhaust particles extracts and their major nitrated polycyclic aromatic hydrocarbon components, 1-nitropyrene and dinitropyrenes, by human cytochrome P450s 1A1, 1A2, and 1B1. Mutat Res 472:129–138
77. Takemoto K, Yamazaki H, Nakajima M, Yokoi T (2002) Genotoxic activation of benzophenone and its two metabolites by human cytochrome P450s in SOS/*umu* assay. Mutat Res 19:199–204
78. Shimada T, Yamazaki H, Shimura H, Tanaka R, Guengerich FP (1990) Metabolic deactivation of furylfuramide by cytochrome P450 in human and rat liver microsomes. Carcinogenesis 11:103–110
79. Shimada T, Guengerich FP (1990) Inactivation of 1,3-,1,6-, and 1,8-dinitropyrene by cytochrome P-450 enzymes in human and rat liver microsomes. Cancer Res 50:2036–2043
80. Ueng Y-F, Shyu C-C, Liu T-Y, Oda Y, Lin Y-L, Liao J-F, Chen C-F (2001) Protective effects of baicalein and wogonin against benzo[*a*]pyrene-and aflatoxin B_1-induced genotoxicities. Biochem Pharmacol 62:1653–1660
81. Rosenkranz HS, Mermelstein R (1983) Mutagenicity and genotoxicity of nitroarenes: all nitro-containing chemicals were not created equal. Mutat Res 114:217–267
82. Tokiwa H, Ohnishi Y (1986) Mutagenicity and carcinogenicity of nitroarenes and their sources in the environment. CRC Crit Rev Toxicol 17:23–60
83. Novak M, Kahley MJ, Eiger E, Helmick JS, Peters HE (1993) Reactivity and selectivity of nitrenium ions derived from ester derivatives of carcinogenic *N*-(4-biphenyllyl)hydroxylamine and the corresponding hydroxamic acid. J Am Chem Soc 115:9453–9460
84. Watanabe M, Nohmi T, Ishidate M Jr (1987) New tester strains of *Salmonella typhimurium* highly sensitive to mutagenic nitroarenes. Biochem Biophys Res Commun 147:974–979
85. Watanabe M, Ishidate M Jr, Nohmi T (1989) A sensitive method for detection of mutagenic nitroarenes: construction of nitroreductase-overproducing derivatives of *Salmonella typhimurium* strains TA98 and TA100. Mutat Res 216:211–220
86. Oda Y, Shimada T, Watanabe M, Nohmi T (1992) A sensitive system (*umu* test) for the detection of mutagenic nitroarenes in *Salmonella typhimurium* strain possessing elevated nitroreductase. Water Sci Technol 25:279–284
87. Oda Y, Shimada T, Watanabe M, Ishidate M Jr, Nohmi T (1992) A sensitive *umu* test system for the detection of mutagenic nitroarenes in *Salmonella typhimurium* NM1011 having a high nitroreductase activity. Mutat Res 272:91–99
88. Oda Y, Yamazaki H, Watanabe M, Nohmi T, Shimada T (1993) Highly sensitive *umu* test system for the detection of mutagenic nitroarenes in *Salmonella typhimurium* NM3009 having high *O*-acetyltransferase and nitroreductase activities. Environ Mol Mutagen 21:357–364
89. Oda Y, Yamazaki H, Watanabe M, Nohmi T, Shimada T (1995) Development of high sensitive *umu* test system: rapid detection of genotoxicity of promutagenic aromatic amines

by *Salmonella typhimurium* strain NM2009 possessing high *O*-acetyltransferase activity. Mutat Res 334:145–156

90. Yamazaki H, Oda Y, Shimada T (1992) Use of a newly developed tester strain *Salmonella typhimurium* NM2009 for the study of metabolic activation of carcinogenic aromatic amines by rat liver microsomal cytochrome P-450 enzymes. Mutat Res 272:183–192
91. Ozturk K, Durusoy M (1999) The detection and comparison of the genotoxic effects of some nitroaromatic compounds by the *umu* and SOS chromotest systems. Toxicol Lett 108:63–68
92. Durusoy M, Kambur S (2003) The application of the *umu* test system for screening mutagenicity of surface water. Turk J Biochem 28:3–7
93. Funasaka K, Kitano M, Nakama A, Yoshikura T, Oda Y (2003) Detection of genotoxicity of atmospheric particles using a high-throughput microplate *umu*-test system. Acta Biochim Pol 50:291–296
94. Ma F, Yuan G, Meng L, Oda Y, Hu J (2012) Contributions of flumequine and nitroarenes to the genotoxicity of river and ground waters. Chemosphere 88:476–483
95. Ono Y, Somiya I, Kawaguchi T (1996) Genotoxicity of substances in the nightsoil and its biologically treated water. Water Res 30:569–577
96. Ono Y, Somiya I, Kawaguchi T, Mohri S (1996) Evaluation of toxic substances in effluents from a wastewater treatment plant. Desalination 106:255–261
97. Ohe T, Nukaya H (1996) Genotoxic activity of 1-nitropyrene in water from the Yodo river, Japan. Sci Total Environ 181:7–12
98. Ohe T (1997) Quantification of mutagenic/carcinogenic heterocyclic amines, MeIQx, Trp-P-1, Trp-P-2, and PhIP, contributing highly to genotoxicity of river water. Mutat Res 393:73–79
99. Shimada T, Gillam EM, Sandhu P, Guo Z, Tukey RH, Guengerich FP (1994) Activation of procarcinogens by human cytochrome P450 enzymes expressed in *Escherichia coli*. Simplified bacterial systems for genotoxicity assays. Carcinogenesis 15:2523–2529
100. Yamazaki H, Oda Y, Funae Y, Imaoka S, Inui Y, Guengerich FP, Shimada T (1992) Participation of rat liver cytochrome P450 2E1 in the activation of *N*-nitrosodimethlamine and *N*-nitrosodiethylamine to products genotoxic in an acetyltransferase-overexpressing *Salmonella typhimurium* strain (NM2009). Carcinogenesis 13:979–985
101. Yamazaki H, Inui Y, Yun C-H, Guengerich FP, Shimada T (1992) Cytochrome P4502E1 and 2A6 enzymes as major catalysts for metabolic activation of *N*-nitrosodialkylamines and tobacco-related nitrosamines in human liver microsomes. Carcinogenesis 13:1789–1794
102. Josephy PD, Lord HL, Snieckus VA (1994) Dimethylnitrosamine genotoxicity: does *N*-acetyltransferase activity play a role? Carcinogenesis 15:479–482
103. Yamazaki H, Shimada T (1992) Activation of 6-aminochrysene to genotoxic products by different forms of rat liver cytochrome P450 in an *O*-acetyltransferase overexpressing *Salmonella typhimurium* strain (NM2009). Biochem Pharmacol 44:913–920
104. Yamazaki H, Mimura M, Oda Y, Inui Y, Shiraga T, Iwasaki K, Guengerich FP, Shimada T (1993) Roles of different form of cytochrome P450 in the activation of the promutagen 6-aminochrysene to genotoxic metabolites in human liver microsomes. Carcinogenesis 14:1271–1278
105. Yamazaki H, Mimura M, Oda Y, Gonzalez FJ, El-Bayoumy K, Chae Y-H, Guengerich FP, Shimada T (1994) Activation of *trans*-1,2-dihydro-1,2-dihydroxy-6-aminochrysene to genotoxic metabolites by rat and human cytochromes P450. Carcinogenesis 15:465–470
106. Yamazaki H, Inui Y, Wrighton SA, Guengerich FP, Shimada T (1995) Procarcinogen activation by cytochrome P450 3A4 and 3A5 expressed in *Escherichia coli* and by human liver microsomes. Carcinogenesis 16:2167–2170
107. Shimada T, Hayes CL, Yamazaki H, Amin S, Hecht SS, Guengerich FP (1996) Activation of chemically diverse procarcinogens by human cytochrome P-450 1B1. Cancer Res 56:2979–2984
108. Shimada T (1999/2000) Human cytochrome P450 1B1 and chemical carcinogenesis. Rev Toxicol 3:139–166. IOS Press

109. Shimada T, Watanabe J, Kawajiri K, Sutter TR, Guengerich FP, Gillam EMJ, Inoue K (1999) Catalytic properties of polymorphic human cytochrome P450 1B1 variants. Carcinogenesis 20:1607–1613
110. Shimada T, Oda Y, Gillam EMJ, Guengerich FP, Inoue K (2001) Metabolic activation of polycyclic aromatic hydrocarbons and other procarcinogens by cytochromes P4501A1and P450 1B1 allelic variants and other human cytochromes P450 in *Salmonella typhimurium* NM2009. Drug Metab Dispos 29:1178–1182
111. Shimada T, Murayama N, Yamazaki H, Tanaka K, Takenaka S, Komori M, Guengerich FP (2013) Metabolic activation of polycyclic aromatic hydrocarbons and aryl and heterocyclic amines by human cytochrome 2A13 and 2A6. Chem Res Toxicol 26:529–537
112. Hatanaka N, Yamazaki H, Oda Y, Guengerich FP, Nakajima M, Yokoi T (2001) Metabolic activation of carcinogenic 1-nitropyrene by human cytochrome P450 1B1 in *Salmonella typhimurium* strain expressing an *O*-acetyltransferase in SOS/*umu* assay. Mutat Res 497:223–233
113. Shimada T, El-Bayoumy K, Upadhyaya P, Sutter TR, Guengerich FP (1997) Inhibition of human cytochrome P450-catalyzed oxidations of xenobiotics and procarcinogens by synthetic organoselenium compounds. Cancer Res 57:4757–4764
114. Shimada T, Guengerich FP (2006) Inhibition of human cytochrome P4501A1-, 1A2-, 1B1-mediated activation of procarcinogens to genotoxic metabolites by polycyclic aromatic hydrocarbons. Chem Res Toxicol 19:288–294
115. Wu Z-L, Sohl CD, Shimada T, Guengerich FP (2006) Recombinant enzymes overexpressed in bacteria show broad catalytic specificity of human cytochrome P450 2W1 and limited activity of human cytochrome P450 2S1. Mol Pharmacol 69:2007–2014
116. Yueh M-F, Nguyen N, Famourzadeh M, Strassburg PS, Oda Y, Guengerich FP, Tukey RH (2001) The contribution of UDP-glucuronosyltransferase 1A9 on CYP1A2-mediated genotoxicity by aromatic and heterocyclic amines. Carcinogenesis 22:943–950
117. Davies R, Crespi CL, Rudo K, Turner TR, Langenbach R (1989) Development of a human cell line by selection and drug-metabolizing gene transfection with increased capacity to activate promutagens. Carcinogenesis 10:885–891
118. Yanagawa Y, Sawada M, Deguchi T, Gonzalez F, Kamataki T (1994) Stable expression of human CYP1A2 and *N*-acetyltransferasesin Chinese hamster CHL cells: mutagenic activation of 2-amino-3,8-dimethylimidazo[4,5-f]quinoxaline. Cancer Res 54:3422–3427
119. Josephy PD, DeBruin LS, Lord HL, Oak JN, Evans DH, Guo Z, Dong MS, Guengerich FP (1995) Bioactivation of aromatic amines by recombinant human cytochrome P4501A2 expressed in Ames tester strain bacteria: a substrate for activation by mammalian tissue preparations. Cancer Res 55:799–802
120. Kranendonk M, Mesquita P, Laires A, Vermeulen NP, Rueff J (1998) Expression of human cytochrome P450 1A2 in *Escherichia coli*: a system for biotransformation and genotoxicity studies of chemical carcinogens. Mutagenesis 13:263–269
121. Suzuki A, Kushida H, Iwata H, Watanabe M, Nohmi T, Fujita K-I, Gonzalez FJ, Kamataki T (1998) Establishment of a *Salmonella* tester strain highly sensitive to mutagenic heterocyclic amines. Cancer Res 58:1833–1838
122. Cooper MT, Porter TD (2000) Mutagenicity of nitrosamines in methyltransferase-deficient strains of *Salmonella typhimurium* coexpressing human cytochrome P450 2E1 and reductase. Mutat Res 454:45–52
123. Kushida H, Fujita K, Suzuki A, Yamada M, Nohmi T, Kamataki T (2000) Development of a *Salmonella* tester strain sensitive to promutagenic *N*-nitrosamines: expression of recombinant CYP2A6 and human NADPH-cytochrome P450 reductase in *S. typhimurium* YG7108. Mutat Res 471:135–143
124. Kushida H, Fujita K, Suzuki A, Yamada M, Endo T, Nohmi T, Kamataki T (2000) Metabolic activation of *N*-alkylnitrosamines in genetically engineered *Salmonella typhimurium* expressing CYP2E1 or CYP2A6 together with human NADPH-cytochrome P450 reductase. Carcinogenesis 21:1227–1232

125. Aryal P, Yoshikawa K, Terashita T, Guengerich FP, Shimada T, Oda Y (1999) Development of a new genotoxicity test system with *Salmonella typhimurium* OY1001/1A2 expressing human CYP1A2 and NADPH-P450 reductase. Mutat Res 442:113–120
126. Aryal P, Terashita T, Guengerich FP, Shimada T, Oda Y (2000) Use of genetically engineered *Salmonella typhimurium* OY1002/1A2 strain coexpressing human cytochrome P450 1A2 and NADPH-cytochrome P450 reductase and bacterial *O*-acetyltransferase in SOS/*umu* assay. Environ Mol Mutagen 36:121–126
127. Oda Y, Aryal P, Terashita T, Gillam EMJ, Guengerich FP, Shimada T (2001) Metabolic activation of heterocyclic amines and other procarcinogens in *Salmonella typhimurium umu* tester strains expressing human cytochrome P4501A1, 1A2, 1B1, 2C9, 2D6, 2E1, and 3A4 and human NADPH-P450 reductase and bacterial *O*-acetyltransferase. Mutat Res 492:81–90
128. Oda Y, Totsuka Y, Wakabayashi K, Guengerich FP, Shimada T (2006) Activation of aminophenylnorharman, aminomethylphenylnorharman and aminophenylharman to genotoxic metabolites by human *N*-acetyltransferases and cytochrome P450 enzymes expressed in *Salmonella typhimurium umu* tester strains. Mutagenesis 21:411–416
129. Oda Y, Hirayama T, Watanabe T (2009) Genotoxic activation of the environmental pollutant 3,6-dinitrobenzo[*e*]pyrene in *Salmonella typhimurium umu* strains expressing human cytochrome P450 and *N*-acetyltransferase. Toxicol Lett 188:258–262
130. Oda Y, Yamazaki H, Shimada T (1999) Role of human *N*-acetyltransferases, NAT1 or NAT2, in genotoxicity of nitroarenes and aromatic amines in *Salmonella typhimurium* NM6001 and NM6002. Carcinogenesis 20:1079–1083
131. Oda Y (2004) Analysis of the involvement of human *N*-acetyltransferase 1 in the genotoxic activation of bladder carcinogenic arylamines using a SOS/*umu* assay system. Mutat Res 554:399–406
132. Oda Y, Zhang Y, Buchinger S, Reifferscheid G, Yang M (2012) Roles of human sulfotransferases in genotoxicity of carcinogens using genetically engineered *umu* test strains. Environ Mol Mutagen 53:152–164
133. Oda Y, Yamazaki H, Their R, Ketterer B, Guengerich FP, Shimada T (1996) A new *Salmonella typhimurium* NM5004 strain expressing rat glutathione *S*-transferase 5-5: use in detection of genotoxicity of dihaloalkanes using an SOS/*umu* test system. Carcinogenesis 17:297–302
134. Shimada T, Yamazaki H, Oda Y, Hiratsuka A, Watabe T, Guengerich FP (1996) Activation and inactivation of carcinogenic dihaloalkenes and other compounds by glutathione *S*-transferase 5-5 in *Salmonella typhimurium* tester strain NM5004. Chem Res Toxicol 9:333–340
135. Habig WH, Jakoby WB (1981) Glutathione *S*-transferases (rat and human). Methods Enzymol 77:218–231
136. Meyer DJ, Coles B, Pemble SE, Gilmore KS, Fraser G, Ketterer B (1991) Theta, a new class of glutathione transferases purified from rat and man. Biochem J 274:409–414
137. Chasseaud LF (1979) The role of glutathione and glutathione transferases in the metabolism of chemical carcinogens and other electrophilic reagents. Adv Cancer Res 29:175–274
138. Monks TJ, Anders MW, Dekant W, Stevens JL, Lau SS, van Bladeren PJ (1990) Glutathione conjugate mediated toxicities. Toxicol Appl Pharmacol 106:1–19
139. Pemple SE, Schroeder KR, Spencer SR, Meyer DJ, Hallier E, Bolt HM, Ketterer B, Taylar JB (1994) Human glutathione *S*-transferase theta (GSTT1): cDNA cloning and the characterization of a genetic polymorphism. Biochem J 300:271–276
140. Thier R, Taylar JB, Pemble SE, Humphreys WG, Persmark M, Ketterer B, Guengerich FP (1993) Expression of mammalian glutathione *S*-transferase 5-5 in *Salmonella typhimurium* TA1535 leads to base-pair mutation upon exposure to dihalomethanes. Proc Natl Acad Sci U S A 90:8576–8580
141. Simura TP, Glancey MJ, Wolf CR (1993) Human glutathione *S*-transferase-expressing *Salmonella typhimurium* tester strains to study the activation/detoxification of mutagenic

compounds: studies with halogenated compounds, aromatic amines and aflatoxin B_1. Carcinogenesis 14:1371–1376
142. Rosenkranz HS, Mermelstein R (1985) The genotoxicity, metabolism and carcinogenicity of nitrated polycyclic aromatic hydrocarbons. J Environ Sci Health C3:221–272
143. Weisburger JH (1988) Past, present, and futurerole of carcinogenic and mutagenic *N*-substituted aryl compounds in human cancer causation. In: King CM, Romano LJ, Schuetzele D (eds) Carcinogenic and mutagenic responses to aromatic amines and nitroarenes. Elsevier, Amsterdam, pp 3–19
144. Grant DM, Josephy PD, Lord HL, Morrison LD (1992) *Salmonella typhimurium* strains expressing human arylamine *N*-acetyltransferases metabolism and mutagenic activation of aromatic amines. Cancer Res 52:3961–3964
145. Minchin RF, Reeves PT, Teitel CH, McManus ME, Mojarrabi B, Ilett KF, Kadlubar FF (1992) *N*- and *O*-acetylation of aromatic and heterocyclic amine carcinogens by human monomorphic and polymorphic acetyltransferases expressed in COS-1 cells. Biochem Biophys Res Commun 185:839–844
146. Watanabe M, Matsuoka A, Yamazaki N, Hayashi M, Deguchi T, Nohmi T, Sofuni T (1994) New sublines of Chinese hamster CHL stably expressing human NAT1 or NAT2 *N*-acetyltransferase or *Salmonella typhimurium O*-acetyltransferase: comparison of the sensitivities to nitroarenes and aromatic amines using the *in vitro* micronucleus test. Cancer Res 54:1672–1677
147. Poindexter EH Jr, Carpenter RO (1962) The isolation of harman and norharman from tobacco and cigarette smoke. Phytochemistry 1:215–221
148. Totsuka Y, Ushiyama H, Ishihara J, Sinha R, Goto S, Sugimura T, Wakabayashi K (1999) Quantification of the co-mutagenic b-carbolines, norharman and harman, in cigarette smoke condensates and cooked food. Cancer Lett 143:139–143
149. Oda Y, Watanabe T, Yamazaki H, Hirayama T (2007) Genotoxic activation of the environmental pollutant 3-nitrobenzanthrone by human cytochrome P450 enzymes expressed in *Salmonella typhimurium umu* tester strains. Genes Environ 29:146–152
150. Ohe T, Watanabe T, Wakabayashi K (2004) Mutagens in surface waters: a review. Mutat Res 567:109–149
151. Oda Y, Watanabe T, Terao Y, Nukaya H, Wakabayashi K (2008) Genotoxic activation of 2-phenylbenzotriazole-type compounds by human cytochrome P4501A1 and *N*-acetyltransferase expressed *Salmonella typhimurium umu* strains. Mutat Res 654:52–57
152. Watanabe T, Takahashi K, Konishi E, Hoshino Y, Hasei T, Asanoma M, Hirayama T, Wakabayashi K (2008) Mutagenicity of surface soil from residential areas in Kyoto city, Japan and identification of major mutagens. Mutat Res 649:201–212
153. Coughtrie MW (2002) Sulfation and molecular action. Pharmacogenomics J 2:297–308
154. Strott CA (2002) Sulfonation and molecular action. Endocr Rev 23:703–732
155. Falany CN (1997) Sulfation and sulfotransferases. 3. Enzymology of human cytosolic sulfotransferases. FASEB J 11:206–216
156. Weishiboum RM, Ottemess DM, Akso IA, Wood TC, Her C, Raftgianis RB (1997) Sulfation and sulfotransferases. 1. Sulfotransferase molecular biology: cDNA and genes. FASEB J 11:3–14
157. Miller JA, Surh Y-J (1994) Sulfonation in chemical carcinogenesis. In: Kauffman FC (ed) - Conjugation-deconjugation reactions in drug metabolism and toxicology. Springer, New York, pp 429–457
158. Glatt H (2005) Human cytosolic sulfotransferases (Chap 13). In: Pacifici GM, Coughtrie MWH (eds) Activation and inactivation of carcinogens and mutagens by human sulfotransferases. CRC, Boca Raton, pp 279–304
159. Daly AK (2003) Pharmacogenetics of the major polymorphic metabolizing enzymes. Fundam Clin Pharmacol 17:27–41
160. Gamage N, Barnett A, Hempel N, Duggleby RG, Martin JL, McManus ME (2006) Human sulfotransferases and their roles in chemical metabolism. Toxicol Sci 90:5–22

161. Glatt H, Pabel U, Meinl W, Frederikson H, Frandsen H, Muckel E (2004) Bioactivation of the heterocyclic aromatic amine 2-amino-3-methyl-9*H*-pyrido[2,3-*b*]indole (MeAαC) in recombinant test systems expressing human xenobiotic-metabolizing enzymes. Carcinogenesis 25:801–807
162. Chou H-C, Lang NP, Kadluber FF (1995) Metabolic activation of *N*-hydroxy arylamines and *N*-hydroxy heterocyclic amines by human sulfotransferase(s). Cancer Res 55:525–529
163. Muckel E, Frandsen H, Glatt H (2002) Heterologous expression of human *N*-acetyltransferases 1 and 2 and sulfotransferase 1A1 in *Salmonella typhimurium* for mutagenicity testing of heterocyclic amines. Food Chem Toxicol 40:1063–1068
164. Arlt VM, Glatt H, Muckel E, Pabel U, Sorg BL, Schmeiser HH, Phillip DH (2002) Metabolic activation of the environmental contaminant 3-nitrobenzanthrone by human acetyltransferases and sulfotransferase. Carcinogenesis 23:1937–1945
165. Sodum RS, Sohn OS, Nie G, Fiala ES (1994) Activation of the liver carcinogen 2-nitropropane by arylsulfotransferase. Chem Res Toxicol 7:344–351
166. Herrmann K, Engst W, Appel KE, Monien BH, Glatt HR (2012) Identification of human and murine sulfotransferases able to activative hydroxylated metabolites of methyleugenol to mutagens in *Salmonella typhimurium* and detection of associated DNA adducts using UPLC-MS/MS methods. Mutagenesis 27:453–464

Adv Biochem Eng Biotechnol (2017) 157: 135–158
DOI: 10.1007/10_2016_47

Published online: 8 August 2016

Reporter Gene Assays in Ecotoxicology

Tal Elad and Shimshon Belkin

Abstract The need for simple and rapid means for evaluating the potential toxic effects of environmental samples has prompted the development of reporter gene assays, based on tester cells (bioreporters) genetically engineered to report on sample toxicity by producing a readily quantifiable signal. Bacteria are especially suitable to serve as bioreporters owing to their fast responses, low cost, convenient preservation, ease of handling, and amenability to genetic manipulations. Various bacterial bioreporters have been introduced for general toxicity and genotoxicity assessment, and the monitoring of endocrine disrupting and dioxin-like compounds has been mostly covered by similarly engineered eukaryotic cells. Some reporter gene assays have been validated, standardized, and accredited, and many others are under constant development. Efforts are aimed at broadening detection spectra, lowering detection thresholds, and combining toxicity identification capabilities with characterization of the toxic effects. Taking advantage of bacterial robustness, attempts are also being made to incorporate bacterial bioreporters into field instrumentation for online continuous monitoring or on-site spot checks. However, key hurdles concerning test validation, cell preservation, and regulatory issues related to the use of genetically modified organisms still remain to be overcome.

Keywords Bacterial bioreporter, Biosensor, Dioxin, EDC, Environmental toxicology, Genetic engineering, Genotoxicity, Reporter gene, Stress response, Toxicity, Whole-cell bioassay

Contents

T. Elad and S. Belkin (✉)
Institute of Life Sciences, The Hebrew University of Jerusalem, Jerusalem 91904, Israel
e-mail: shimshon.belkin@mail.huji.ac.il

1 Introduction

When attempting to decipher the effect of a sample of interest on living organisms, an environmental toxicologist has at his disposal a tool kit with a choice of bioassays, which may be broadly categorized into two groups. The first is aimed at quantifying the general toxicity of the studied sample based on its effect on a test organism. Common general toxicity bioassays are based on higher organisms, mainly fish and crustaceans, including the standardized zebrafish [1, 2] and *Daphnia magna* acute toxicity tests [3]. The second group of bioassays allows for the detection of compounds with a defined effect at the molecular level, as pioneered by the *Salmonella* Mutagenicity Test [4]. Integral to effect testing are reporter gene toxicity bioassays, which utilize cells that respond to environmental stimuli through the synthesis of reporter proteins. In this chapter we explain the concept of reporter gene toxicity bioassays, review past and current advances, and discuss future prospects, mainly focusing on bacteria-based systems.

2 Microbes as Toxicity Test Organisms

An interesting characteristic of ecotoxicology is that whereas, on the one hand, there exists a set of long-established toxicity testing procedures, standardized, validated and in routine use, on the other hand, there is a continuous stream of proposed new bioassays. The latter are designed to address new classes of chemical challenges, employ different test organisms, or generally improve our ability to protect both the environment and human health from the presence of deleterious compounds. Although many of these assays remain as unrealized proposals in the scientific literature, others proceed all the way to full standardization, validation, and regulatory acceptance.

An important group of test organisms often found among such publications is bacteria. Although a direct relevance to human health does not always exist, bacteria

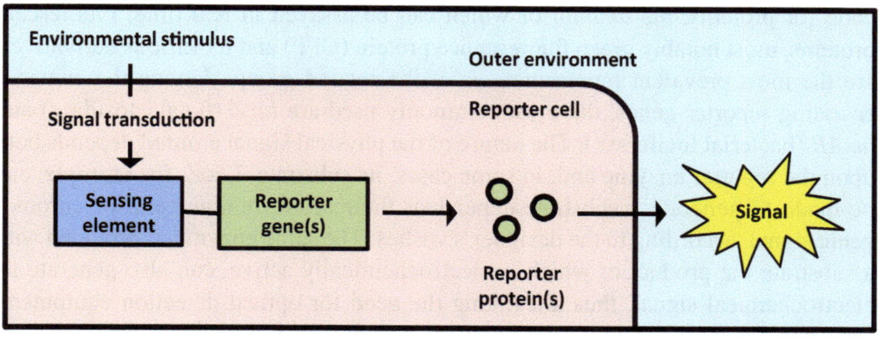

Fig. 1 Schematic depiction of a microbial bioreporter. A host cell is transformed with a fusion of a reporter gene(s) to a gene promoter (sensing element) known to be activated by a specific stress or chemical. Under the relevant environmental conditions, the resulting construct synthesizes the reporter protein(s), producing a readily quantifiable dose-dependent signal

are nevertheless endowed with several characteristics that render them highly attractive test organisms: they are fast to respond, they can easily and cost-effectively be obtained as large and homogeneous populations of live cells, and they can be preserved for prolonged periods of time and made available at short notice. These advantages, realized several decades ago, has prompted the development of the now widely accepted luminescent bacteria test, in which the short-term decrease in light emission by the naturally bioluminescent marine bacterium *Aliivibrio fischeri* (previously known as *Vibrio fischeri*) serves as an indication of sample toxicity [1, 2].

Another important advantage of bacteria as test organisms is their easy amenability to molecular engineering. The fusion of inducible gene promoters to diverse reporter genes, originally employed as molecular laboratory tools for studying gene regulation and expression, has found a different niche in the detection of biological effects exerted by the environment and sensed by the cells. The combination of a sensing element (in most cases a gene promoter), known to respond to a specific environmental condition, and a reporting element, the activity (or presence) of the product(s) of which can be easily monitored in real time, opens a broad spectrum of options for the design and construction of what are now often referred to as bacterial bioreporters. These can be tailored to recognize and detect specific chemicals, groups of chemicals, or deleterious global bio-effects such as toxicity or genotoxicity. The latter group is at the focus of the present review (Fig. 1).

3 Molecular Engineering of General Toxicity Bioreporters

As indicated above, the design of a toxicity bioreporter calls for an a priori selection of two major building blocks: a sensing and a reporting element. The reporting elements available comprise a relatively short list of options; most of these genes encode enzymes *the activity* of which produces a physical signal, whereas a few code for proteins, *the amount* of which can be assayed in real time. Fluorescent proteins, most notably green fluorescence protein (GFP) and its various derivatives, are the most prevalent representatives of the second group. Among the enzyme-encoding reporter genes, the most commonly used are *lacZ* (β-galactosidase) and *luxAB* (bacterial luciferase). The nature of the physical signal emitted depends both upon the reporter enzyme and, in some cases, its substrate. LacZ, for example, can be made to generate a substrate-dependent fluorescent, luminescent, or chromogenic signal, according to the designer's wishes. The same enzyme, if provided with a substrate the product of which is electrochemically active, can also generate an electrochemical signal, thus alleviating the need for optical detection equipment. Reporter genes commonly employed in bioreporter design have been listed in several comprehensive reviews [5–7].

Whereas, as noted above, the available selection of reporter genes is relatively limited, the options for gene promoters acting as the sensing element of the bioreporter are practically unlimited, and it may be stated with confidence that a sensing element can be found for any chemical or group of chemicals exerting a biological effect on live bacterial cells. Bacterial bioreporters that can sensitively and specifically detect polycyclic aromatic hydrocarbons such as naphthalene [8] or heavy metals such as mercury [9] were first described over 20 years ago. Since then many reports have described the construction and characterization of bacterial bioreporters for the detection of diverse targets including additional heavy metals [10, 11] and aromatic compounds [12, 13], nutrients [14, 15], and traces of explosives [16, 17].

For the detection of general toxicity, however, the issue becomes more challenging as gene promoters need to be identified that will respond not to specific chemicals but rather to general toxicity. Such genes should be induced in response to an imbalance in cellular homeostasis, and their degree of activation should represent the overall toxicity level. Over the years, several bacterial constructs have been reported to be characterized by relatively broad response spectra, among them those based on the promoters of the *Escherichia coli* genes *grpE* [18] and *fabA* [19] as sensing elements. These genes, involved in the regulation of the bacterial heat shock response and fatty acid synthesis, respectively, were shown to be induced by diverse toxic compounds belonging to different chemical groups. Other gene promoters reported to be of a relatively broad response spectrum are *uspA* [20] and *micF*. The latter, demonstrated to be induced by superoxide radicals [21], serves as an excellent indication of the involvement of oxidative stress in the toxic effects of numerous chemicals.

Regardless of how sophisticated is the molecular engineering, it is unreasonable to expect that any single bioreporter will be able to detect the deleterious effects of the entire spectrum of toxic compounds of interest, which may be of highly varied chemical natures and toxicological characteristics. The obvious solution to this limitation is the use of a panel of reporter strains, each responsive to a different class of compounds, which together provide a much better coverage of the relevant threat lists. Several such panels have been reported, varying in the number of their members from 5–14 [21–28] to several dozen [29] to genome-wide collections covering the entire *E. coli* promoter set [30]. Importantly, in these cases, advantage is taken of the fact that a panel's response not only indicates the existence of a toxic threat but also provides information concerning the toxic and chemical nature of the sample contents, as may be derived from its response pattern.

Moreover, comprehensive libraries of reporter gene fusions can offer an alternative to the more laborious and more expensive DNA microarray and RNA sequencing (RNA-seq) techniques for genome-wide transcriptional analysis [31, 32]. In particular, they can be used for assessing the toxicity and for determining the mode of action of known and potential environmental pollutants through gene expression, as proposed in a series of publications, in which mercury [33], nanomaterials [34], naphthenic acids [35], triphenyltin chloride [36], hydroxylated polybrominated diphenyl ethers (OH-PBDEs; [37]), bisphenol A [38], and flame retardants [39, 40] were tested against arrays of dozens to thousands of different bacterial bioreporters. Also explored in this context was the opportunity to account for the cells' dynamic response and to look beyond predefined endpoints [41].

4 Genotoxicity Bacterial Bioreporters

A special position among toxicity testing constructs is held by those tailored to detect the potential DNA damage hazards inherent in the sample, and thus assess its mutagenic and possibly carcinogenic risks. In the selection of sensing elements to be used for the construction of genotoxicity reporter systems, the most promising candidates are promoters of genes involved in DNA repair. Such genes, induced in response to either actual DNA damage or to the presence of DNA damaging agents, mostly belong to one of two inducible bacterial systems: the *recA*-dependent, *lexA*-controlled SOS response and the *recA*-independent, *ada*-controlled adaptive system induced in response to alkylation damage of DNA.

4.1 SOS-Based Assays

The bacterial SOS response is negatively controlled by the LexA protein that binds to the SOS box in the promoter region of the regulon genes, repressing their expression. De-repression occurs when the RecA protein binds to single-stranded DNA at replication forks blocked by DNA damage. Once bound to DNA, the RecA protein changes conformation and acts as a co-protease in the cleavage of LexA, thus allowing transcription of the SOS genes [42, 43].

The first commercial bacteria-based genotoxicity test kit, the SOS chromotest, was proposed in 1982 [44]. This test employs an *E. coli* strain with a transcriptional fusion between the SOS cell division inhibitor gene *sfiA* (also known as *sulA*) and the reporter enzyme β-galactosidase (LacZ). Oda and co-workers introduced the *umu* test, based on a *Salmonella typhimurium* strain harboring a *lacZ* fusion to the promoter of another SOS response gene, *umuC* [45]. Results from the *umu* test correlated well with those from both the Ames test and the SOS chromotest, and were highly representative of rodent carcinogenicity [46, 47]. Following several modifications, the *umu* test achieved ISO standardization [48]. Another SOS gene promoter that has been integrated into a commercial product is that of *recN*, which is involved in double stranded DNA break repair. Known as the VITOTOX® test, this assay comprises *E. coli* and *S. typhimurium* strains, which harbor a fusion of the *recN* promoter to the *A. fischeri luxCDABE* gene cassette [49]. The VITOTOX® test strains were evaluated with a variety of chemicals and environmental samples [49–52].

Additional notable SOS genes, on the basis of which genotoxicity bacterial bioreporters were constructed, are *recA* and *cda*. As described, the RecA recombinase gene plays a key role in the SOS response by its co-protease activity on the LexA repressor. Nunoshiba and Nishioka [53] described, and tested against an array of compounds, the colorimetric Rec-lac test, based on *E. coli* strain GE94 that carries a *recA::lacZ* fusion and its DNA repair-deficient derivative strains such as KY946 (*uvrA*), KY945 (*recA*), and KY943 (*lexA*). A different reporter system, the *A. fischeri luxCDABE* operon, was used by Vollmer et al. [54] to generate several *E. coli* reporter strains, one of them (DPD2794) carrying a *recA::luxCDABE* fusion. Bacterial bioreporter constructs of *recA* have since been widely used alone or with other reporter strains in diverse studies for toxicity assessment (e.g., [24, 27, 55, 56]). The colicin D gene *cda*, a constituent of the ColD plasmid [57], served as a basis for a bioluminescent genotoxicity sensor using *Photobacterium leiognathi luxCDABE* as a reporter. This SOS *lux* test responded to diverse genotoxic agents and was comparable in sensitivity to the SOS chromotest, the *umu* test, and the Ames test [58]. In their GenoTox version, *cda*-based genotoxicity bacterial bioreporters use *gfp* as a reporter gene [59]. The signal is measured by means of flow cytometry, which was reported to enhance the sensitivity of fluorescent bioreporters [60, 61]. Furthermore, the *cda-gfp* GenoTox assay exhibited lower detection thresholds than other genotoxicity reporter gene assays that use different promoter-reporter combinations [62].

4.2 Non-SOS-Based Assays

Several bacterial DNA protection and repair systems that are independent of RecA-LexA control have been described; one of these, most efficiently induced by alkylating agents, has been generally termed the adaptive response. The adaptive system responds to the presence of methylated phosphotriesters generated by DNA alkylation that activate the *ada* gene product which, in turn, triggers the transcription of genes such as *ada*, *alkA*, *alkB*, and *aidB* [63]. Vollmer et al. [54] used *alkA*, which encodes a repair glycosylase (*N*3-methyladenine DNA glycosylase II; [63]) for the construction of a bioluminescent reporter strain. The construct displayed a very strong response to the alkylating agent methylnitronitrosoguanidine (MNNG), the magnitude of which was enhanced by very low background bioluminescence.

Another non-SOS gene employed was *nrdA*, which encodes for a ribonucleoside diphosphate reductase and is strongly affected by DNA damage. An *E. coli* strain carrying a fusion between the *nrdA* promoter and the *Photorhabdus luminescens luxCDABE* operon responded to the DNA damaging agents nalidixic acid, mitomycin C, MNNG, 4-nitroquinoline 1-oxide (4-NQO), and hydrogen peroxide, but not to other oxidants or phenolic compounds [64]. Combined together, *sulA*-, *recN*-, *recA*-, *alkA*-, and *nrdA*-based luminescent bacterial bioreporters were shown to be able not only to detect genotoxic agents but also to classify their potential mode of action [65, 66].

5 EDCs and Dioxins

Endocrine disrupting compounds (EDCs), as well as dioxin and dioxin-like compounds (hereafter dioxins), are given special attention by monitoring agencies because of their potent negative impact on both human and environmental health. The mode of action of EDCs and dioxins is based on the translocation of an activated receptor to the nucleus. By definition, such a sequence of events can be directly detected only by use of eukaryotic bioreporters. Indeed, yeast and mammalian cells are central in the design of either EDCs or dioxins reporter gene assays.

The yeast estrogen and androgen screens (also known as YES and YAS, respectively) are prominent among yeast-based bioreporter assays for EDC detection [67, 68]. The *Saccharomyces cerevisiae* YES and YAS bioreporter strains are genetically modified to express the human estrogen receptor (hER-α) and the human androgen receptor, respectively. They are also inserted with a fusion of the *lacZ* reporter gene and human estrogen/androgen response elements, and, supplemented with a chromogenic substrate, yield a color product once a recognition event between receptor and compound occurs. The YES and YAS were tested with, and adapted to, various matrices, including dairy wastes [69] and aquifer samples [70], and are commercially available. The coupling of high-performance thin-layer chromatography with the yeast estrogen screen (planar-YES, p-YES) can serve as a fast and robust screening tool in effect-directed analysis, which links chemical analysis with effect testing [71].

Following the same guiding principle as the YES and YAS, similar yeast-based bioreporter assays have been developed by employing *gfp* [72, 73] and firefly luciferase [74] reporter genes. A *luxAB* bioluminescent modification was shown to be applicable for the United States Environmental Protection Agency's Endocrine Disruptor Screening and Testing Program [75], and most recently facilitated the comparison of treated wastewater from a traditional activated sludge facility and a membrane bioreactor [76]. Human cell line-based reporter gene assays are also at the forefront of EDC detection. Such are the transactivation in vitro assays using the hERα-HeLa-9903 human cervical tumor and the BG1Luc-4E2 human ovarian adenocarcinoma cell lines. These two cell lines express ER and have been stably transfected with an ER-responsive luciferase reporter gene, and the two assays are accredited in the OECD Guidelines for the Testing of Chemicals [77]. Widely used are the ER CALUX® strains, which include human bone marrow and breast cancer cell lines incorporating the firefly luciferase gene coupled to estrogen responsive elements [78].

Dioxins exert their adverse effect by activating the intracellular aryl hydrocarbon receptor and transcription factor (AhR) of all vertebrates, including humans [79]. The major reporter gene assays for dioxin detection accordingly utilize mammalian cells and measure AhR-dependent gene expression. The DR CALUX ® bioassay, for example, consists of a genetically modified H4IIe rat hepatoma cell line that contains the firefly luciferase gene fused downstream to AhR response elements [80, 81]. The DR CALUX® was used for screening of feed and food samples in official control programs [82], and was also validated against chromatographic analytical methods [83] and calibrated in intra- and inter-laboratory studies [84]. A somewhat different variation of the reporter gene assay concept is manifested in the H4IIe rat hepatoma ethoxyresorufin-*O*-deethylase (EROD) bioassay, in which the cytochrome P450 enzyme CYP1A1, the expression of which is AhR-dependent, serves as an endogenous reporter gene, de-ethylating its substrate 7-ethoxyresorufin to an easily measured fluorescent product, resorufin [85]. Although the EROD bioassay is considered rapid and simple, its sensitivity might be hampered by the inhibiting effect of AhR ligands on CYP1A1 [86]. In addition to H4IIe, a diverse range of cell lines has been employed for the development of AhR-based bioassays by use of either transfected reporter genes (e.g., luciferase, *gfp*, *lacZ*) or CYP1A1 as an endogenous reporter (see [87] for a review). To circumvent the difficulty inherent in harnessing prokaryotes for the detection of dioxin-like compounds, Turner et al. [88] have introduced an *E. coli* luminescent recombinant strain for the detection of hydroxylated polychlorinated biphenyls (OH-PCBs) by taking advantage of the recognition of a group of related compounds, namely hydroxylated biphenyls, by the product of the *hbpR* gene in the *hbp* operon from *Pseudomonas azelaica* HBP1.

6 From Bioassays to Biosensors

A biosensor is commonly defined as a device that combines a biological entity and an electronic system for the detection of various substances in water, air, or body fluids [6]. A capacity to withstand environmental conditions, easy handling, rapid responses, and readily measurable signals make bacterial bioreporters highly suitable for the development of biosensors for field applications. A bacterial bioreporter field biosensor should provide the conditions for reporter cell long term preservation, and be comprised of a signal transducer, signal processing and analysis algorithms, and, if required, a communication unit. Although a fully operational bacterial bioreporter biosensor embodying this entire wish list has not yet been introduced commercially, substantial steps towards its introduction have been taken. Some examples are described below, organized into two categories: online continuous monitoring and portable field kits.

6.1 Online Continuous Monitoring

Online continuous water quality monitoring is normally based on the continuous measurement of physicochemical parameters such as temperature, pH, conductivity, and oxygen, and is often supplemented by automated sampling for specific analysis of organic and inorganic pollutants, performed using laboratory-based chromatographic techniques. Because of the elaborate sample preparation procedures and high costs, the time resolution of the data provided by chromatography-based continuous monitoring is low; in most monitoring programs a daily sampling frequency is considered high [89]. Furthermore, it is difficult for this approach, geared towards the analysis of specific chemicals, to detect unknown compounds. As a result, there is a risk that sporadic contamination is not identified on time. As a partial solution to this need, online continuous biomonitoring systems have been developed to provide a rapid alarm against the presence of contaminants. These biological early warning systems are based on the monitoring of behavioral or physiological stress responses of test organisms that are continuously exposed to the tested water either in bypass systems or directly in-stream. The test organisms range from fish [90], daphnia, and mussels [89, 91, 92] to algae and bacteria [93].

Bacterial bioreporters offer simple and rapid online continuous biomonitoring solutions that can indicate not only the occurrence of toxic events but also the specific effect of the chemicals involved. Such designs mostly include luminescent reporting elements, which require neither the addition of exogenous reagents nor the hardware for optical excitation. Gu and co-workers [94, 95] have developed a multichannel system featuring stress-responsive reporter strains kept in suspension in a steady physiological state. Each channel was assigned with a reporter strain responsive to a specific kind of stress, and the response profile was used for pollution classification. To avoid the complexity inherent in maintaining a continuous culture [96], cells have been immobilized on the tips of a liquid light guide [97] or an optical fiber [98, 99]. Immobilized reporters have also been loaded onto a disposable card, which was employed in the assembly of a multi-strain biosensor for online detection of metal pollution and its classification [100]. Another report described the construction of a biosensor for online continuous water monitoring, at the heart of which was a panel of luminescent bacterial reporter strains immobilized in specialized flow-through chambers and probed by sensitive light detectors [101]. The living cell panel consisted of a general toxicity reporter (*micF-lux*), a genotoxicity reporter (*recA-lux*), and a heavy metal reporter (*arsR-lux*). The biosensor was shown to operate for at least 10 consecutive days under continuous tap water flow, detecting and classifying spikes of both model chemicals and industrial wastewater within 0.5–2.5 h, and requiring minimal maintenance. The detection was facilitated by a signal processing procedure laid out to determine whether an observed increase in the light intensity is significant enough for raising the alarm on a possible pollution event. The procedure was centered on the rate of change of the light signal, which was found to be a better indicator of reporter induction than the signal's absolute value; basal light signal fluctuations were also considered to eliminate false positives (Fig. 2).

6.2 Portable Field Kits

Easy-to-operate field kits for on-site monitoring allowing real-time analysis is beneficial when a rapid response is required or in remote areas where equipped laboratories or skilled personal are scarce. In an attempt to realize the substantial potential of microbial bioreporters to be incorporated in to such instrumentation, researchers have proposed and demonstrated the feasibility of various portable biosensor configurations, adopting diverse reporting systems and preservation techniques. Representative examples follow.

Nivens et al. [102] pioneered the field by describing a Bioluminescent Bioreporter Integrated Circuit (BBIC), a setup designated for field deployment consisting of bioluminescent bacterial bioreporters sustained within a micro-environment, a complementary metal oxide semiconductor (CMOS) integrated circuit microluminometer, and a light-tight enclosure. Equipped with the bacterial bioreporter strain *Pseudomonas fluorescens* 5RL, the BBIC setup was reported to detect low concentrations of model compounds salicylate and naphthalene in liquid and gas environments, respectively [102, 103]. Eltzov et al. [104] assembled a portable biosensor integrating a PMT detector and a liquid light guide. For testing, the biosensor was equipped with disposable calcium alginate pads containing three stress-specific recombinant reporter strains, harboring a plasmid-borne fusion of *recA*, *grpE*, or *fabA* to the *luxCDABE* gene cassette. The biosensor was then exposed to organic solvents, heavy metals, and endocrine disrupting compounds, and to environmental water samples, and its capability to detect the presence of toxicants was confirmed. Notably, the group demonstrated that the same biosensor can also be used for indoor air toxicity monitoring [105].

Using β-galactosidase reporter protein and *E. coli* hosts, Stocker et al. [106] have developed a paper strip test. According to the described protocol, reporter cells are suspended in drying protectant solution and spotted on paper strips, which are in turn vacuum-dried and stored. Upon use, the strips are placed inside the sample for 30 min. A substrate (X-gal) solution is then spotted on the bacterial spot and the color that develops is compared to that of a reference test. Originally demonstrated for rapid and quantitative measurements of arsenic in potable water, this paper strip whole-cell biosensor was later also successfully tested with quorum signaling molecules [107].

Daunert and co-workers have incorporated spore-forming bioreporters into a centrifugal compact disk (CD) microfluidic platform towards the development of a

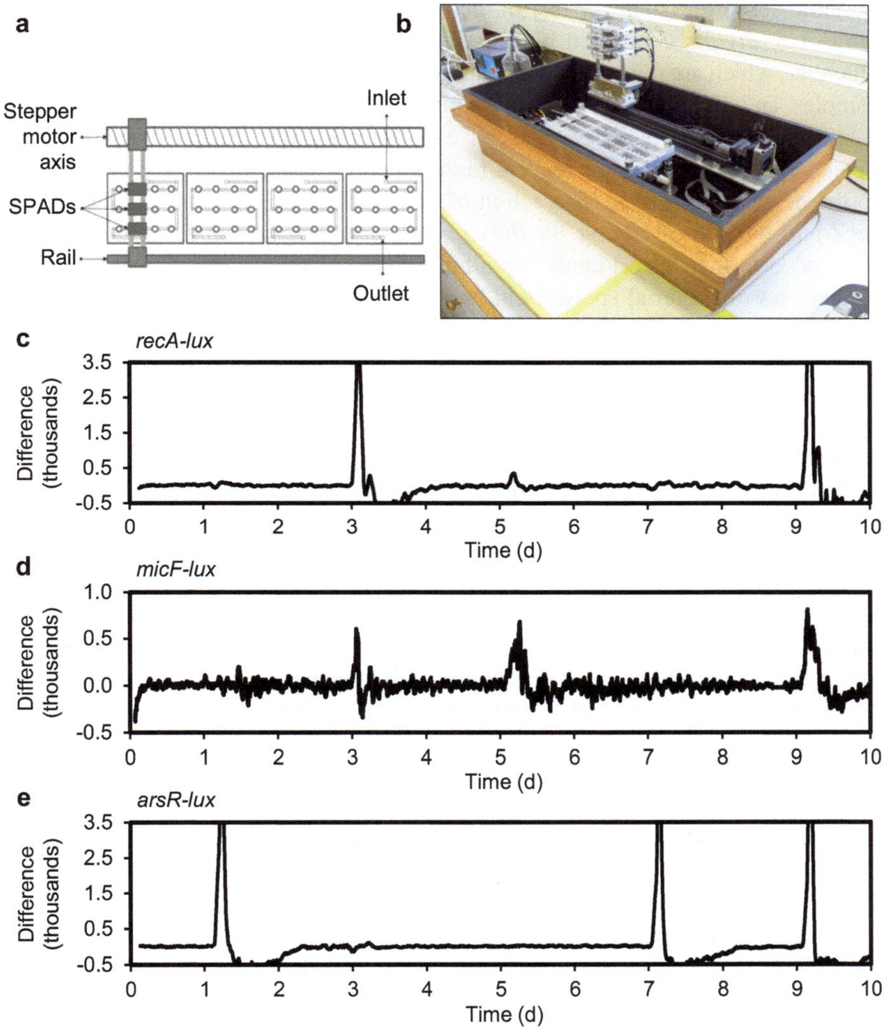

Fig. 2 A bacterial bioreporter-based biosensor for online continuous monitoring of water toxicity. The biosensor consists of four separate flow-through chambers. Each chamber comprises 12 wells, in which reporter cells are immobilized in agar. Three aligned single-photon avalanche diodes (SPADs), connected to a single-axis linear stepper motor, measure the light signal at 1-min intervals. Each of *recA*-, *micF*-, and *arsR-lux E. coli* reporter strains was immobilized in an individual flow-through chamber in 12 replicates. Tap water was pumped continuously through the system for 10 days, in the course of which five simulations of pollution events were carried out by 2-h pulses of tap water spiked with arsenite (day 1 and day 7), nalidixic acid (a genotoxic agent; day 3), paraquat (an oxidative herbicide; day 5), and a mix of all three (day 9). (**a**) A partial scheme of the four biochips and the photodetectors. (**b**) A photograph of the system. (**c–e**) The recorded signals of *recA-lux*, *micF-lux*, and *arsR-lux* reporters, respectively; the y-axis represents the difference between two consecutive photon counts [101]

portable sensing system for fast on-field analysis [108]. The platform consists of reagent reservoirs controlled by burst valves, a mixing channel, and a detection chamber, which are microfabricated on a circular disk. Bacterial bioreporter spores, samples, and, if needed, a substrate solution, are placed in the reagent reservoirs. At a specific rotation frequency, the valves controlling the reagent reservoirs burst open and the fluids from the reagent reservoirs flow to the mixing channel and then to the detection chamber. The action of the platform was demonstrated by use of *lacZ* and *gfp* spore-forming *Bacillus spp.* reporter strains, which germinated and responded to their target analytes within 2.5–3 h. The advantages inherent in the ruggedness of bacterial spores hold promise in providing a biosensor with a long shelf life. Indeed, the same group reported that the sensing spores could be effectively stored for over 2 years at room temperature and for at least 1 year under extreme conditions [108–110]. Centrifugal CD microfluidics is an active area of research and development; for more information on the state-of-the-art thereof the reader is directed to a recent critical review [111].

Roda et al. [112] have developed a portable device comprising a disposable microwell cartridge with immobilized microbial bioreporters placed in contact with a compact CCD sensor through a fiber optic taper. Cells were immobilized in an aqueous mixture of agarose, PVP, and collagen and kept their vitality for at least 1 month at 4 °C. To demonstrate the biosensor's capability, the authors used, among others, a new yeast bioreporter, which emits both green light in response to androgens and red light as vitality control. Light signal separation was carried out by using printed filters, and testosterone was quantified after correction for cell vitality. The same group has also developed a cell phone-based biosensor by transforming a smart-phone, low-cost 3D-printed adapter, cell cartridges, and a specially-programmed application for bioluminescence detection into a standalone user-friendly platform for quantitative toxicity assessment [113]. A proof-of-concept was provided by the use of genetically engineered human cells that constitutively express a powerful green-light-emitting luciferase mutant.

Biran et al. [114] have constructed a bacterial bioreporter in which the tightly controlled strong promoter of the *E. coli* SOS response gene *sulA* was fused to the alkaline phosphatase-coding *phoA* reporter gene. Together with the design of a portable electrochemical detection platform [115], a chip able to detect DNA-damaging agents when dipped in water is envisaged. Tsai et al. [116] have introduced LumiSense, a portable system combining luminescent bacterial bioreporters loaded on a biochip and a linear charge-coupled device (CCD), the protective window of which was removed for high light collection efficiency. The system picked up a dose-dependent signal from a reporter strain harboring a fusion of the *recA* promoter and the *luxCDABE* gene cassette as the latter was exposed to various concentrations of nalidixic acid.

Choi and Gu [24] combined into a portable field kit a small light-proof test chamber, an optic-fiber, a luminometer, and stress-responsive bacterial bioreporters, harboring *fabA*::, *grpE*::, *recA*::, and *katG::luxCDABE* fusions, freeze-dried within vials. When the kit was challenged with phenolic compounds, general toxicity was reported by the *fabA* and *grpE* bioreporters. The ARSOlux biosensor test kit, developed by scientists at the Helmholtz Centre for Environmental Research—UFZ (Germany) and the Université de Lausanne (Switzerland), is based on lyophilized arsenic-specific luminescent reporter cells. Bioreporter vials are injected on site with the sampled water, incubated for 2 h, and arsenic concentrations are inferred by measuring the bacterial responses using a portable luminometer. The system, successfully tested in Bangladesh and Argentina, requires no preparatory steps except for filtration of turbid samples [117, 118].

7 Performance Enhancement

Bacterial bioreporters are not only carriers of a simple promoter-reporter fusion but can also be subjects of additional genetic manipulations aimed at enhancing their performance. Such manipulations have taken several forms, including the introduction of mutations into the tester strains, the modification of the sensing or reporting elements, and, for genotoxicity detection, the incorporation of metabolic activation capabilities.

7.1 Introduction of Mutations into Host Tester Strains

Several mutations introduced into bacterial bioreporters have been reported to lower their detection thresholds. Mutations of *uvrAB* and *rfa*, which hinder DNA repair and increase membrane permeability, respectively, render genotoxicity bacterial bioreporters responsive to lower concentrations of certain genotoxic agents [4, 44, 45, 51, 53, 59, 114, 119, 120]. A similar effect on reporter gene assays for genotoxicity detection is achieved by *tag* and *oxyR* mutations; the former inactivates 3-methyl-adenine DNA glycosylase I, an enzyme involved in the removal of alkylated bases from DNA, and thereby lowers the tester strain's detection thresholds towards alkylating agents, and the latter suppresses the oxidative stress response under the control of OxyR transcription regulator, increasing sensitivity to peroxide-generating compounds [46]. Last in this set of examples, limiting the host cell's efflux capacity by a *tolC* mutation lowered the detection thresholds of *cda*-, *recA*-, and *umuD*-based genotoxicity bioreporters [120–123] and those of *grpE*- and *fabA*-based general toxicity sensors [18, 19].

7.2 Modifying the Sensing or Reporting Elements

Molecular manipulations of the DNA fragment harboring the sensing promoter element to improve the bacterial response to target chemicals have been described. The influence of addition/subtraction of *lexA* binding sites on reporter gene expression in the SOS promoter::*uidA* fusion under genotoxic stress was examined with several SOS promoters (*umuD*, *sulA*, *recA*, and *recN*; [124]). The highest signals were produced by constructs that either harbored an additional *lexA* binding site in the *sulA* promoter that overlapped the −35 promoter region or lacked one of the two *recN* LexA binding sites. Replacing the wild-type −35 promoter sequence with the −35 sequence of the highly active *trp* gene promoter improved the performance of a *umuDC*::*gfp* construct [125].

Changing specific nucleotides in specific positions using site-directed techniques is also efficient in improving bioreporter abilities. The VITOTOX® *S. typhimurium* bioluminescent genotoxicity tester strain was enhanced by an up promoter mutation where a consensus nucleotide was introduced in the *recN* promoter −35 region [49]. Similarly, the −35 element of the *sulA* promoter in a *sulA::luxCDABE* fusion was changed to a consensus sequence, improving light signal intensity, response time, and detection threshold [126].

An opposite approach to the precisely planned point mutations targeting nucleotides at specific positions is random mutagenesis in a directed evolution process. This approach is generally based on error-prone polymerase chain reaction (PCR), with or without the combination of a DNA shuffling procedure, which can be performed on the target DNA sequence; the resulting library of variants is then screened for the desired feature and selected isolates are reprocessed in the same manner. This procedure is often applied for modifying protein sequence and performance, but it may just as well be applied to regulatory areas of the bacterial genome. In one cycle of an error-prone PCR performed on the *sulA* promoter in a *sulA::luxCDABE* reporter, an improved mutant was isolated, harboring three point mutations relative to the wild-type sequence [126]. Interestingly, these mutations were not located within well-defined regulatory sequences. This finding demonstrates the potential power of the directed evolution process, which introduces mutations into random locations that would not necessarily be identified as obvious targets by bioinformatic tools.

Reporting elements have also been modified with the aim of enhancing bacterial bioreporter performance. The *Photorhabdus luminescens luxCDABE* cassette was rearranged and split into two independent functional units: *luxAB*, coding for the luciferase enzyme, and *luxCDE*, coding for the reductase complex, which catalyzes the synthesis of the aldehyde necessary for the reaction [127]. The expression of each subunit was put under the control of either an inducible stress-responsive promoter or a synthetic constitutive promoter in both general toxicity and genotoxicity *E. coli* reporter strains. In all cases, the split combinations proved to be superior to the native *luxCDABE* configuration. The best combination was that of an inducible *luxAB* and a constitutive *luxCDE*, resulting in dramatically improved assay parameters and, at the same time, indicating that aldehyde availability may be limited when the five genes are expressed together. Another modification of the *P. luminescens lux* system reduced the half-life of these proteins, allowing for the monitoring of both the initiation and the termination of transcription in real time, thus upgrading circuit performance [128].

7.3 Incorporating Metabolic Activation Capabilities

Bacterial-based assays cannot carry out the complex biochemical reactions collectively known as metabolic activation, which take place mainly in mammalian liver cells, and in which some xenobiotics are transformed into genotoxic forms [129]. To overcome this limitation, rodent-derived cytochrome P450 (S9) preparations, which can convert susceptible compounds into active genotoxic entities, are included in the procedures of, for example, the Ames test and the *umu* test [4, 48]. To reduce at least partially the dependency of reporter gene assays on S9 extracts, several attempts have been made to genetically engineer bacterial cells to incorporate some of the enzymatic activities involved in the activation process of xenobiotics. These include the development of a series of *S. typhimurium* tester strains with high *O*-acetyltransferase (O-AT) and nitroreductase (NR) activities, which displayed increased sensitivity towards nitroarenes and aromatic amines [59, 130–133]. Such successful efforts clearly demonstrate the viability and potential of this approach; nevertheless, it should be remembered that the cyt P-450 complex is composed of a plurality of activities, the integration of which into a single reporter strain is highly challenging.

8 Outlook

The development of reporter gene assays and their application as ecotoxicological assessment tools have been prompted over the last decades by the need for simple and rapid methods for toxicity assessment and pollutant detection. New and innovative assays are constantly being proposed and tested in many institutions worldwide. The new reporter gene assays show great promise in broadening the detection spectrum and increasing the sensitivity offered by presently available techniques. Moreover, microbial bioreporters have been successfully integrated into biosensor hardware for both laboratory and field use. Based on this progress, it is tempting to envision a validated set of bioreporters, covering a range of biological effects and indicative as to the nature of toxic samples, routinely employed either as a part of an online early warning system or for spot checks, and serving as first-line safeguards of human and ecosystem health. However, for such assays and biosensor devices to be implementable in real-world applications, several key hurdles still need to be overcome.

First, most proposed reporter gene assays have not yet been validated in large-scale comparative studies against standard methods, nor have they been the subjects of inter-laboratory round-robin tests to determine their reproducibility. Second, additional progress needs to be made in keeping microbial bioreporters alive and active for prolonged periods of time; lack of long term preservation is often a major factor limiting the application of whole-cell biosensors. Third, strict regulations prohibit the use of microbial bioreporters outside laboratories because of concerns regarding the risks involved in the use of genetically modified organisms (GMOs). Although highly efficient measures to prevent the escape of bioreporters from portable field devices or continuous monitoring units can be taken, they might not suffice under current legislation.

Acknowledgements T.E. acknowledges with gratitude the Eshkol fellowship from the Israeli Ministry of Science and Technology. Research was supported in part by the Minerva Center for Bio-Hybrid Complex Systems and by the EU FP7 project BRAAVOO.

References

1. ISO (2007) ISO 11348:2007 Water quality – determination of the inhibitory effect of water samples on the light emission of *Vibrio fischeri* (luminescent bacteria test). ISO, Geneva
2. ISO (2007) ISO 15088:2007 Water quality – determination of the acute toxicity of waste water to zebrafish eggs (*Danio rerio*). ISO, Geneva
3. ISO (2012) ISO 6341:2012 Water quality – determination of the inhibition of the mobility of *Daphnia magna* Straus (*Cladocera, Crustacea*) – acute toxicity test. ISO, Geneva
4. Ames BN, McCann J, Yamasaki E (1975) Methods for detecting carcinogens and mutagens with the salmonella/mammalian-microsome mutagenicity test. Mutat Res 31:347–364
5. Daunert S, Barrett G, Feliciano JS, Shetty RS, Shrestha S, Smith-Spencer W (2000) Genetically engineered whole-cell sensing systems: coupling biological recognition with reporter genes. Chem Rev 100(7):2705–2738
6. van der Meer JR, Belkin S (2010) Where microbiology meets microengineering: design and applications of reporter bacteria. Nat Rev Microbiol 8(7):511–522
7. Yagi K (2007) Applications of whole-cell bacterial sensors in biotechnology and environmental science. Appl Microbiol Biotechnol 73(6):1251–1258
8. King JM, Digrazia PM, Applegate B, Burlage R, Sanseverino J, Dunbar P, Larimer F, Sayler GS (1990) Rapid, sensitive bioluminescent reporter technology for naphthalene exposure and biodegradation. Science 249(4970):778–781
9. Selifonova O, Burlage R, Barkay T (1993) Bioluminescent sensors for detection of bioavailable Hg(II) in the environment. Appl Environ Microbiol 59(9):3083–3090
10. Corbisier P, van der Lelie D, Borremans B, Provoost A, Lorenzo V, Brown NL, Lloyd JR, Hobman JL, Csöregi E, Johansson G, Mattiasson B (1999) Whole cell- and protein-based biosensors for the detection of bioavailable heavy metals in environmental samples. Anal Chim Acta 387(3):235–244
11. Magrisso S, Erel Y, Belkin S (2008) Microbial reporters of metal bioavailability. Microb Biotechnol 1:320–330
12. Applegate BM, Kehrmeyer SR, Sayler GS (1998) A chromosomally based *tod-luxCDABE* whole-cell reporter for benzene, toluene, ethybenzene, and xylene (BTEX) sensing. Appl Environ Microbiol 64(7):2730–2735
13. Tecon R, Beggah S, Czechowska K, Sentchilo V, Chronopoulou PM, McGenity TJ, van der Meer JR (2010) Development of a multistrain bacterial bioreporter platform for the monitoring of hydrocarbon contaminants in marine environments. Environ Sci Technol 44 (3):1049–1055
14. Gillor O, Hadas O, Post FA, Belkin S (2002) Phosphorus bioavailability monitoring by a luminescent cyanobacterial sensor strain. J Phycol 38:107–115
15. Mbeunkui F, Richaud C, Etienne AL, Schmid RD, Bachmann TT (2002) Bioavailable nitrate detection in water by an immobilized luminescent cyanobacterial reporter strain. Appl Microbiol Biotechnol 60:306–312

16. Garmendia J, De Las Heras A, Galvão TC, De Lorenzo V (2008) Tracing explosives in soil with transcriptional regulators of *Pseudomonas putida* evolved for responding to nitrotoluenes. Microb Biotechnol 1:236–246
17. Yagur-Kroll S, Lalush C, Rosen R, Bachar N, Moskovitz Y, Belkin S (2014) *Escherichia coli* bioreporters for the detection of 2,4-dinitrotoluene and 2,4,6-trinitrotoluene. Appl Microbiol Biotechnol 98(2):885–895
18. Van Dyk TK, Majarian WR, Konstantinov KB, Young RM, Dhurjati PS, LaRossa RA (1994) Rapid and sensitive pollutant detection by induction of heat shock gene-bioluminescence gene fusions. Appl Environ Microbiol 60(5):1414–1420
19. Bechor O, Smulski DR, Van Dyk TK, LaRossa RA, Belkin S (2002) Recombinant microorganisms as environmental biosensors: pollutants detection by Escherichia coli bearing *fabA'::lux* fusions. J Biotechnol 94(1):125–132
20. Van Dyk TK, Smulski DR, Reed TR, Belkin S, Vollmer AC, LaRossa RA (1995) Responses to toxicants of an *Escherichia coli* strain carrying a *uspA'::lux* genetic fusion and an *E. coli* strain carrying a *grpE'::lux* fusion are similar. Appl Environ Microbiol 61(11):4124–4127
21. Belkin S, Smulski DR, Dadon S, Vollmer AC, Van Dyk TK, LaRossa RA (1997) A panel of stress-responsive luminous bacteria for the detection of selected classes of toxicants. Water Res 31(12):3009–3016
22. Ahn J-M, Kim JH, Kim JH, Gu MB (2010) Randomly distributed arrays of optically coded functional microbeads for toxicity screening and monitoring. Lab Chip 10(20):2695–2701
23. Ben-Israel O, Ben-Israel H, Ulitzur S (1998) Identification and quantification of toxic chemicals by use of *Escherichia coli* carrying *lux* genes fused to stress promoters. Appl Environ Microbiol 64(11):4346–4352
24. Choi SH, Gu MB (2002) A portable toxicity biosensor using freeze-dried recombinant bioluminescent bacteria. Biosens Bioelectron 17(5):433–440
25. Dardenne F, Smolders R, De Coen W, Blust R (2007) Prokaryotic gene profiling assays to detect sediment toxicity: evaluating the ecotoxicological relevance of a cell-based assay. Environ Sci Technol 41(5):1790–1796
26. Elad T, Benovich E, Magrisso S, Belkin S (2008) Toxicant identification by a luminescent bacterial bioreporter panel: application of pattern classification algorithms. Environ Sci Technol 42(22):8486–8491
27. Pedahzur R, Polyak B, Marks RS, Belkin S (2004) Water toxicity detection by a panel of stress-responsive luminescent bacteria. J Appl Toxicol 24(5):343–348
28. Sazykin IS, Sazykina MA, Khmelevtsova LE, Mirina EA, Kudeevskaya EM, Rogulin EA, Rakin AV (2016) Biosensor-based comparison of the ecotoxicological contamination of the wastewaters of Southern Russia and Southern Germany. Int J Environ Sci Technol 13 (3):945–954
29. Hug C, Zhang X, Guan M, Krauss M, Bloch R, Schulze T, Reinecke T, Hollert H, Brack W (2015) Microbial reporter gene assay as a diagnostic and early warning tool for the detection and characterization of toxic pollution in surface waters. Environ Toxicol Chem 34:2523–2532
30. Elad T, Belkin S (2013) Broad spectrum detection and "barcoding" of water pollutants by a genome-wide bacterial sensor array. Water Res 47(11):3782–3790
31. Van Dyk TK, DeRose EJ, Gonye GE (2001) LuxArray, a high-density, genomewide transcription analysis of *Escherichia coli* using bioluminescent reporter strains. J Bacteriol 183 (19):5496–5505
32. Zaslaver A, Bren A, Ronen M, Itzkovitz S, Kikoin I, Shavit S, Liebermeister W, Surette MG, Alon U (2006) A comprehensive library of fluorescent transcriptional reporters for *Escherichia coli*. Nat Methods 3(8):623–628
33. Onnis-Hayden A, Weng H, He M, Hansen S, Ilyin V, Lewis K, Gu AZ (2009) Prokaryotic real-time gene expression profiling for toxicity assessment. Environ Sci Technol 43 (12):4574–4581

34. Gou N, Onnis-Hayden A, Gu AZ (2010) Mechanistic toxicity assessment of nanomaterials by whole-cell-array stress genes expression analysis. Environ Sci Technol 44(15):5964–5970
35. Zhang X, Wiseman S, Yu H, Liu H, Giesy JP, Hecker M (2011) Assessing the toxicity of naphthenic acids using a microbial genome wide live cell reporter array system. Environ Sci Technol 45(4):1984–1991
36. Su G, Zhang X, Raine JC, Xing L, Higley E, Hecker M, Giesy JP, Yu H (2013) Mechanisms of toxicity of triphenyltin chloride (TPTC) determined by a live cell reporter array. Environ Sci Pollut Res Int 20(2):803–811
37. Su G, Yu H, Lam MHW, Giesy JP, Zhang X (2014) Mechanisms of toxicity of hydroxylated polybrominated diphenyl ethers (HO-PBDEs) determined by toxicogenomic analysis with a live cell array coupled with mutagenesis in *Escherichia coli*. Environ Sci Technol 48 (10):5929–5937
38. Gou N, Yuan S, Lan J, Gao C, Alshawabkeh AN, Gu AZ (2014) A quantitative toxicogenomics assay reveals the evolution and nature of toxicity during the transformation of environmental pollutants. Environ Sci Technol 48(15):8855–8863
39. Guan M, Su G, Giesy JP, Zhang X (2016) Classification and toxicity mechanisms of novel flame retardants (NFRs) based on whole genome expression profiling. Chemosphere 144:2150–2157
40. Krivoshiev BV, Dardenne F, Blust R, Covaci A, Husson SJ (2015) Elucidating toxicological mechanisms of current flame retardants using a bacterial gene profiling assay. Toxicol Vitro 29(8):2124–2132
41. Gao C, Weisman D, Lan J, Gou N, Gu AZ (2015) Toxicity mechanisms identification via gene set enrichment analysis of time-series toxicogenomics data: impact of time and concentration. Environ Sci Technol 49(7):4618–4626
42. Janion C (2008) Inducible SOS response system of DNA repair and mutagenesis in *Escherichia coli*. Int J Biol Sci 4(6):338–344
43. Little JW, Edmiston SH, Pacelli LZ, Mount DW (1980) Cleavage of the *Escherichia coli lexA* protein by the *recA* protease. Proc Natl Acad Sci U S A 77(6):3225–3229
44. Quillardet P, Huisman O, D'Ari R, Hofnung M (1982) SOS chromotest, a direct assay of induction of an SOS function in *Escherichia coli* K-12 to measure genotoxicity. Proc Natl Acad Sci U S A 79(19):5971–5975
45. Oda Y, Nakamura S, Oki I, Kato T, Shinagawa H (1985) Evaluation of the new system (*umu* test) for the detection of environmental mutagens and carcinogens. Mutat Res 147 (5):219–229
46. Quillardet P, Hofnung M (1993) The SOS chromotest: a review. Mutat Res 297(3):235–279
47. Reifferscheid G, Heil J (1996) Validation of the SOS/*umu* test using test results of 486 chemicals and comparison with the Ames test and carcinogenicity data. Mutat Res 369:129–145
48. ISO (2000) ISO 13829:2000 Water quality – determination of the genotoxicity of water and waste water using the *umu*-test. ISO, Geneva
49. van der Lelie D, Regniers L, Borremans B, Provoost A, Verschaeve L (1997) The VITOTOX® test, an SOS bioluminescence *Salmonella typhimurium* test to measure genotoxicity kinetics. Mutat Res 389(2–3):279–290
50. Verschaeve L (2002) Genotoxicity studies in groundwater, surface waters, and contaminated soil. ScientificWorldJournal 2:1247–1253
51. Verschaeve L, Van Gompel J, Thilemans L, Regniers L, Vanparys P, van der Lelie D (1999) VITOTOX® bacterial genotoxicity and toxicity test for the rapid screening of chemicals. Environ Mol Mutagen 33(3):240–248
52. Westerink WMA, Stevenson JCR, Lauwers A, Griffioen G, Horbach GJ, Schoonen WGEJ (2009) Evaluation of the Vitotox™ and RadarScreen assays for the rapid assessment of genotoxicity in the early research phase of drug development. Mutat Res 676(1–2):113–130
53. Nunoshiba T, Nishioka H (1991) 'Rec-lac test' for detecting SOS-inducing activity of environmental genotoxic substances. Mutat Res 254(1):71–77

54. Vollmer AC, Belkin S, Smulski DR, Van Dyk TK, LaRossa RA (1997) Detection of DNA damage by use of *Escherichia coli* carrying *recA'::lux*, *uvrA'::lux*, or *alkA'::lux* reporter plasmids. Appl Environ Microbiol 63(7):2566–2571
55. Jia K, Marks RS, Ionescu RE (2014) Influence of carbon-based nanomaterials on *lux*-bioreporter *Escherichia coli*. Talanta 126:208–213
56. Song Y, Jiang B, Tian S, Tang H, Liu Z, Li C, Jia J, Huang WE, Zhang X, Li G (2014) A whole-cell bioreporter approach for the genotoxicity assessment of bioavailability of toxic compounds in contaminated soil in China. Environ Pollut 195:178–184
57. Frey J, Ghersa P, Palacios PG, Belet M (1986) Physical and genetic analysis of the ColD plasmid. J Bacteriol 166(1):15–19
58. Ptitsyn LR, Horneck G, Komova O, Kozubek S, Krasavin EA, Bonev M, Rettberg P (1997) A biosensor for environmental genotoxin screening based on an SOS *lux* assay in recombinant *Escherichia coli* cells. Appl Environ Microbiol 63(11):4377–4384
59. Østergaard TG, Hansen LH, Binderup M-L, Norman A, Sørensen SJ (2007) The *cda* GenoTox assay: a new and sensitive method for detection of environmental genotoxins, including nitroarenes and aromatic amines. Mutat Res 631(2):77–84
60. Busam S, McNabb M, Wackwitz A, Senevirathna W, Beggah S, van der Meer JR, Wells M, Breuer U, Harms H (2007) Artificial neural network study of whole-cell bacterial bioreporter response determined using fluorescence flow cytometry. Anal Chem 79(23):9107–9114
61. Norman A, Hansen LH, Sørensen SJ (2006) A flow cytometry-optimized assay using an SOS–green fluorescent protein (SOS–GFP) whole-cell biosensor for the detection of genotoxins in complex environments. Mutat Res 603(2):164–172
62. Biran A, Yagur-Kroll S, Pedahzur R, Buchinger S, Reifferscheid G, Ben-Yoav H, Shacham-Diamand Y, Belkin S (2010) Bacterial genotoxicity bioreporters. Microb Biotechnol 3:412–427
63. Volkert MR (1988) Adaptive response of *Escherichia coli* to alkylation damage. Environ Mol Mutagen 11(2):241–255
64. Hwang ET, Ahn J-M, Kim BC, Gu MB (2008) Construction of a *nrdA::luxCDABE* fusion and its use in *Escherichia coli* as a DNA damage biosensor. Sensors 8(2):1297–1307
65. Ahn J-M, Gu MB (2012) Geno-Tox: cell array biochip for genotoxicity monitoring and classification. Appl Biochem Biotechnol 168(4):752–760
66. Ahn J-M, Hwang ET, Youn C-H, Banu DL, Kim BC, Niazi JH, Gu MB (2009) Prediction and classification of the modes of genotoxic actions using bacterial biosensors specific for DNA damages. Biosens Bioelectron 25(4):767–772
67. Purvis IJ, Chotai D, Dykes CW, Lubahn DB, French FS, Wilson EM, Hobden AN (1991) An androgen-inducible expression system for *Saccharomyces cerevisiae*. Gene 106(1):35–42
68. Routledge EJ, Sumpter JP (1996) Estrogenic activity of surfactants and some of their degradation products assessed using a recombinant yeast screen. Environ Toxicol Chem 15 (3):241–248
69. Raman DR, Williams EL, Layton AC, Burns RT, Easter JP, Daugherty AS, Mullen MD, Sayler GS (2004) Estrogen content of dairy and swine wastes. Environ Sci Technol 38 (13):3567–3573
70. Conroy O, Quanrud DM, Ela WP, Wicke D, Lansey KE, Arnold RG (2005) Fate of wastewater effluent hER-agonists and hER-antagonists during soil aquifer treatment. Environ Sci Technol 39(7):2287–2293
71. Buchinger S, Spira D, Bröder K, Schlüsener M, Ternes T, Reifferscheid G (2013) Direct coupling of thin-layer chromatography with a bioassay for the detection of estrogenic compounds: applications for effect-directed analysis. Anal Chem 85(15):7248–7256
72. Bovee TFH, Helsdingen RJR, Koks PD, Kuiper HA, Hoogenboom RLAP, Keijer J (2004) Development of a rapid yeast estrogen bioassay, based on the expression of green fluorescent protein. Gene 325:187–200

73. Bovee TFH, Helsdingen RJR, Hamers ARM, van Duursen MBM, Nielen MWF, Hoogenboom RLAP (2007) A new highly specific and robust yeast androgen bioassay for the detection of agonists and antagonists. Anal Bioanal Chem 389(5):1549–1558
74. Michelini E, Leskinen P, Virta M, Karp M, Roda A (2005) A new recombinant cell-based bioluminescent assay for sensitive androgen-like compound detection. Biosens Bioelectron 20(11):2261–2267
75. Sanseverino J, Eldridge ML, Layton AC, Easter JP, Yarbrough J, Schultz TW, Sayler GS (2009) Screening of potentially hormonally active chemicals using bioluminescent yeast bioreporters. Toxicol Sci 107(1):122–134
76. Wang J, Eldridge M, Menn FM, Dykes T, Sayler G (2015) Standardized application of yeast bioluminescent reporters as endocrine disruptor screen for comparative analysis of wastewater effluents from membrane bioreactor and traditional activated sludge. Ecotoxicology 24 (10):2088–2099
77. OECD (2015) Test no. 455: performance-based test guideline for stably transfected transactivation in vitro assays to detect estrogen receptor agonists and antagonists, OECD Guidelines for the Testing of Chemicals, Section 4. OECD, Paris
78. van der Burg B, van der Linden S, Man H-y, Winter R, Jonker L, van Vugt-Lussenburg B, Brouwer A (2013) A panel of quantitative Calux® reporter gene assays for reliable high-throughput toxicity screening of chemicals and complex mixtures. In: Steinberg P (ed) High-throughput screening methods in toxicity testing. Wiley, Hoboken, pp 519–532
79. White SS, Birnbaum LS (2009) An overview of the effects of dioxins and dioxin-like compounds on vertebrates, as documented in human and ecological epidemiology. J Environ Sci Health C Environ Carcinog Ecotoxicol Rev 27(4):197–211
80. Aarts JMMJG, Denison MS, De Haan LHJ, Schalk JAC, Cox MA, Brouwer A (1993) Ah receptor-mediated luciferase expression: a tool for monitoring dioxin-like toxicity. Organohalogen Compd 13:361–364
81. Denison MS, El-Fouly MH, Aarts JMMJG, Brouwer A, Richter C, Giesy JP (1993) Production of novel recombinant cell line bioassay systems for detection of 2,3,7,8-tetrachlorodibenzo-p-dioxin-like chemicals. Organohalogen Compd 13:365–368
82. Hoogenboom R, Bovee T, Traag W, Hoogerbrugge R, Baumann B, Portier L, van de Weg G, de Vries J (2006) The use of the DR CALUX bioassay and indicator polychlorinated biphenyls for screening of elevated levels of dioxins and dioxin-like polychlorinated biphenyls in eel. Mol Nutr Food Res 50(10):945–957
83. Besselink H, Leonards P, Felzel E, Brouwer B (2002) Analysis of polychlorinated dibenzo-p-dioxins (PCDD), dibenzofurans (PCDF) and biphenyls (PCB) in fish using DR CALUX® and GC/MS: a comparison. Organohalogen Compd 58:412–415
84. Besselink HT, Schipper C, Klamer H, Leonards P, Verhaar H, Felzel E, Murk AJ, Thain J, Hosoe K, Schoeters G, Legler J, Brouwer B (2004) Intra- and interlaboratory calibration of the DR CALUX® bioassay for the analysis of dioxins and dioxin-like chemicals in sediments. Environ Toxicol Chem 23:2781–2789
85. Whyte JJ, Schmitt CJ, Tillitt DE (2004) The H4IIE cell bioassay as an indicator of dioxin-like chemicals in wildlife and the environment. Crit Rev Toxicol 34(1):1–83
86. Petrulis JR, Bunce NJ (1999) Competitive inhibition by inducer as a confounding factor in the use of the ethoxyresorufin-O-deethylase (EROD) assay to estimate exposure to dioxin-like compounds. Toxicol Lett 105(3):251–260
87. Denison MS, Zhao B, Baston DS, Clark GC, Murata H, Han D (2004) Recombinant cell bioassay systems for the detection and relative quantitation of halogenated dioxins and related chemicals. Talanta 63(5):1123–1133
88. Turner K, Xu S, Pasini P, Deo S, Bachas L, Daunert S (2007) Hydroxylated polychlorinated biphenyl detection based on a genetically engineered bioluminescent whole-cell sensing system. Anal Chem 79(15):5740–5745
89. Kramer KJM, Foekema EM (2000) The "Musselmonitor®" as a biological early warning system. In: Butterworth FM, Gunatilaka A, Gonsebatt ME (eds) Biomonitors and biomarkers

as indicators of environmental change 2. Kluwer Academic/Plenum Publishers, New York, pp 59–88
90. van der Schalie WH, Shedd TR, Knechtges PL, Widder MW (2001) Using higher organisms in biological early warning systems for real-time toxicity detection. Biosens Bioelectron 16:457–465
91. Gerhardt A, Janssens de Bisthoven L, Penders E (2003) Quality control of drinking water from the River Rhine (Netherlands) with the multispecies freshwater biomonitor. Aquat Ecosyst Health Manag 6(2):159–166
92. Gunatilaka A, Diehl P, Puzicha H (2000) The evaluation of the "Dynamic Daphnia Test" after a decade of use. In: Butterworth FM, Gunatilaka A, Gonsebatt ME (eds) Biomonitors and biomarkers as indicators of environmental change 2. Kluwer Academic/Plenum Publishers, New York, pp 29–58
93. Gerhardt A, Ingram MK, Kang IJ, Ulitzur S (2006) In situ on-line toxicity biomonitoring in water: recent developments. Environ Toxicol Chem 25(9):2263–2271
94. Kim BC, Gu MB (2005) A multi-channel continuous water toxicity monitoring system: its evaluation and application to water discharged from a power plant. Environ Monit Assess 109:123–133
95. Lee JH, Mitchell RJ, Kim BC, Cullen DC, Gu MB (2005) A cell array biosensor for environmental toxicity analysis. Biosens Bioelectron 21(3):500–507
96. Bjerketorp J, Håkansson S, Belkin S, Jansson JK (2006) Advances in preservation methods: keeping biosensor microorganisms alive and active. Curr Opin Biotechnol 17(1):43–49
97. Heitzer A, Malachowsky K, Thonnard JE, Bienkowski PR, White DC, Sayler GS (1994) Optical biosensor for environmental on-line monitoring of naphthalene and salicylate bioavailability with an immobilized bioluminescent catabolic reporter bacterium. Appl Environ Microbiol 60:1487–1494
98. Eltzov E, Marks RS, Voost S, Wullings BA, Heringa MB (2009) Flow-through real time bacterial biosensor for toxic compounds in water. Sens Actuat B Chem 142:11–18
99. Eltzov E, Slobodnik V, Ionescu RE, Marks RS (2015) On-line biosensor for the detection of putative toxicity in water contaminants. Talanta 132:583–590
100. Charrier T, Chapeau C, Bendria L, Picart P, Daniel P, Thouand G (2011) A multichannel bioluminescent bacterial biosensor for the on-line detection of metals and toxicity. Part II: technical development and proof of concept of the biosensor. Anal Bioanal Chem 400:1061–1070
101. Elad T, Almog R, Yagur-Kroll S, Levkov K, Melamed S, Shacham-Diamand Y, Belkin S (2011) Online monitoring of water toxicity by use of bioluminescent reporter bacterial biochips. Environ Sci Technol 45(19):8536–8544
102. Nivens DE, McKnight TE, Moser SA, Osbourn SJ, Simpson ML, Sayler GS (2004) Bioluminescent bioreporter integrated circuits: potentially small, rugged and inexpensive whole-cell biosensors for remote environmental monitoring. J Appl Microbiol 96(1):33–46
103. Vijayaraghavan R, Islam SK, Zhang M, Ripp S, Caylor S, Bull ND, Moser S, Terry SC, Blalock BJ, Sayler GS (2007) A bioreporter bioluminescent integrated circuit for very low-level chemical sensing in both gas and liquid environments. Sens Actuat B Chem 123 (2):922–928
104. Eltzov E, Yehuda A, Marks RS (2015) Creation of a new portable biosensor for water toxicity determination. Sens Actuat B Chem 221:1044–1054
105. Eltzov E, Cohen A, Marks RS (2015) Bioluminescent liquid light guide pad biosensor for indoor air toxicity monitoring. Anal Chem 87(7):3655–3661
106. Stocker J, Balluch D, Gsell M, Harms H, Feliciano J, Daunert S, Malik KA, van der Meer JR (2003) Development of a set of simple bacterial biosensors for quantitative and rapid measurements of arsenite and arsenate in potable water. Environ Sci Technol 37 (20):4743–4750

107. Struss A, Pasini P, Ensor CM, Raut N, Daunert S (2010) Paper strip whole cell biosensors: a portable test for the semiquantitative detection of bacterial quorum signaling molecules. Anal Chem 82(11):4457–4463
108. Date A, Pasini P, Daunert S (2010) Integration of spore-based genetically engineered whole-cell sensing systems into portable centrifugal microfluidic platforms. Anal Bioanal Chem 398 (1):349–356
109. Date A, Pasini P, Sangal A, Daunert S (2010) Packaging sensing cells in spores for long-term preservation of sensors: a tool for biomedical and environmental analysis. Anal Chem 82 (14):6098–6103
110. Sangal A, Pasini P, Daunert S (2011) Stability of spore-based biosensing systems under extreme conditions. Sens Actuat Chem 158(1):377–382
111. Gilmore J, Islam M, Martinez-Duarte R (2016) Challenges in the use of compact disc-based centrifugal microfluidics for healthcare diagnostics at the extreme point of care. Micromachines 7(4):52
112. Roda A, Cevenini L, Michelini E, Branchini BR (2011) A portable bioluminescence engineered cell-based biosensor for on-site applications. Biosens Bioelectron 26 (8):3647–3653
113. Cevenini L, Calabretta MM, Tarantino G, Michelini E, Roda A (2016) Smartphone-interfaced 3D printed toxicity biosensor integrating bioluminescent "sentinel cells". Sens Actuat B Chem 225:249–257
114. Biran A, Ben Yoav H, Yagur-Kroll S, Pedahzur R, Buchinger S, Shacham-Diamand Y, Reifferscheid G, Belkin S (2011) Microbial genotoxicity bioreporters based on *sulA* activation. Anal Bioanal Chem 400(9):3013–3024
115. Ben-Yoav H, Biran A, Pedahzur R, Belkin S, Buchinger S, Reifferscheid G, Shacham-Diamand Y (2009) A whole cell electrochemical biosensor for water genotoxicity bio-detection. Electrochim Acta 54(25):6113–6118
116. Tsai H-F, Tsai Y-C, Yagur-Kroll S, Palevsky N, Belkin S, Cheng J-Y (2015) Water pollutant monitoring by a whole cell array through lens-free detection on CCD. Lab Chip 15 (6):1472–1480
117. Siegfried K, Endes C, Bhuiyan AF, Kuppardt A, Mattusch J, van der Meer JR, Chatzinotas A, Harms H (2012) Field testing of arsenic in groundwater samples of Bangladesh using a test kit based on lyophilized bioreporter bacteria. Environ Sci Technol 46(6):3281–3287
118. Siegfried K, Hahn-Tomer S, Koelsch A, Osterwalder E, Mattusch J, Staerk H-J, Meichtry JM, De Seta GE, Reina FD, Panigatti C, Litter MI, Harms H (2015) Introducing simple detection of bioavailable arsenic at Rafaela (Santa Fe Province, Argentina) using the ARSOlux biosensor. Int J Environ Res Public Health 12(5):5465–5482
119. Baumstark-Khan C, Rode A, Rettberg P, Horneck G (2001) Application of the Lux-Fluoro test as bioassay for combined genotoxicity and cytotoxicity measurements by means of recombinant *Salmonella typhimurium* TA1535 cells. Anal Chim Acta 437(1):23–30
120. Rettberg P, Bandel K, Baumstark-Khan C, Horneck G (2001) Increased sensitivity of the SOS-LUX-Test for the detection of hydrophobic genotoxic substances with *Salmonella typhimurium* TA1535 as host strain. Anal Chim Acta 426(2):167–173
121. Davidov Y, Rozen R, Smulski DR, Van Dyk TK, Vollmer AC, Elsemore DA, LaRossa RA, Belkin S (2000) Improved bacterial SOS promoter::*lux* fusions for genotoxicity detection. Mutat Res 466(1):97–107
122. Maehana K, Tani H, Shiba T, Kamidate T (2004) Effects of using a low-copy plasmid and controlling membrane permeability in SOS-based genotoxic bioassay. Anal Chim Acta 522 (2):189–195
123. Norman A, Hestbjerg Hansen L, Sørensen SJ (2005) Construction of a ColD cda promoter-based SOS-green fluorescent protein whole-cell biosensor with higher sensitivity toward genotoxic compounds than constructs based on *recA*, *umuDC*, or *sulA* promoters. Appl Environ Microbiol 71(5):2338–2346

124. Dreier J, Breitmaier EB, Gocke E, Apfel CM, Page MGP (2002) Direct influence of S9 liver homogenate on fluorescence signals: impact on practical applications in a bacterial genotoxicity assay. Mutat Res 513(1–2):169–182
125. Arai R, Makita Y, Oda Y, Nagamune T (2001) Construction of green fluorescent protein reporter genes for genotoxicity test (SOS/*umu*-test) and improvement of mutagen-sensitivity. J Biosci Bioeng 92(3):301–304
126. Yagur-Kroll S, Bilic B, Belkin S (2010) Strategies for enhancing bioluminescent bacterial sensor performance by promoter region manipulation. Microb Biotechnol 3:300–310
127. Yagur-Kroll S, Belkin S (2011) Upgrading bioluminescent bacterial bioreporter performance by splitting the *lux* operon. Anal Bioanal Chem 400(4):1071–1082
128. Allen MS, Wilgus JR, Chewning CS, Sayler GS, Simpson ML (2007) A destabilized bacterial luciferase for dynamic gene expression studies. Syst Synth Biol 1(1):3–9
129. Malling HV (1971) Dimethylnitrosamine: formation of mutagenic compounds by interaction with mouse liver microsomes. Mutat Res 13(4):425–429
130. Oda Y, Shimada T, Watanabe M, Ishidate M, Nohmi T (1992) A sensitive umu test system for the detection of mutagenic nitroarenes in *Salmonella typhimurium* NM1011 having a high nitroreductase activity. Mutat Res 272(2):91–99
131. Oda Y, Yamazaki H, Watanabe M, Nohmi T, Shimada T (1993) Highly sensitive *umu* test system for the detection of mutagenic nitroarenes in *Salmonella typhimurium* NM3009 having high O-acetyltransferase and nitroreductase activities. Environ Mol Mutagen 21 (4):357–364
132. Oda Y, Yamazaki H, Watanabe M, Nohmi T, Shimada T (1995) Development of high sensitive *umu* test system: rapid detection of genotoxicity of promutagenic aromatic amines by *Salmonella typhimurium* strain NM2009 possessing high O-acetyltransferase activity. Mutat Res 334(2):145–156
133. Oda Y, Kunihiro F, Masaaki K, Akihiko N, Taro Y (2004) Use of a high-throughput *umu*-microplate test system for rapid detection of genotoxicity produced by mutagenic carcinogens and airborne particulate matter. Environ Mol Mutagen 43(1):10–19

Adv Biochem Eng Biotechnol (2017) 157: 159–186
DOI: 10.1007/10_2016_15

Published online: 9 November 2016

Toxicogenomics in Environmental Science

Alexandra Brinke and Sebastian Buchinger

Abstract This chapter reviews the current knowledge and recent progress in the field of environmental, aquatic ecotoxicogenomics with a focus on transcriptomic methods. In ecotoxicogenomics the omics technologies are applied for the detection and assessment of adverse effects in the environment, and thus are to be distinguished from omics used in human toxicology [Snape et al., Aquat Toxicol 67:143–154, 2004]. Transcriptomic methods in ecotoxicology are applied to gain a mechanistic understanding of toxic effects on organisms or populations, and thus aim to bridge the gap between cause and effect. A worthwhile effect-based interpretation of stressor induced changes on the transcriptome is based on the principle of phenotypic-anchoring [Paules, Environ Health Perspect 111:A338–A339, 2003]. Thereby, changes on the transcriptomic level can only be identified as effects if they are clearly linked to a specific stressor-induced effect on the macroscopic level. By integrating those macroscopic and transcriptomic effects, conclusions on the effect-inducing type of the stressor can be drawn. Stressor-specific effects on the transcriptomic level can be identified as stressor-specific induced pathways, transcriptomic patterns, or stressors-specific genetic biomarkers. In this chapter, examples of the combined application of macroscopic and transcriptional effects for the identification of environmental stressors, such as aquatic pollutants, are given and discussed. By means of these examples, challenges on the way to a standardized application of transcriptomics in ecotoxicology are discussed. This is also done against the background of the application of transcriptomic methods in environmental regulation such as the EU regulation Registration, Evaluation, Authorisation and Restriction of Chemicals (REACH).

A. Brinke (✉) and S. Buchinger
Division of Biochemistry and Ecotoxicology, Federal Institute of Hydrology (BfG), Koblenz 56068, Germany
e-mail: alexandra.brinke@bafg.de

Keywords (Eco-)toxicogenomics, omics in regulation, Pathway analysis, Transcriptional pattern, Transcriptomic biomarker

Contents

1 Introduction

At the end of the last century "toxicogenomics", which "is concerned with the identification of potential human and environmental toxicants, and their putative mechanisms of action, through the use of genomics resources" [1], entered the stage of environmental science for the first time. Most published articles agree on the high potential of this approach because of the possibility of comprehensively capturing a cellular response, for example on the transcript level, to a toxic stressor.

The fundamental idea of toxicogenomics is formulated, for example, by Nuwaysir et al. [1]: "Almost without exception, gene expression is altered during toxicity, as either a direct or indirect result of toxicant exposure." Later on this concept was broadened to the analysis of proteins or metabolites by means of proteomics [2, 3] and metabolomics [4, 5]. Regardless of the biological level on which effects occur, the European Food Safety Authority defines a toxicogenomic approach as a study of the response of a genome to hazardous substances using omics technologies such as genomic-scale expression of messenger-RNA (mRNA, transcriptomics), cell- and tissue-wide protein expression (proteomics), and metabolite profiling (metabolomics), in combination with bioinformatic methods and conventional toxicology [6]. The term ecotoxicogenomics for the application of transcriptomics in ecotoxicology was proposed by Snape et al. [7] to distinguish clearly between these types of studies and the use of transcriptomics in human toxicology.

Virtually each toxic noxa affects at least one biological level directly, that is, gene expression, protein function, or metabolite levels (Fig. 1). Because of regulatory feedback loops it is likely that the effect of a toxicant is somehow reflected at other biological levels as well. A change in a metabolite level caused by an

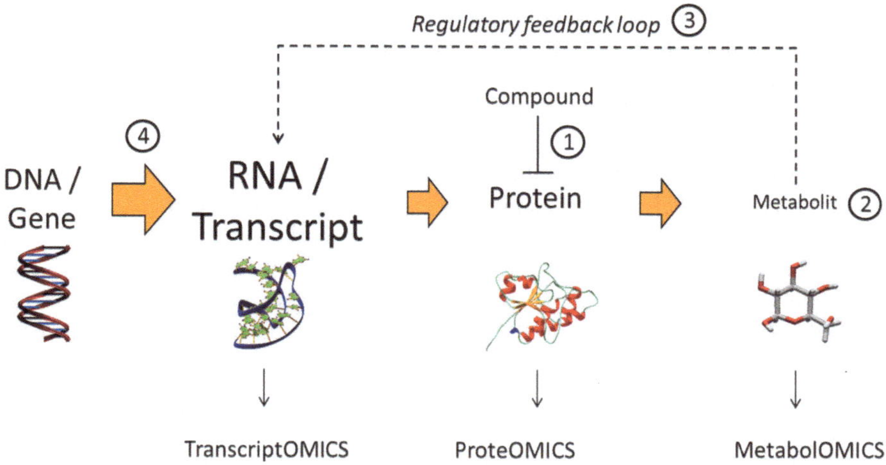

Fig. 1 Toxic effects are reflected on various molecular levels. (*1*) A substance might act as a stressor by inhibiting the activity of an enzyme. (*2*) Because of the lower activity of the enzyme, the concentration of a metabolite generated by the enzyme falls. (*3*) The decrease in the metabolite concentration might act as a signal, influencing the transcription of the gene encoding the inhibited enzyme. (*4*) The transcription of the respective gene is activated to compensate for the decreased metabolite concentration. The initial mode of action of the compound is the inhibition of the enzyme, but because of regulatory feedback loops – a very simple one is illustrated here – other molecular levels are also affected. The omics technologies aim to capture the concentration of all molecules in a cell, tissue, or organism belonging to a certain type of biomolecule under given conditions, that is, transcript concentrations (mRNA) are targeted by transcriptomics, protein concentrations by proteomics, and metabolite concentrations by metabolomics

inhibition of an enzyme can induce the expression of certain genes to compensate the alteration in the cellular physiology. The analysis of changes on the molecular level can thus help to bridge the knowledge gap between a toxic effect and its cause by providing a better understanding of the toxicants mode of action (MOA). Snape et al. [7] stated in a publication in 2004 that "the challenge in ecotoxicology for the 21st century is to understand the mechanisms of toxicity to different wildlife species," underlining the importance and the expected potential of ecotoxicogenomics. However, there are several aspects leading to very complex challenges when toxic effects are analyzed by means of toxicogenomics.

– "The good into the pot, the bad into the crop": It is highly important to stress the fact that not all alterations that might be detectable on the molecular level after exposure to a toxic compound are toxic effects per se. Only if patterns in gene expression or the expression of distinct biomarker genes are linked to effects on higher biological levels [8] can transcriptomics be meaningfully integrated in ecotoxicology. Therefore, it is essential to differentiate between adaptive and toxic effects as is done, for example, by using the concept of "phenotypic anchoring" [9] which correlates changes on the molecular level to adverse biological effects visible on the macroscopic level. Environmental science can only benefit from toxicogenomic approaches if there is a proven relationship – either mechanistic or at least statistically – between the molecular event and the adverse outcome.

- *Multicellular "noise"*: The response of a cell to a toxicant can differ with respect to the cell type [10]. In a multicellular organism, cells are differentiated and organized in tissues. Even within one tissue there can be a number of different cell types [11, 12]. If whole animals are exposed to a toxic compound, specific molecular effects in a certain cell type might be masked by the response of other cells in the exposed organism. Even if all the other cells are just not reacting to the toxic compound, the specific molecular effects in the target cells are "diluted" by the non-target [10]. For larger animals the analysis can be focused on specific tissues, but in the case of small animals the preparation of tissues is extremely laborious or even impossible. Alternatively, cell cultures can be used, but they suffer from certain restrictions, in particular the lack of all aspects of toxicokinetics such as distribution and metabolization of compounds.
- "We cannot observe without disturbing the observed phenomenon" [13]: The advantage of effects on the molecular level, namely their rapidness, becomes an obstacle if molecular effects are investigated by means of toxicogenomics. Organismic biotests may take some hours or even days to weeks in the case of chronic tests. However, changes on the molecular level, for example, alteration of metabolite pools, can occur within minutes or even seconds, necessitating quenching methods for metabolome analysis [14, 15]. Alterations in mRNA or protein concentrations take place within minutes to hours. The consequence is that, in contrast to macroscopic effects, molecular endpoints (mRNA, protein, or metabolite levels) are unstable and may change quickly after the end of the test during the harvesting of exposed cells and in particular during the preparation of tissues from whole animal experiments. A frequently used approach to circumvent this problem is shock freezing of the exposed organisms or cells, but a bias in results caused by the instability of the biomolecules under investigation cannot be excluded completely, in particular with respect to the analysis of metabolites.

Nevertheless, it is expected that the characterization of (toxic) effects on the molecular levels can facilitate (1) identification of toxic compounds in environmental samples, (2) predictions of adverse in vivo effects that occur at later time points after exposure and/or at higher concentrations of the toxicant or sample under investigation, (3) substance evaluation by comparing effects on the molecular level between well characterized model compounds and the substance to be assessed, and (4) investigation of mixture toxicity.

This chapter focuses on transcriptomic approaches, that is, the analysis of mRNA levels, because the respective techniques for mRNA quantification and data interpretation are more developed and broadly used compared to proteomics and metabolomics. Furthermore, the claim of omics to cover virtually all molecular entities of the respective molecular level is best met by transcriptomics because of the possibility of using highly specific sequences of nucleotides within the RNA to differentiate between RNA molecules. On the other hand, the physicochemical properties between different RNA molecules are very similar, allowing a less biased extraction of "all" RNA molecules from a cell or a tissue. In contrast to this, the physicochemical properties of proteins and especially metabolites are quite diverse – for example, comparing fatty acids with carbohydrates, which make it virtually impossible to use a single method for a complete extraction and analysis of proteins and metabolites. In terms of specificity, metabolites within a compounds class, for example, monosaccharides, are especially hard to differentiate by mass spectrometry – a common technique in metabolomics and proteomics for the identification of molecules. For this reason, approaches combining different analytical methods for a comprehensive analysis of metabolites are used [16]. The described obstacles of proteomics and metabolomics are counterbalanced to some extent because, in any case, it is impossible to extrapolate from an RNA level or a change in RNA level to a change in protein levels, activities, or metabolite pools – in other words, to extrapolate from information about the mRNA to the actual "doing" of the cell. This is because of further cellular regulatory mechanisms such as the regulation of translation, posttranslational modifications of proteins, protein degradation, allosteric regulation of enzyme activities, etc. A further advantage of metabolomics is the universal language of metabolites which is independent of the species under investigation. The general architecture of central metabolic pathways, for example, glycolysis or the citric acid cycle, are conserved from bacteria to humans – a fact making the metabolomics approach a highly useful tool in studying non-model organisms.

This chapter aims to inform about methodical approaches, applications, and potentials of transcriptomic analysis, especially in aquatic ecotoxicology. The choice of suitable techniques for the quantification of mRNA and strategies for experimental design and data evaluation depend strongly on the underlying question or purpose of the experiment. Therefore, after briefly describing available techniques for mRNA analysis, this chapter is structured according to the general objectives frequently addressed by studies such as (1) identification of biomarkers and use of expression patterns for classification and (2) studies aiming for a mechanistic understanding of toxic effects. In each section, examples from the literature are shown for illustration. In the final section of this chapter, the use and requirements for the implementation of toxicogenomics in regulation are discussed.

2 Techniques for Quantification of mRNA

The choice of the right technology in the wide field of transcriptomics is the first and a crucial step in the experimental design. As there are several excellent reviews on the technical aspects of transcriptome analysis available [2, 17–21], and the most common methods are only superficially presented in the following section for a better understanding.

Methods for transcriptomic analysis are mainly divided into methods for the measuring of global gene expression and methods for the expression of distinct genes that might serve as biomarkers for specific questions. The most common methods for global gene expression involve DNA microarrays and RNA sequencing, whereas single gene expression is commonly detected by means of quantitative real time PCR (qPCR). DNA microarrays and RNA sequencing are high-throughput methods able to measure the gene expression of a high number of genes or even the total genome of a test organism in parallel. In contrast, qPCR focuses on distinct genes (target genes), but is much cheaper compared to DNA microarrays and RNA sequencing. Furthermore, qPCR offers a more quantitative measurement, as the quantification is done in real time during amplification and in a linear range [22]. The advantage of RNA sequencing lies in its independence from previous genomic sequence information, which is why non-model organisms can also be sequenced.

Independent from the subsequent technology, the first step of a gene expression analysis is the isolation of total RNA from the cell culture, tissue, compartment, or even a whole organism (e.g., *Caenorhabditis elegans*) under investigation. Because the quality of the isolated RNA is crucial for the analysis, the RNA has to be characterized in terms of purity, quantity, and integrity. The purity and quantity of RNA can be determined spectrophotometrically [23] and the RNA integrity can be measured by means of capillary electrophoresis and the calculation of an RNA integrity number based on the signals of the ribosomal RNA [24–26].

2.1 DNA Microarray

A DNA microarray consists of several thousand single stranded DNA oligomers – termed probes – that are immobilized in distinct spots on the surface of a solid slide. Each spot contains one type of DNA oligomer representing one gene from the genome of a given test organism by its specific DNA sequence. For the gene expression analysis, a fluorescent-labeled copy DNA (cDNA) or copy RNA (cRNA, a retranscribed cDNA) is synthesized by reverse transcription of the messenger RNA (mRNA) by making use of either a labeled poly-T primer and/or a mixture of labeled random primers. The cDNA (or cRNA) reflects the mRNA composition of the sample under investigation and is hybridized against the DNA microarray. This results in the immobilization of the labeled cRNA or cDNA at the probe which has a complementary sequence to a specific cRNA or cDNA species. Spots on the DNA microarray that represent highly expressed genes capture a higher number of labeled cRNA or cDNA compared to genes that are expressed at lower levels. Consequently, the expression level of a gene is reflected by the amount of fluorescent dye that is immobilized at a specific probe, that is, a specific area on the microarray at which the DNA spot is located. By image scanning analysis the relative gene expression of – for example – a treated sample against a control is obtained for each gene. The hybridization of the control and the treated sample can be done on the same DNA microarray if the respective cRNA or cDNA is labeled differently. This approach is termed "double labeling." Alternatively, the same labeling can be used for the control and the treated sample (single labeling approach) but the hybridization is performed on two different DNA microarrays. The choice of these two different strategies depends on the design of the overall study (see Lettieri [19] for detailed information).

Unless cDNA is used for the construction of the microarray, an essential prerequisite for this method is the knowledge of at least parts of the genome of the test organism in order to design specific probes spotted on the array. Therefore, this method is popular for global gene analysis of completely sequenced test organisms, such as *Oryza sativa* ssp. *indica* or *japonica*, *Arabidopsis thaliana*, *Brassica oleracea*, *Mus musculus*, *Rattus norvegicus*, *C. elegans*, *Danio rerio*, *Salmon salar*, *Xenopus leavis*, and others. For these test organisms, commercial platforms are available that offer a high quality standard. Customized arrays with a defined group of genes are also available.

2.2 RNA Sequencing

RNA sequencing is an evaluation methodology in the field of next-generation sequencing (NGS – also known as high-throughput sequencing) and is a collective term describing a number of modern sequencing methods, such as Illumina sequencing and Roches 454 pyrosequencing, Ion torrent: Proton/PGM sequencing, or SOLiD sequencing [27]. These methods are enhanced versions of the Sanger dideoxy sequencing method, because it is possible to sequence DNA or RNA more quickly and much cheaper. Further advantages are the relative long read length, the possibility of high-throughput measurements, and the robust performance. The underlying principle of the four above-mentioned methods is similar. The isolated DNA or RNA is fragmented and immobilized as a single strand. For each immobilized fragment the complementary strand is synthesized by a DNA polymerase. For each elongation step only one of the four DNA nucleotides is offered. The integration of the given DNA nucleotide can be recorded by different means, for example, by the generation of pyrophosphate (pyrosequencing), the release of fluorescent molecules (e.g., Illumina sequencing), or even the formation of a proton during the reaction (Ion torrent). Thus, if a base was integrated in the complementary strand, a signal (e.g., light impulse, change of pH) is detected and captured in an image. The images from NGS are converted into sequence reads. This is done by means of a base-calling system, corresponding to the respective method. The reads are filtered and mapped to either single expressed genes or a transcriptome in a further step. These data are used for the profiling of gene expression of target genes or the entire genome by means of statistical analysis. These methods are described as highly quantitative over a broad range of expression levels with detection limits down to a few transcripts per cell [8] and are thus more sensitive in detecting differentially expressed genes compared to DNA microarrays [27]. The described methods allow the generation of large amounts of genomic data and the study of toxicogenomic responses without previous sequence information. Therefore this method is also highly suitable for non-model organisms, such as many common test species in ecotoxicology. For de novo assembly, which is used, for example, to create customized DNA microarrays for non-model organisms, mainly 454 pyrosequencing is used, whereas the other named methods are usually applied for resequencing [27].

2.3 Quantitative PCR (qPCR)

The expression of selected target genes can be achieved by a quantitative polymerase chain amplification of DNA fragments, flanked by primers that are highly specific for the gene under investigation. A copy DNA (cDNA) is synthesized by reverse transcription of the messenger RNA (mRNA). This cDNA is used as a template for the subsequent amplification of one or more target genes by polymerase chain reaction (PCR). The amplification of a cDNA of interest can be monitored in real time either by using a fluorescent dye that intercalates nonspecifically into double stranded DNA, resulting in an increase of the fluorescence signal or by making use of fluorescent reporter probes which specifically bind only to DNA containing the probe sequence. In the former case the fluorescence signal increases with time if DNA is amplified and the amount of double stranded DNA is doubled during each cycle of the PCR. In the latter case the fluorescence signal increases by the release of the fluorescent dye from the probe sequence which is depolymerized by the DNA polymerase in the course of the synthesis of the new DNA strand. Nowadays there are many technical variations available for the realization of a qPCR and statistical analyses; for more information see [28–34].

In terms of a gene expression analysis, the expression of a gene of interest is quantified in relation to one or more reference genes, usually housekeeping genes, which are used as internal controls [34].

3 Use of Toxicogenomics in Ecotoxicology

The following sections describe the use of toxicogenomics in ecotoxicology on three specific scientific questions with increasing complexity:

- Identification of biomarkers for chemical stressors and/or toxic effects
- Analysis of transcriptional expression patterns for a sample classification
- Understanding of toxicological mechanisms and pathways

Each section first provides general aspects and definitions about the respective topic and gives some recommendations for suitable experimental techniques followed by more detailed descriptions showing selected case studies. An overview of various studies published in the field of aquatic ecotoxicology is given in Table 1.

Table 1 List of case studies on transcriptomics in ecotoxicology

Level of analysis	Technique	Compound/sample	Organism	Reference
MB	A	Bisphenol A	*Daphnia magna*	[69]
MB	A	Cadmium, lead[a]	*Mytilus edulis*	[70]
MB	A/B	17-β Estradiol, nonylphenol, chlorophenol	*Oryzias latipes*	[71]
MB	B	Acetaminophen, atenolol, carbamazepine, oxytetracycline, sulfamethoxazole, trimethoprim, chlorpyrifos	*Brachionus koreanus*	[72]
MB	B	Temperature shift, cadmium chloride, butyl benzyl phthalate, diethylhexyl phthalate, bisphenol A, 4-nonylphenol, ethinylestradiol, pentachlorophenol, tributyltin	*Chironomus riparios*	[73]
MB	B	Natural fresh water, field study	*Chironomus riparius*	[74]
MB	B	3,4- and 3,5-Dichloroaniline, 1,4-dichlorobenzene, 2,4-dinitrophenol, atrazine, parathion-ethyl, chlorotoluron, genistein, 4-nitroquinoline-1-oxide, cadmium	*Danio rerio*	[64]
MB	B	Water-soluble fraction of weathered crude oil	*Danio rerio*	[75]
MB	B	Rubber wastewater (zinc)	*Daphnia magna*	[40]
MB	B	Nitrofurazone	*Euplotes vannus*	[76]
MB	B	Natural sediments (metals)	*Melita plumulosa*	[37]
MB	B	Cadmium, copper, zinc	*Perca flavescens*	[77]
MB	B	Bifenthrin	*Pimephales promelas*	[78]
MB	B	Heavy fuel oil no. 6, styrene	*Scophthalmus maximus*	[79]
MB	C	Metal/diesel contaminated sediments	*Melita plumulosa*	[36]
MB/TP	A/B	Natural sediments/water (copper)	*Melita plumulosa*	[80]
MB/TP	B	Cadmium, copper, lead	*Chlamydomonas rheinhardtii*	[81]
MB/PA	A/B	Arsenate, arsenite	*Oryza sativa*	[82]
MB/PA	A/B	Benzo(a)pyrene	*Thalassiosira pseudonana*	[83–85]
PA	A	Benzene	*Caenorhabditis elegans*	[86]
PA	A	Drospirenone, progesterone	*Danio rerio*	[87]
PA	A	17-α Ethinylestradiol and flutamide	*Danio rerio*	[88]
PA	A	Dieldrin	*Micropterus salmoides*	[89]
PA	A	Finasteride	*Silurana tropicalis*	[90]

(continued)

Table 1 (continued)

Level of analysis	Technique	Compound/sample	Organism	Reference
PA	A/B	Mixture of tributyltin, benzo(a)pyrene[a]	*Danio rerio*	[91]
PA	A/B	Munition pollutants (2,4,6-trinitrotoluene, 2,4-dinitrotoluene, 2,6-dinitrotoluene, trinitrobenzene, dinitrobenzene, 1,3,5-trinitro-1,3,5-triazacyclohexane)	*Daphnia magna*	[92] based on [93, 94]
PA	A/B	Munition pollutants single substance, mixture, contaminated groundwater (2,4,6-trinitrotoluene, 2,4-dinitrotoluene, 2,6-dinitrotoluene, trinitrobenzene, dinitrobenzene, 1,3,5-trinitro-1,3,5-triazacyclohexane)[a]	*Daphnia magna*	[93]
PA	A/B	Ibuprofen	*Daphnia magna*	[51]
PA	C	Tributyltin	*Nucella lapillus*	[95]
TP		Methylmercury chloride, $CdCl_2$, PbCl2, As_2O_3, aroclor 1254, acrylamide, *tert*-butylhydroquinone, 4-chloroaniline, 1,1-bis-(4-chlorophenyl)2,2,2-trichloroethane, 2,3,7,8-tetrachlorodibenzo-*p*-dioxin, valproic acid	Danio rerio	[96]
TP	A	Natural sediments	*Caenorhabditis elegans*	[97]
TP	A	Benzo(a)pyrene	*Danion rerio*	[98]
TP	A	Polymetallic solution	*Escherischia coli*	[99]
TP	A/B	3,4-Dichloroaniline	*Danio rerio*	[100]
TP	A/B	Estrogenic wastewater treatment plant effluent	*Pimephales promelas*	[101]
MB	B	Atrazine, diurone, isoproturone, glyphosate (mixture)	*Platichthys flesus*	[102]
MB	B	3,4-Dichloroaniline, 3,5-dichloroaniline, pentachloroaniline, 1,4-dichlorobenzene, pentachlorobenzene, 2,4-dinitrophenol, atrazine, lindane, parathion-ethyl, chlorotoluron, genistein, 4-nitroquinoline-1-oxide, cadmium, ivermectin	*Danio rerio*	[64]
TP	A	Fadrozole	*Danio rerio*	[103]
TP/PA	A	Wastewater treatment plant effluent	*Pimephales promelas*	[104]

The table is structured after the level of analysis and the applied techniques
MB molecular biomarker, *TP* transcriptional pattern, *PA* pathway analysis
A = microarray platform, B = RT-qPCR, C = RNA-Seq
[a]Mixture toxicity

In this table the level of analysis (biomarker, expression pattern and pathway analysis) and experimental techniques used are indicated together with investigated sample types or compounds and the organisms used. Because of the large number of studies already published, the table is necessarily incomplete. The literature search is complicated because of the numerous synonymous search terms for transcriptomics such as "transcriptional analysis," "transcriptomics," or "expression analysis." Despite this limitation, Table 1 shows a broad selection of instructive examples for the use of toxicogenomics in aquatic ecotoxicology. Most case studies were found with a main focus on molecular biomarkers and pathway analysis, whereas the analysis of transcriptomic patterns were mostly used as a basis for biomarker analysis. Microarray analysis and qPCR for validation were mainly used for transcriptomic analysis of all types of data analysis.

3.1 Biomarker

Genes that are differentially expressed in response to chemical stressors (e.g., cadmium, heavy metals, compound with estrogenic activity) or are in correlation to a specific apical endpoint may serve as molecular biomarkers. The identification of biomarkers indicative for environmental contaminations and/or adverse effects is the "easiest" and most practical approach to use.

Molecular biomarkers may be used for risk identification if they are predictive for classical ecotoxicological endpoints such as survival, growth, and reproduction. Following the principle of phenotypic anchoring, to guarantee a better interpretation of data the expression of molecular biomarkers should correlate with these classical endpoints [9]. If those biomarkers can detect effects on the molecular level earlier than the manifestation of adverse macroscopic effects and if they are clearly correlated to these macroscopic effects, they might be used as predictive tools for the assessment of environmental risks. Furthermore, long-term chronic exposures of animals could be avoided.

The two main approaches to select candidate transcriptomic biomarkers are presented below. These two approaches are explained in more detail by case studies.

3.1.1 The Usual Suspects

If distinct candidate biomarker genes are already known from previous studies or if stressor-inducing well-described pathways are in focus of a study, the expression of biomarker genes can be measured by means of qPCR. Of course, it is also possible to measure the expression of your candidate biomarker genes using DNA microarrays or RNA sequencing. However, this would be taking a sledge hammer to crack a nut in the sense of price and effort.

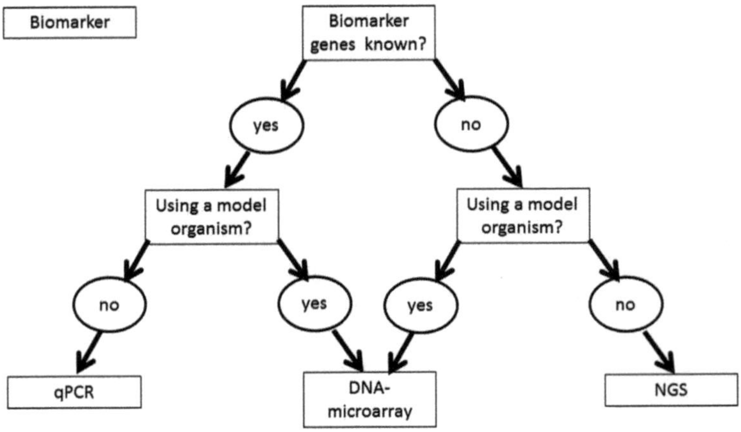

Fig. 2 Scheme for the selection of a suitable technique for the identification and analysis of a transcriptional biomarker. *NGS* next generation sequencing

3.1.2 Biomarker Discovery

In the case where there is no information on suitable candidate biomarkers for a stressor of interest and the selected test organism, a more unbiased identification of candidate biomarker genes has to be made (Fig. 2). A global gene expression analysis can be conducted to identify genes that are differentially expressed in response to a selected stressor. This may also comprise non-annotated genes, that is, genes of which the function is still unknown. By means of RNA sequencing, both, model organisms and non-model organisms can be targeted whereas the use of DNA microarrays is restricted mainly to sequenced model organisms. Furthermore, it is possible to design customized microarrays for non-model organisms based on data obtained by RNA sequencing or by making use of cDNA libraries. Taking into account price and time for the design and the production of a customized microarray, the use of RNA sequencing is more efficient. Candidate biomarkers identified by means of a "global" analysis of transcript levels should be further validated by means of qPCR, as this is the analytical method of choice once a biomarker is identified. On the one hand qPCR is considered to produce more accurate results, and on the other the analysis of single biomarker genes by qPCR is less cost-intensive compared to the application of global gene expression analyses.

3.1.3 Case Studies Using Biomarker Genes

The following case study is exemplary for the identification of candidate biomarker genes from global gene expression analyses, with a focus on the subsequent application and evaluation of natural samples. Hook et al. [35] used the results from an RNA sequencing analysis [36] to create a customized microarray for the non-model amphipod *Melita plumulosa* and subsequently conducted a pattern analysis to identify unique gene expression profiles in response to the exposure of *M. plumulosa* to different chemical groups. Observed patterns were assessed to be sufficiently unique to allow the identification of dominant stressors in contaminated sediment samples. A set of six biomarker genes for aquatic pollutants, such as heavy metals, was identified based on the observed expression patterns. The selected candidate biomarker genes were transcripts encoding two digestive proteases (chymotrypsin, carboxypeptidase), chitinase, hemocyanin, alpha tubulin, and long-form myosin. Based on these results, it was hypothesized that an identification of the most relevant stressor in field-contaminated sediments should be possible by making use of a battery of qPCR analyses. To confirm the specificity of the selected candidate biomarkers, a qPCR analysis was performed. Biomarker expression changes in *M. plumolosa* were surveyed on natural sediments spiked with the metals copper (Cu^{2+}), nickel (Ni^{2+}), and zinc (Zn^{2+}). The concentrations were selected to affect the reproduction of *M. plumulosa* but not its survival. The genes encoding chymotrypsin, the carboxypeptidase, and the chitinase were consistently down-regulated under metal exposure, whereas the other candidate genes showed a high inter-individual variability. Despite the high variability, a metal induced up-regulation of the hemocyanin transcript level was assumed but was not statistically significant.

In a further evaluation step, *M. plumulosa* was exposed to complex toxicant mixtures in field-collected natural sediments. On the macroscopic level, the three tested sediments induced toxic effects that were roughly equivalent to those measured in single-substance exposures. The results obtained from two of the three sediments investigated were at least partly in accordance with the transcript levels expected from the spiking experiments. Although the metal concentrations in the three sediments were comparable, the results from one sediment investigated showed an opposite expression profile [37]. The reasons were assumed to be uncharacterized toxicants or non-chemical stressors, and a lack of knowledge concerning the MOA of the test substances in crustaceans [37]. In general, it was recommended that more candidate biomarkers be included in the test battery, which should be better characterized in terms of the influence of non-contaminant stressors. To reduce the high inter-individual variance, it was recommended that more replicates be used. Furthermore, only one reference gene was used for the normalization of the qPCR, which might be an insufficient number of reference genes for satisfactory elimination of technical variation possibly masking true biological variation. The authors further recommended using a broader range of test substances with which to compare the changes in transcript abundance [37]. To summarize, the results of this study show exemplarily the main challenge of biomarker discovery on the molecular level, namely the transfer of results obtained from single compound exposures to field samples. This task is likely to be more challenging in the case of studies with soil and sediment compared to an aquatic exposure because structure and composition of soil or sediment may also influence transcript levels of biomarkers and thus add a layer of further complexity to the analysis.

A larger battery of transcriptional biomarkers for aquatic pollutants was identified by Regier et al. [38] using RNA sequencing on the Illumina platform. In the first part of the study, the transcriptome of the macrophyte *Elodea nutallii* was characterized after exposure to mercury and cadmium under laboratory and field conditions. Based on a pattern analysis, two subsets or pattern of differentially expressed genes induced by either mercury or cadmium were identified. It was assumed they were applicable to the detection of toxic metal exposure in fresh water or sediments. The transcriptomic analysis for the subset of genes was conducted by means of the NanoString nCounter technology, which hybridizes probes specifically labeled with several fluorescent dyes to the target molecules of interest. As the probes are "barcoded" by the fluorescent dyes the number and identity of hybridized probes reflecting the number of target molecules in the sample can be detected by fluorescence microscopy [39].

The application of the distinct subsets of 170 mercury-responsive genes and the subset of 212 cadmium-responsive genes showed a distinctive response of the test plants to elevated metal concentrations spiked to fresh water. This was assessed in comparison to lower but environmental relevant concentrations in different exposure scenarios such as the exposure to spiked fresh water or to natural contaminated sediments, both under laboratory conditions, and the exposure to contaminated fresh water under field conditions in three different depths. In addition, the influence of abiotic stressors, such as darkness, lower temperature (10 °C), and higher salinity on the transcriptome, were assessed to identify potential masking effects. Darkness induced an expression pattern similar to the exposure to low metal concentrations, whereas increased salinity induced an expression pattern that was distinct from patterns which are induced by low and high metal concentrations. This shows that factors such as light regime and salinity may interfere with gene expression patterns induced by chemical stressors. However, under laboratory conditions, a clear dose–response relationship of changes in the gene expression pattern in *E. nutalli* exposed to gradually increasing mercury concentrations in an aquatic test system was detected. This can be considered an example for the importance and the difficulties in defining reference conditions for biotest approaches. It was concluded that the subset of genes used as a biomarker battery for the detection of metal-responsive expression patterns might still have been small and a higher number of genes could have facilitated the detection of metal contamination in natural water or sediment samples by means of the presented bioassay. The presented study did not encompass macroscopic endpoints, to which the results on the transcriptomic level could be linked, and thus are not applicable in this form to a toxicological risk assessment.

Candidate transcript biomarkers are quite frequently selected based on the biological function of their gene product. Despite the undoubted advantage of such a targeted approach, it neglects one of the main features of the omics technologies, namely the possibility to analyze the genome-wide response to a stressor. One example to utilize genes with unknown annotation as biomarkers for the assessment of environmental samples was published by Jo and Jung [40] who used three genes without known biological function in *Daphnia magna* as biomarkers for rubber wastewater contamination that caused immobility on the macroscopic level. For two candidate biomarker a dose-dependent induction by rubber wastewater was detected. One of the genes was inducible as well by zinc which was shown in a single-substance biotest with *D. magna*. In comparison to the macroscopic endpoint immobility, the candidate biomarker proved more sensitive. It was therefore assessed as a sensitive and possible zinc-specific biomarker in water and wastewater.

A lack of knowledge about the biological function of a gene product should not lead to its exclusion from a biomarker battery if its expression level is clearly correlated to the occurrence of a stressor and/or the induction of an apical endpoint. Such a finding should rather initiate further research to elucidate the molecular and cellular function of the respective gene.

3.2 Transcriptional Pattern

Both the analysis of transcriptional patterns and the pathway analysis described below are performed on large transcriptional data sets which can be obtained by means of global gene expression. Of course, smaller gene sets can also be assessed by these methods, but commonly they are applied to datasets from whole genome microarrays or RNA sequencing analysis – the latter technique is especially useful for non-model organisms without genome-wide sequence information.

Transcriptional patterns or transcriptional signatures are groups or classes of genes which are significantly up- or down-regulated compared to the control state of the organism in response to a specific stressor or group of stressors. Therefore, they can be used as biomarkers of exposure or effect if they allow the classification of biological samples that were exposed to different conditions or (groups of) stressors [41]. Machine learning algorithms such as Support Vector Machines (SVM) or K-nearest neighbor are frequently used to achieve a classification of transcriptional patterns [41]. Because of the large public available transcriptomic data sets in the area of medicine, especially with respect to cancer, the mentioned classification approaches are widely used in clinical studies, for example, for the classification of various cancer types or to predict the therapeutic success of a cancer treatment [42].

In principle, such methods could also be more widely used in environmental toxicology for a qualitative classification of aquatic samples and biota. The general approach is as follows. After the definition of classes reflecting the underlying scientific question, for example, exposed vs not exposed to a given stressor or occurrence vs absence of an apical endpoint which might be manifested at later time points, the data set is divided into a training set and a validation set. For the training set the classes are indicated and the underlying algorithm maximizes the selectivity and specificity of the classification using the data in the training set. Support Vector Machines (SVM), for example, try to find "functions" that separate the defined classes in an n-dimensional space with a minimum number of false negative

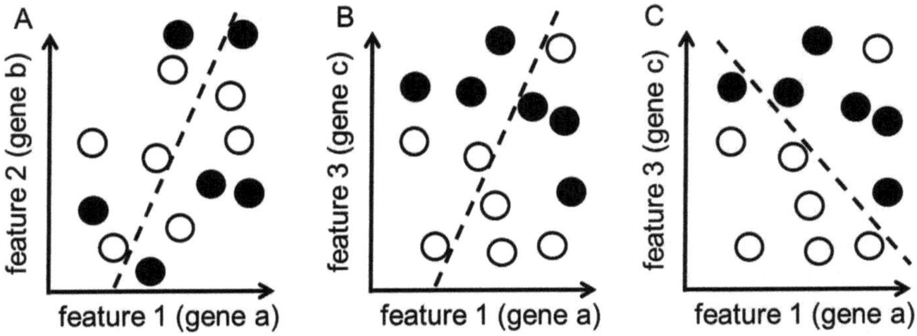

Fig. 3 Schematic illustration of the general approach of support vector machines. The *hollow circles* represent a control class, for example, non-exposed and the *black circles* represent the treated class. Features 1–3 are the expression levels of the genes a, b, and c, respectively. (**a**) Using the features 1 and 2, that is, expression data from gene 1 and 2, it is impossible to find a linear function that allows a clear distinction between the two classes defined. (**b**) The combination of the selected features 1 and 3 – the expression data of the genes a and c – provides a better data basis for a classification, although the selected linear function fails to distinguish exposed from non-exposed samples. (**c**) With the linear function indicated in the figure, it is possible to distinguish control samples from treated samples. In each group there is one misclassified, that is, wrong, sample

and false positive classifications. This approach is illustrated in Fig. 3 in two-dimensional space.

Each feature used for the SVM adds a dimension for the classification of the samples. The first important step is the feature selection answering the question of which set of genes has to be chosen for a classification of the samples. The second step is then the definition of a function that separates the defined classes most accurately. It is very important to validate the trained classification with data that was not used for the training of the SVM. Such approaches make use of the multidimensional nature of transcriptomics data. It is a practical approach to use gene expression data to assess a contamination of an aquatic sample or the exposure of animals from the field. Furthermore, compounds with an unknown mode of action can be compared to model compounds to find whether the compound to be assessed acts in a similar way to the model compound – see the case study below. A deeper understanding of the molecular functions of the selected features (i.e., genes) is not necessary per se. Despite the power of such approaches, an obstacle might be a limited data basis, especially in case of environmental toxicology. Furthermore, a high number of features might possibly be required for a classification with sufficient accuracy, a clear drawback for routine analysis, because expression data for all genes selected as features for the classification have to be measured. In several open source software tools, such as the Multiple Experimental Viewer (MEV), classification algorithms are implemented [43].

3.2.1 Case Study

The following study shows the applicability of global transcriptomic analyses, in particular pattern analysis, to classify substances with an unknown MOA for, for example, the risk assessment of chemicals [44]. This was done by comparing substances with an unknown MOA to reference toxicants with a known MOA regarding their effects on the macroscopic and the molecular level. A pattern analysis on transcriptomic data from a microarray analysis on zebra fish embryo was conducted by Hawliczek et al. [45]. The authors hypothesized that the PAHs benzo(*b*)fluorene (BBF) and 4-azapyrene (AP) induce a dioxin-like developmental toxicity in zebrafish. Therefore, effects on the macroscopic and molecular levels were compared to those induced by the known teratogene retene.

On the macroscopic level, the expected dioxin-like phenotype was induced by the reference substance retene and also under BBF, but not under AP stress which showed distinct developmental toxicity effects. A similar behavior of retene and BBF was also detected in global gene expression. A microarray analysis was applied at sublethal concentrations (EC_{10}) of the tested PAHs. Data were analyzed by means of Ingenuity Pathway Analysis, to identify networks of genes that can characterize each tested substance. Gene networks were identified that were most significant to the set of differentially expressed genes induced by each PAH. In the gene set of BBF and retene, one distinct network each was found. In contrast, the gene set of AP included two distinct networks. As substance specific patterns were found in the gene sets, tests were carried out to determine whether machine learning methods can discriminate between the three different PAHs as single substances and in a synthetic mixture. By means of diagonal linear discriminant analysis and k-nearest-neighbor machine learning, a set of genes was found that could almost in all cases clearly predict the three PAHs and the mixture based on their unique expression pattern. This gene set was also used for discrimination between the test substances and the mixture by means of PCA and hierarchical clustering (HCL). PCA and HCL indicated a closer relationship of BBF and retene. In conclusion, the two previously uncharacterized PAHs BBF and AP induced developmental toxicity. However, by means of pattern analysis, differences in the underlying MOA between BBF, as well as retene, and AP were detected. Thus, the presented approach is an example of how to categorize chemicals concerning the hazard they might pose by means of toxicogenomics.

3.3 Pathway Analysis

Pathway analysis, similar to pattern analysis, is performed on large gene sets from transcriptional analysis. By means of gene ontology (GO) term enrichment, genes with known annotation can be linked to corresponding GO terms. Thus, underlying biological, molecular, or cellular functions induced by a stressor in comparison to the physiological reference state of the test organism can be identified. As indicated, GO consists of three hierarchically structured vocabularies (ontologies) that describe gene products in terms of their associated biological processes, cellular components, and molecular functions [46]. Overrepresented annotations are identified by statistical analysis to identify predominant effects on the molecular level in the tested sample. Various web applications based on diverse algorithms can be used for GO term enrichment; examples are AmiGO [47], Gorilla for ranked lists of genes [46], GoMiner [48], FatiGO [49], and DAVID [50]. Unfortunately, many genes from non-model organisms are not yet annotated and therefore cannot be included in the analysis. Furthermore, there are large differences concerning the integration of organisms in the GO term databases, as not all model species are included in each application and the formats of demanded gene identifiers vary between applications. However, GO term analysis is a valuable tool to gain insight into the MOA of specific stressors. For mixtures of stressors, induced GO terms might highlight processes of mixture toxicity as demonstrated in the following case study.

3.3.1 Case Study

In an early system biology study, Heckmann et al. [51] aimed to bridge the gap between transcriptomic and phenotypic stress responses in the aquatic model organism *D. magna* exposed to ibuprofen. Ibuprofen was designed to interrupt the eicanoseid metabolism in mammals, but little was known of its effects on non-target organisms. Therefore, chronic population experiments were linked to acute effects on the transcriptomic level to obtain an insight into ibuprofen's MOA in *D. magna* and to study its effects on life history traits and population dynamics. On the macroscopic level, a strong reduction in reproduction resulting in a population decline for *D. magna* was detected. Effects on the transcriptomic level were assessed by means of global gene expression analysis using DNA microarrays with a subsequent GO term analysis. Several interlinked pathways and biological processes were suggested in *D. magna* under ibuprofen stress. These were the interruption of the eicosanoid metabolism, resulting in the impairment of signal transduction, finally leading to the disruption of the endocrine system related to juvenile hormone esterase metabolism and oogenesis. The results show a strong link between acute effects on the transcriptomic level and chronic effects on the organismic and population level.

4 Toxicogenomics in Regulation

The following section focus on the questions of whether and how toxicogenomics is or should be integrated into environmental regulation.

(Eco-)"toxicology is defined as the study of stressors and their adverse effects. One sub-discipline deals with hazard identification, mechanistic toxicology, and risk assessment" [22]. (Eco-)toxicogenomics integrates omics methods that offer mechanistic information on stress responses, with adverse effects on the organ, organism, or population level [9, 22]. Methods defined as omics technologies measure biochemical changes on the level of RNA/DNA (transcriptomics), proteins (proteomics), and the whole metabolome (metabolomics) [22]. omics technologies in (eco-)toxicology are expected to serve as methods used to create stepping stones (mechanistic information on stressor related response) that facilitate crossing the river that separates cause ((environmental-)stressor) and effect (assessed on organ, organism or population level). By consolidating many single stepping stones, the river between cause and effect can be crossed on a causeway. Regarding a risk assessment of chemicals, for example, this causeway can support the elucidation of MOA or AOPs [6]. "An AOP is a conceptual construct that portrays existing knowledge concerning the linkage between a direct molecular initiating event and an adverse outcome at a biological level of organization relevant to risk assessment" [52].

Several scientific programs address, among other things, the question of whether new alternative technologies such as omics are worth integrating into hazard assessment of chemical, cosmetic, or food safety, such as:

- OECD – Adverse Outcome Pathways, Molecular Screening, and Toxicogenomics [53]
- US-EPA – TOX-21, a quantitative high-through put screening for chemical toxicity on humans, including the acquisition of gene/protein expression data [54]
- EU initiative SEURAT-1, a research cluster addressing the issue of alternatives to animal testing for prediction of repeated dose systemic toxicity [55]

Hopes and expectations regarding the integration of omics in (environmental) risk assessment, specifically in hazard assessment, both hazard identification and hazard characterization, are based on diverse studies, and some of these are discussed in the following. By elucidating biological mechanisms around critical events [56], several existing toxicity prediction strategies for the risk assessment of stressors, mainly chemical substances show the potential to be enhanced by toxicogenomic technologies. Every study on effects induced by a specific stressor (e.g., chemicals), ideally analyzed on different levels of effect (e.g., molecular and macroscopic) and organization (e.g., cell, organ, organism, population) resembles a thin thread of information. Of course mostly there are more studies on a given stressor focusing possibly on different levels of effect. Taking those thin, stressor specific, threads together, a thicker and more robust thread can be woven (e.g., MOA, Weight of Evidence, Toxicity identification Evaluation, Sediment Quality guidelines). All existing threads for diverse stressors can then be the basis for knots in an ever growing net of knowledge (e.g., database). With each knot, more meshes are formed and the net of knowledge grows denser until it can be used for fishing for specific information on the substances that formed the threads or substances with unknown toxicity (e.g., read-across in chemical registration). Some "robust threads" to be supported by omics are:

- AOP/MOA: The main focus of discussion lies in the support or elucidation of MOA or AOPs by means of omics technologies. The OECD appreciates the applicability of mechanistic information obtained from studies using omics technologies regarding the creation of AOP [53], especially for skin sensitization [57]. Thereby, information obtained with omics technologies can serve "to identify further information needs specific to the target chemical or chemical category" [57]. This can be done by combining datasets on in vivo and in vitro methods for distinct compounds in large databases with knowledge of MOA or AOPs. Results from omics methods can be the identification of distinct genes or transcriptomic pattern working as biomarkers of effect that are related to specific MOAs/AOPs.
  Whereas the OECD focuses on human toxicology, Ankley et al. [52] suggests the incorporation of the AOP approach into a broader system's biology framework.
- Weight of Evidence: The international consortium on fish toxicogenomics sponsored by the U.K. Natural Environment Research Council (Fish Toxicogenomics – Moving into Regulation and Monitoring, held 21–23 April 2008 at the Pacific Environmental Science Centre, Vancouver, BC, Canada, discussed omics technologies as "already useful for elucidating modes of action of toxicants and can contribute to the risk assessment process as part of a weight-of-evidence approach" [44].
- Toxicity identification evaluations: Ankley et al. [58] proposed omics as a promising approach to assess the biologically relevant toxicity pathways in organisms exposed to complex mixtures of contaminants that would greatly assist the TIE process.
- Sediment quality guidelines: During the International Conference on Deriving Environmental Quality Standards for the Protection of Aquatic Ecosystems held in Hong Kong in December 2011, an expert group, focusing on sediment management, demanded new toxicity study approaches such as biomarkers and high throughput OMIC-based toxicity screening for sediment risk assessment [59]

Today, toxicogenomic technologies are not integrated as standard methods in regulation. However, some regulatory agencies either already consider the use of submitted data on a case-by-case basis for assessment purposes, such as the U.S. Environment Protection Agency, the U.S. Food and Drug Administration [56], or, in the case of the European Chemical Agency, discuss whether toxicogenomic data should be included in data submissions for regulated substances and how this could be implemented in the standard procedures. This discussion is mainly focused on REACH [60] but also addresses the Water Framework Directive [61]. Specifically, ANNEX XI of the REACH regulation suggests the possibility of substantially improving the hazard assessment of substances by making use of alternative methods, especially by focusing on read-across methods.

Alternative methods in this sense could be omics technologies, such as transcriptomics, as they can offer mechanistic information on the MOA of chemicals. In the registration process, omics are thought to support a read-across/ grouping of target substances with missing information on hazard [62, 63]. This means, for example, that registrants for substances with missing information, which show structural similarities to registered substances, could support the indication of comparable MOAs or even AOPs by means of omics technologies to facilitate their classification.

Important factors to integrate omics technologies in regulation, such as chemical registration, are:

- A shorter test duration compared to common test systems, as effects on the molecular level are often measurable long before adverse effects become manifest on the macroscopic level [1]. Therefore, omics can be further regarded as a new tool for predictive toxicology that can assess chronic effects which are not assessed in most biotests.
- Effects at low doses are better detectable on a molecular scale [6]
- In REACH [60], sub-organismic testing is demanded instead of animal testing [64]. Therefore, an integral part of ongoing research initiatives such as TOX-21 and SEURAT-1 deals with the use of inter alia omics technologies in terms of alternative methods to animal testing [6, 65, 66].
  The substitution of animal testing in REACH is demanded for several reasons:
  - The implementation of REACH leads to higher numbers of test animals. The substitution of animal tests is a question of ethics [6].
  - Alternative methods to animal testing, for example, using omics technologies, are also expected to offer a higher inter- and intra-species comparability and to omit the extrapolation from high to low doses, and from short- to long-term exposure. Thus, they are suggested to strengthen the predictability in human hazard assessment [65].

Despite the high expectations for the integration of omics technologies in regulation, there are some factors that need to be respected and some technical hurdles that must be overcome first:

- One has to respect that phenotypic anchoring is needed for a worthwhile interpretation of omics readouts. This means that only if results from transcriptomic analysis are integrated in AOPs and thus are linked to adverse effects on higher biological or ecological levels can transcriptomics be meaningfully integrated in ecotoxicological hazard identification and characterization [8, 56, 61].
- The integration of omics technologies in general in hazard assessment is still limited by their state of validation and the level of their standardization. A better standardization was demanded, for example, to minimize technical variation for DNA microarray experiments by Brazma et al. [67] and for RT-qPCR by Bustin et al. [68]. Although these guidelines were important steps toward standardization and comparability between studies, they are not yet obligatory for

publication in all scientific journals. One reason among others for this situation might be higher costs by abiding to these guidelines

- omics technologies might suffer from a high level of biological variation [10]. The biological variation reflects changes in molecular expression caused not only by a specific stressor but also by other factors such as sex, cell type, etc. Specifically with respect to field studies, differing abiotic factors can further complicate the interpretation of data [10].
- Another problem is the type of results obtained by omics analysis. Regulators deal with readouts that can be translated in a quantitative risk assessment, that is, no effect concentrations of compounds for various organisms and endpoints. A comparable type of results is not yet provided by omics methods. For an application in regulation, omics results must be transformed into guidance values or a margin of exposure [6]. Uncertainties and degrees of confidence have to be understandable by decision makers [58]. Data have to reflect "adequately quantitative dose- and time-dependent (e.g., threshold) responses that are the hallmark of toxicological evaluation, making their immediate acceptance in regulatory arenas circumspect" [56].
- Last but not least, for routine use, omics technologies have to be less cost intensive.

5 Conclusion

(Eco-)toxicogenomics is regarded as an opportunity to bridge the gap between cause and effect. omics technologies span different molecular levels by the application of transcriptomics, proteomics, and metabolomics. These technologies offer information on biological and molecular responses. This chapter is focused on transcriptomics that aims to quantify stressor-induced changes on the transcript, that is, mRNA level. Global transcriptomic changes are generally assessed by means of DNA microarray or RNA sequencing. methods. Individual, specific (e.g., MOA related) genes, can be assessed by means of qPCR. Data obtained by these techniques have to be correlated to ecotoxicological data capturing apical endpoints. This idea of phenotypic anchoring was first mentioned by Paules [2]. In the case of clinical studies, for example, this strategy allowed the classification of different cancer types with respect to the diagnostic outcome. The basis of this achievement is a sound and broad database providing both transcriptomics and the respective clinical data. It would be highly attractive to translate this success to the field of ecotoxicogenomics. Regarding the need for the required amount of data, it becomes evident that scientists have to join forces in collaborative studied. The collection of large data sets across research groups necessitates the further development of standards for both, experimental protocols and data storage. The amount of data requires intelligent strategies to make use of this "big data," avoiding not seeing the wood for the trees.

Ecotoxicological data obtained by transcriptomics are considered to be worthwhile integrated in risk assessment, as part of MOAs or AOPs for specific stressors, but also in WoE or TIER approaches.

Transcriptomics is expected to be an added value in ecotoxicology or risk assessment, because it can elucidate the mechanistic understanding of toxic effects, can predict chronic toxicity at a shorter test-duration, and is assumed to differentiate between severe and less severe outcome of effects. Despite the need for further developments data obtained by transcriptomics, the data are considered to be worthwhile integrated in ecotoxicological risk assessment, for example, as part of MOAs or AOPs for specific stressors, but also in WoE or TIER approaches. Specifically, the application of transcriptomic or omics in general, for read-across or classifications in the process of the registration of chemicals, are in focus [62, 63].

In summary, there is a lot of evidence that omics approaches can support environmental toxicology in a multitude of ways, although major challenges have still to be solved to implement these techniques in regulatory risk assessment.

References

1. Nuwaysir EF, Bittner M, Trent J et al (1999) Microarrays and toxicology: the advent of toxicogenomics. Mol Carcinog 24:153–159. doi:10.1002/(SICI)1098-2744(199903)24:3<153::AID-MC1>3.0.CO;2-P
2. Calzolai L, Ansorge W, Calabrese E et al (2007) Transcriptomics and proteomics. Applications to ecotoxicology. Comp Biochem Physiol Part D Genomics Proteomics 2:245–249. doi:10.1016/j.cbd.2007.04.007
3. Monsinjon T, Knigge T (2007) Proteomic applications in ecotoxicology. Proteomics 7:2997–3009. doi:10.1002/pmic.200700101
4. Bundy JG, Davey MP, Viant MR (2009) Environmental metabolomics: a critical review and future perspectives. Metabolomics 5:3–21. doi:10.1007/s11306-008-0152-0
5. Lankadurai BP, Nagato EG, Simpson MJ (2013) Environmental metabolomics: an emerging approach to study organism responses to environmental stressors. Environ Rev 21:180–205. doi:10.1139/er-2013-0011
6. European Food Safety Authority (2014) Modern methodologies and tools for human hazard assessment of chemicals. EFSA J 12(4):3638:87 pp. doi:10.2903/j.efsa.2014.3638
7. Snape JR, Maund SJ, Pickford DB, Hutchinson TH (2004) Ecotoxicogenomics: the challenge of integrating genomics into aquatic and terrestrial ecotoxicology. Aquat Toxicol 67:143–154. doi:10.1016/j.aquatox.2003.11.011
8. Schirmer K, Fischer B, Madureira D, Pillai S (2010) Transcriptomics in ecotoxicology. Anal Bioanal Chem 397:917–923. doi:10.1007/s00216-010-3662-3
9. Paules R (2003) Phenotypic anchoring: linking cause and effect. Environ Health Perspect 111:A338–A339
10. Simmons DBD, Benskin JP, Cosgrove JR et al (2015) Omics for aquatic ecotoxicology; control of extraneous variability to enhance the analysis of environmental effects. Environ Toxicol Chem 34:1693–1704. doi:10.1002/etc.3002
11. Grisham JW (1983) Cell types in rat liver cultures: their identification and isolation. In: Najjar VA (ed) Enzyme induction modules. Springer, Boston, pp 23–33

12. Marceau N, Blouin M-J, Germain L, Noel M (1989) Role of different epithelial cell types in liver ontogenesis, regeneration and neoplasia. In Vitro Cell Dev Biol 25:336–341. doi:10.1007/BF02624596
13. Heisenberg W (1971) Physics and beyond: encounters and conversations. G. Allen & Unwin, London
14. Canelas AB, Ras C, ten Pierick A et al (2008) Leakage-free rapid quenching technique for yeast metabolomics. Metabolomics 4:226–239. doi:10.1007/s11306-008-0116-4
15. Teng Q, Huang W, Collette TW et al (2009) A direct cell quenching method for cell-culture based metabolomics. Metabolomics 5:199–208. doi:10.1007/s11306-008-0137-z
16. Tufi S, Lamoree MH, De Boer J, Leonards PEG (2015) Cross-platform metabolic profiling: application to the aquatic model organism Lymnaea stagnalis. Anal Bioanal Chem 407:1901–1912. doi:10.1007/s00216-014-8431-2
17. Afshari CA, Hamadeh HK, Bushel PR (2011) The evolution of bioinformatics in toxicology: advancing toxicogenomics. Toxicol Sci 120:S225–S237. doi:10.1093/toxsci/kfq373
18. Ju Z, Wells MC, Walter RB (2006) DNA microarray technology in toxicogenomics of aquatic models: methods and applications. Comp Biochem Physiol Part C Toxicol Pharmacol 145:5–14. doi:10.1016/j.cbpc.2006.04.017
19. Lettieri T (2006) Recent applications of DNA microarray technology to toxicology and ecotoxicology. Environ Health Perspect 114:4–9. doi:10.1289/ehp.8194
20. Slonim DK, Yanai I (2009) Getting started in gene expression microarray analysis. PLos Comput Biol 5, E1000543. doi:10.1371/journal.pcbi.1000543
21. Slonim D (2002) From patterns to pathways: gene expression data analysis comes of age. Nat Genet 32:502–508. doi:10.1038/ng1033
22. Hamadeh HK, Amin RP, Paules RS, Afshari CA (2002) An overview of toxicogenomics. Curr Issues Mol Biol 4:45–56
23. Gayral P, Weinert L, Chiari Y et al (2011) Next-generation sequencing of transcriptomes: a guide to RNA isolation in nonmodel animals. Mol Ecol Resour 11:650–661. doi:10.1111/j.1755-0998.2011.03010.x
24. Fleige S, Walf V, Huch S et al (2006) Comparison of relative mRNA quantification models and the impact of RNA integrity in quantitative real-time RT-PCR. Biotechnol Lett 28:1601–1613
25. Mueller O, Lightfoot S, Schroder A (2004) RNA integrity number (RIN) standardization of RNA quality control. Tech Rep 5989-1165EN Agilent Application Note
26. Schroeder A, Mueller O, Stocker S et al (2006) The RIN: an RNA integrity number for assigning integrity values to RNA measurements. BMC Mol Biol 7:3. doi:10.1186/1471-2199-7-3
27. Su Z, Ning B, Fang H, Hong H, Perkins R, Tong W and Shi L (2011) Next-generation sequencing: a revolutionary tool for toxicogenomics. In: General, applied and systems toxicology. doi:10.1002/9780470744307.gat232
28. Heid C, Stevens J, Livak K, Williams P (1996) Real time quantitative PCR. Genome Res 6:986–994. doi:10.1101/gr.6.10.986
29. Karlen Y, McNair A, Perseguers S et al (2007) Statistical significance of quantitative PCR. BMC Bioinf 8:131. doi:10.1186/1471-2105-8-131
30. Livak K, Schmittgen T (2001) Analysis of relative gene expression data using real-time quantitative PCR and the 2-DeltaDeltaCt method. Methods 25:402–408. doi:10.1006/meth.2001.1262
31. Nolan T, Hands RE, Bustin SA (2006) Quantification of mRNA using real-time RT-PCR. Nat Protoc 1:1559–1582. doi:10.1038/nprot.2006.236
32. Pfaffl M (2001) A new mathematical model for relative quantification in real-time RT-PCR. Nucleic Acids Res 29, e45
33. Pfaffl M (2004) Quantification strategies in real-time PCR. In: Bustin SA (ed) A-Z of quantitative PCR 1. International University Line, La Jolla, pp 89–113

34. Regier N, Frey B (2010) Experimental comparison of relative RT-qPCR quantification approaches for gene expression studies in poplar. BMC Mol Biol 11:57. doi:10.1186/1471-2199-11-57
35. Hook SE, Osborn HL, Spadaro DA, Simpson SL (2014) Assessing mechanisms of toxicant response in the amphipod Melita plumulosa through transcriptomic profiling. Aquat Toxicol 146:247–257. doi:10.1016/j.aquatox.2013.11.001
36. Hook SE, Twine NA, Simpson SL et al (2014) 454 pyrosequencing-based analysis of gene expression profiles in the amphipod Melita plumulosa: transcriptome assembly and toxicant induced changes. Aquat Toxicol 153:73–88. doi:10.1016/j.aquatox.2013.11.022
37. Hook S, Osborn H, Spadaro D, Simpson S (2014) Challenges for using quantitative PCR test batteries as a TIE-type approach to identify metal exposure in benthic invertebrates. Environ Sci Pollut Res 22:17280–17289. doi:10.1007/s11356-014-3234-8
38. Regier N, Baerlocher L, Munsterkotter M et al (2013) Analysis of the Elodea nuttallii transcriptome in response to mercury and cadmium pollution: development of sensitive tools for rapid ecotoxicological testing. Environ Sci Technol 26:26. doi:10.1021/es401082h
39. Geiss GK, Bumgarner RE, Birditt B et al (2008) Direct multiplexed measurement of gene expression with color-coded probe pairs. Nat Biotechnol 26:317–325. doi:10.1038/nbt1385
40. Jo HJ, Jung J (2008) Quantification of differentially expressed genes in Daphnia magna exposed to rubber wastewater. Chemosphere 73:261–266. doi:10.1016/j.chemosphere.2008.06.038
41. Wang R-L, Bencic D, Biales A et al (2008) DNA microarray-based ecotoxicological biomarker discovery in a small fish model species. Environ Toxicol Chem 27:664–675. doi:10.1897/07-192.1
42. van't Veer LJ, Dai H, van de Vijver MJ et al (2002) Gene expression profiling predicts clinical outcome of breast cancer. Nature 415:530–536. doi:10.1038/415530a
43. Saeed AI, Sharov V, White J et al (2003) TM4: a free, open-source system for microarray data management and analysis. Biotechniques 34:374–378
44. Van Aggelen G, Ankley GT, Baldwin WS et al (2010) Integrating omic technologies into aquatic ecological risk assessment and environmental monitoring: hurdles, achievements, and future outlook. Environ Health Perspect 118:1–5. doi:10.1289/ehp.0900985
45. Hawliczek A, Nota B, Cenijn P et al (2012) Developmental toxicity and endocrine disrupting potency of 4-azapyrene, benzo[b]fluorene and retene in the zebrafish Danio rerio. Reprod Toxicol 33:213–223. doi:10.1016/j.reprotox.2011.11.001
46. Eden E, Navon R, Steinfeld I et al (2009) GOrilla: a tool for discovery and visualization of enriched GO terms in ranked gene lists. BMC Bioinf 10:1–7. doi:10.1186/1471-2105-10-48
47. Carbon S, Ireland A, Mungall CJ et al (2009) AmiGO: online access to ontology and annotation data. Bioinformatics 25:288–289. doi:10.1093/bioinformatics/btn615
48. Zeeberg BR, Feng W, Wang G et al (2003) GoMiner: a resource for biological interpretation of genomic and proteomic data. Genome Biol 4:R28. doi:10.1186/gb-2003-4-4-r28
49. Al-Shahrour F, Diaz-Uriarte R, Dopazo J (2004) FatiGO: a web tool for finding significant associations of Gene Ontology terms with groups of genes. Bioinformatics 20:578–580. doi:10.1093/bioinformatics/btg455
50. Dennis G Jr, Sherman BT, Hosack DA et al (2003) DAVID: database for annotation, visualization, and integrated discovery. Genome Biol 4:P3. doi:10.1186/gb-2003-4-9-r60
51. Heckmann L-H, Sibly RM, Connon R et al (2008) Systems biology meets stress ecology: linking molecular and organismal stress responses in Daphnia magna. Genome Biol 9:R40. doi:10.1186/gb-2008-9-2-r40
52. Ankley GT, Bennett RS, Erickson RJ et al (2010) Adverse outcome pathways: a conceptual framework to support ecotoxicology research and risk assessment. Environ Toxicol Chem 29:730–741. doi:10.1002/etc.34
53. OECD (2013) Guidance document on developing and assessing adverse outcome pathways. Series on Testing and Assessment No. 184, ENV/JM/MONO(2013)6

54. Schmidt CW (2009) TOX 21: new dimensions of toxicity testing. Environ Health Perspect 117:A348–A353. doi:10.1289/ehp.117-a348
55. Gocht T (2014) The SEURAT-1 approach towards animal free human safety assessment. ALTEX 32:9–24. doi:http://dx.doi.org/10.14573/altex.1408041
56. Corvi R, Ahr H-J, Albertini S et al (2005) Meeting report: validation of toxicogenomics-based test systems: ECVAM–ICCVAM/NICEATM considerations for regulatory use. Environ Health Perspect 114:420–429. doi:10.1289/ehp.8247
57. OECD (2012) The adverse outcome pathway for skin sensitisation initiated by covalent binding to proteins. Part 2: Use of the AOP to develop chemical categories and integrated assessment and testing approaches. Series on Testing and Assessment No. 168, ENV/JM/MONO(2012)10/PART2
58. Ankley GT, Daston GP, Degitz SJ et al (2006) Toxicogenomics in regulatory ecotoxicology. Environ Sci Technol 40:4055–4065. doi:10.1021/es0630184
59. Kwok KH, Batley G, Wenning R et al (2014) Sediment quality guidelines: challenges and opportunities for improving sediment management. Environ Sci Pollut Res 21:17–27. doi:10.1007/s11356-013-1778-7
60. Regulation (EC) No 1907/2006 of the European Parliament and of the Council of 18 December 2006 concerning the Registration, Evaluation, Authorisation and Restriction of Chemicals (REACH), establishing a European Chemicals Agency, amending Directive 1999/45/EC and repealing Council Regulation (EEC) No 793/93 and Commission Regulation (EC) No 1488/94 as well as Council Directive 76/769/EEC and Commission Directives 91/155/EEC, 93/67/EEC, 93/105/EC and 2000/21/EC
61. Wernersson A-S, Carere M, Maggi C et al (2015) The European technical report on aquatic effect-based monitoring tools under the water framework directive. Environ Sci Eur. doi:10.1186/s12302-015-0039-4
62. Berggren E, Amcoff P, Benigni R et al (2015) Chemical safety assessment using read-across: assessing the use of novel testing methods to strengthen the evidence base for decision making. Environ Health Perspect 123(12):1232–1240. doi:10.1289/ehp.1409342
63. Schultz TW, Amcoff P, Berggren E et al (2015) A strategy for structuring and reporting a read-across prediction of toxicity. Regul Toxicol Pharmacol 72:586–601. doi:10.1016/j.yrtph.2015.05.016
64. Weil M, Scholz S, Zimmer M et al (2009) Gene expression analysis in zebrafish embryos: a potential approach to predict effect concentrations in the fish early life stage test. Environ Toxicol Chem 28:1970–1978. doi:10.1897/08-627.1
65. Tralau T, Luch A (2015) Moving from rats to cellular omics in regulatory toxicology: great challenge toward sustainability or "up-shit-creek without a paddle"? Arch Toxicol 89:819–821. doi:10.1007/s00204-015-1511-z
66. Tralau T, Oelgeschläger M, Gürtler R et al (2015) Regulatory toxicology in the twenty-first century: challenges, perspectives and possible solutions. Arch Toxicol 89:823–850. doi:10.1007/s00204-015-1510-0
67. Brazma A, Hingamp P, Quackenbush J et al (2001) Minimum information about a microarray experiment (MIAME) toward standards for microarray data. Nat Genet 29:365–371. doi:10.1038/ng1201-365
68. Bustin SA, Benes V, Garson JA et al (2009) The MIQE guidelines: minimum information for publication of quantitative real-time PCR experiments. Clin Chem 55:611–622. doi:10.1373/clinchem.2008.112797
69. Jeong SW, Lee SM, Yum SS et al (2013) Genomic expression responses toward bisphenol-A toxicity in Daphnia magna in terms of reproductive activity. Mol Cell Toxicol 9:149–158. doi:10.1007/s13273-013-0019-y
70. Poynton HC, Robinson WE, Blalock BJ, Hannigan RE (2014) Correlation of transcriptomic responses and metal bioaccumulation in Mytilus edulis L. reveals early indicators of stress. Aquat Toxicol 155:129–141. doi:10.1016/j.aquatox.2014.06.015

71. Chi Hoa P, Park KS, Kim BC et al (2011) Construction and characterization of Japanese medaka (Oryzias latipes) hepatic cDNA library and its implementation to biomarker screening in aquatic toxicology. Aquat Toxicol 105:569–575. doi:10.1016/j.aquatox.2011.08.019
72. Rhee JS, Kim BM, Jeong CB et al (2013) Effect of pharmaceuticals exposure on acetylcholinesterase (AchE) activity and on the expression of AchE gene in the monogonont rotifer, Brachionus koreanus. Comp Biochem Physiol Part C Toxicol Pharmacol 158:216–224. doi:10.1016/j.cbpc.2013.08.005
73. Morales M, Planello R, Martinez-Paz P et al (2011) Characterization of Hsp70 gene in Chironomus riparius: expression in response to endocrine disrupting pollutants as a marker of ecotoxicological stress. Comp Biochem Physiol Part C Toxicol Pharmacol 153:150–158. doi:10.1016/j.cbpc.2010.10.003
74. Planello R, Servia MJ, Gomez-Sande P et al (2015) Transcriptional responses, metabolic activity and mouthpart deformities in natural populations of Chironomus riparius larvae exposed to environmental pollutants. Environ Toxicol 30:383–395. doi:10.1002/tox.21893
75. Salaberria I, Brakstad OG, Olsen AJ et al (2014) Endocrine and AhR-CYP1A pathway responses to the water-soluble fraction of oil in zebrafish (Danio rerio Hamilton). J Toxicol Environ Health A 77:506–515. doi:10.1080/15287394.2014.886983
76. Hong Y, Liu S, Lin X et al (2015) Recognizing the importance of exposure–dose–response dynamics for ecotoxicity assessment: nitrofurazone-induced antioxidase activity and mRNA expression in model protozoan Euplotes vannus. Environ Sci Pollut Res Int 22 (12):9544–9553. doi:10.1007/s11356-015-4096-4
77. Defo MA, Bernatchez L, Campbell PGC, Couture P (2015) Transcriptional and biochemical markers in transplanted Perca flavescens to characterize cadmium- and copper-induced oxidative stress in the field. Aquat Toxicol 162:39–53. doi:10.1016/j.aquatox.2015.02.014
78. Beggel S, Connon R, Werner I, Geist J (2011) Changes in gene transcription and whole organism responses in larval fathead minnow (Pimephales promelas) following short-term exposure to the synthetic pyrethroid bifenthrin. Aquat Toxicol 105:180–188. doi:10.1016/j.aquatox.2011.06.004
79. Ruiz P, Ortiz-Zarragoitia M, Orbea A et al (2012) Responses of conventional and molecular biomarkers in turbot Scophthalmus maximus exposed to heavy fuel oil no. 6 and styrene. Aquat Toxicol 116–117:116–128. doi:10.1016/j.aquatox.2012.02.004
80. Hook SE, Osborn HL, Golding LA et al (2014) Dissolved and particulate copper exposure induces differing gene expression profiles and mechanisms of toxicity in the deposit feeding amphipod Melita plumulosa. Environ Sci Technol 48:3504–3512. doi:10.1021/es405322s
81. Hutchins CM, Simon DF, Zerges W, Wilkinson KJ (2010) Transcriptomic signatures in Chlamydomonas reinhardtii as Cd biomarkers in metal mixtures. Aquat Toxicol 100:120–127. doi:10.1016/j.aquatox.2010.07.017
82. Chakrabarty D, Trivedi PK, Misra P et al (2009) Comparative transcriptome analysis of arsenate and arsenite stresses in rice seedlings. Chemosphere 74:688–702. doi:10.1016/j.chemosphere.2008.09.082
83. Carvalho RN, Bopp SK, Lettieri T (2010) Gene and protein regulation in the marine diatom Thalassiosira pseudonana upon exposure to polycyclic aromatic hydrocarbons. Comp Biochem Physiol Mol Integr Physiol 157:S51–S52. doi:10.1016/j.cbpa.2010.06.146
84. Carvalho RN, Bopp SK, Lettieri T (2011) Transcriptomics responses in marine diatom Thalassiosira pseudonana exposed to the polycyclic aromatic hydrocarbon benzo(a)pyrene. PLos One 6:9. doi:10.1371/journal.pone.0026985
85. Carvalho RN, Burchardt AD, Sena F et al (2011) Gene biomarkers in diatom Thalassiosira pseudonana exposed to polycyclic aromatic hydrocarbons from contaminated marine surface sediments. Aquat Toxicol 101:244–253. doi:10.1016/j.aquatox.2010.10.004
86. Eom HJ, Kim H, Kim BM et al (2014) Integrative assessment of benzene exposure to Caenorhabditis elegans using computational behavior and toxicogenomic analyses. Environ Sci Technol 48:8143–8151. doi:10.1021/es500608e

87. Zucchi S, Mirbahai L, Castiglioni S, Fent K (2014) Transcriptional and physiological responses induced by binary mixtures of drospirenone and progesterone in zebrafish (Danio rerio). Environ Sci Technol 48:3523–3531. doi:10.1021/es405584f
88. Schiller V, Wichmann A, Kriehuber R et al (2013) Transcriptome alterations in zebrafish embryos after exposure to environmental estrogens and anti-androgens can reveal endocrine disruption. Reprod Toxicol 42:210–223. doi:10.1016/j.reprotox.2013.09.003
89. Martyniuk CJ, Doperalski NJ, Kroll KJ et al (2013) Sexually dimorphic transcriptomic responses in the teleostean hypothalamus: a case study with the organochlorine pesticide dieldrin. Neurotoxicology 34:105–117. doi:10.1016/j.neuro.2012.09.012
90. Bissegger S, Martyniuk CJ, Langlois VS (2014) Transcriptomic profiling in Silurana tropicalis testes exposed to finasteride. Gen Comp Endocrinol 203:137–145. doi:10.1016/j.ygcen.2014.01.018
91. Huang L, Zuo Z, Zhang Y, Wang C (2015) Toxicogenomic analysis in the combined effect of tributyltin and benzo[a]pyrene on the development of zebrafish embryos. Aquat Toxicol 158:157–164. doi:10.1016/j.aquatox.2014.10.024
92. Cassese A, Guindani M, Antczak P et al (2015) A Bayesian model for the identification of differentially expressed genes in Daphnia magna exposed to munition pollutants. Biometrics 71:803–811. doi:10.1111/biom.12303
93. Garcia-Reyero N, Escalon BL, Loh P et al (2012) Assessment of chemical mixtures and groundwater effects on Daphnia magna transcriptomics. Environ Sci Technol 46:42–50. doi:10.1021/es201245b
94. Garcia-Reyero N (2015) Are adverse outcome pathways here to stay? Environ Sci Technol 49:3–9. doi:10.1021/es504976d
95. Pascoal S, Carvalho G, Vasieva O et al (2013) Transcriptomics and in vivo tests reveal novel mechanisms underlying endocrine disruption in an ecological sentinel, Nucella lapillus. Mol Ecol 22:1589–1608. doi:10.1111/mec.12137
96. Yang L, Kemadjou J, Zinsmeister C et al (2007) Transcriptional profiling reveals barcode-like toxicogenomic responses in the zebrafish embryo. Genome Biol 8:R227. doi:10.1186/gb-2007-8-10-r227
97. Menzel R, Swain S, Hoess S et al (2009) Gene expression profiling to characterize sediment toxicity - a pilot study using Caenorhabditis elegans whole genome microarrays. BMC Genomics 10:160. doi:10.1186/1471-2164-10-160
98. Huang LX, Zuo ZH, Zhang YY et al (2014) Use of toxicogenomics to predict the potential toxic effect of benzo(a)pyrene on zebrafish embryos: ocular developmental toxicity. Chemosphere 108:55–61. doi:10.1016/j.chemosphere.2014.02.078
99. Gomez-Sagasti MT, Becerril JM, Martin I et al (2014) cDNA microarray assessment of early gene expression profiles in Escherichia coli cells exposed to a mixture of heavy metals. Cell Biol Toxicol 30:207–232. doi:10.1007/s10565-014-9281-6
100. Voelker D, Vess C, Tillmann M et al (2007) Differential gene expression as a toxicant-sensitive endpoint in zebrafish embryos and larvae. Aquat Toxicol 81:355–364. doi:10.1016/j.aquatox.2006.12.013
101. Filby AL, Santos EM, Thorpe KL et al (2007) Gene expression profiling for understanding chemical causation of biological effects for complex mixtures: a case study on estrogens. Environ Sci Technol 41:8187–8194. doi:10.1021/es071278v
102. Marchand J, Tanguy A, Charrier G et al (2006) Molecular identification and expression of differentially regulated genes of the European flounder, Platichthys flesus, submitted to pesticide exposure. Mar Biotechnol 8:275–294. doi:10.1007/s10126-005-0099-3
103. Villeneuve DL, Wang R-L, Bencic DC et al (2009) Altered gene expression in the brain and ovaries of zebrafish (Danio rerio) exposed to the aromatase inhibitor fadrozole: microarray analysis and hypothesis generation. Environ Toxicol Chem 28:1767–1782. doi:10.1897/08-653.1
104. Garcia-Reyero N, Adelman I, Liu L, Denslow N (2008) Gene expression profiles of fathead minnows exposed to surface waters above and below a sewage treatment plant in Minnesota. Mar Environ Res 66:134–136. doi:10.1016/j.marenvres.2008.02.073

Adv Biochem Eng Biotechnol (2017) 157: 187–224
DOI: 10.1007/10_2016_16

Published online: 19 October 2016

Coupling of In Vitro Bioassays with Planar Chromatography in Effect-Directed Analysis

Stefan C. Weiss, Nicole Egetenmeyer, and Wolfgang Schulz

Abstract Modern analytical test methods increasingly detect anthropogenic organic substances and their transformation products in water samples and in the environment. The presence of these compounds might pose a risk to the aquatic environment. To determine a possible (eco)toxicological risk, aquatic samples are tested using various bioassays, including sub-organismic assays such as the luminescent bacteria inhibition test, the acetylcholinesterase inhibition test, and the umu-test. The effect-directed analysis (EDA) combines physicochemical separation methods with biological (in vitro) tests. High-performance thin-layer chromatography (HPTLC) has proved to be particularly well suited for the separation of organic compounds and the subsequent analysis of effects by the application of the biotests directly on the surface of the HPTLC plate. The advantage of using HPTLC in comparison to high-performance liquid chromatography (HPLC) for EDA is that the solvent which is used as a mobile phase during chromatography is completely evaporated after the separation and therefore can no longer influence the applied bioassays.

A prioritization during the complex identification process can be achieved when observed effects are associated with the separated zones in HPTLC. This increases the probability of identifying the substance responsible for an adverse effect from the multitude of organic trace substances in environmental samples. Furthermore, by comparing the pattern of biological effects of a separated sample, it is possible to track and assess changes in biological activity over time, over space, or in the course of a process, even without identifying the substance. HPTLC has already been coupled with various bioassays.

Because HPTLC is a very flexible system, various detection techniques can be used and combined. In addition to the UV/Vis absorption and fluorescence

S.C. Weiss (✉), N. Egetenmeyer, and W. Schulz
Betriebs und Forschungslaboratorium, Zweckverband Landeswasserversorgung (LW), Am Spitzigen Berg 1, 89129 Langenau, Germany
e-mail: weiss.s@lw-online.de

measurements, TLC can also be coupled with a mass spectrometer (MS) for compound identification. In addition, detection of functional groups by means of derivatization reagents can support this identification. It is also possible to combine derivatization and HPLC-MS.

Two case studies are used to illustrate the significance of HPTLC-EDA in investigating water quality:

- Study on a wastewater treatment plant
- Possible influence of an artificial turf surface on ground water

Keywords Antibacterial effect, Bioautography, Bioluminescence inhibition, Derivatization, Environmental analysis, Estrogenic effect, Fungicidal effect, HPTLC/AMD, Neurotoxicity, Photosynthesis inhibition, TLC-MS

Contents

1 Introduction

Because of their frequent use, a wide variety of organic compounds, such as pharmaceuticals, pesticides, industrial, or household chemicals, are emitted into the environment via, for example, wastewater, agriculture, contaminated landfills, or street run-off [1–4]. By transformation processes in the environment and during wastewater treatment, the number of compounds is further increased [1, 5]. These compounds of anthropogenic origin (often referred to as anthropogenic micropollution) represent a potential risk to the aquatic environment and to drinking water resources [2, 5]. For the toxicological assessment of contaminants in a given sample, biological test methods are used [6–8]. However, biological tests do not provide any structural information about the bioactive compounds, which is needed to identify the origin of substances causing adverse biological effects and to take measures for their reduction.

To record the wide variety of substances in the environment, in addition to the target analysis of individual substances, non-target screening by means of high-performance liquid chromatography coupled with high-resolution mass spectrometry (HPLC-HRMS) is used. The non-target screening allows the detection of yet unidentified substances. An overview of the use of non-target screening is provided by Bletsou et al. [9] or Leendert et al. [10]. However, this physicochemical analytical method does not provide any information on the effect of detected substances, which hampers a toxicological assessment of the detected contaminant.

Even if a toxicological test is carried out in parallel with a non-target screening, matching a detected compound to an effect is virtually impossible. A solution that utilizes the combined approach of physicochemical analysis and in vitro bioassay is effect-directed analysis (EDA). In EDA, the sample, which usually has a complex composition, is first fractionated in a separation process and then examined further in a biological testing system [11]. Matching detected effects with respective HPLC fractions supports the identification of the substance(s) that trigger(s) the measured effects [11, 12]. The initial separation of the sample by HPLC provides additional information on the characteristics of the active substances (e.g., K_{OW}) [11]. The effects of the individual fractions result in a pattern of biological activity. Based on the changes in the effect pattern during a process (e.g., wastewater treatment), the pattern can be assessed along with observed changes in compound composition.

Because of the high separation performance and extensive automation, high-performance liquid chromatography (HPLC) is a frequently described separation technique for EDA in the literature. Despite these often-mentioned advantages, HPLC has some significant disadvantages for application in EDA. The time differences between fractions containing the separated compounds are usually far shorter than the exposure time required by many bioassays. In addition, the organic solvents needed for HPLC might affect the performance of the subsequent bioassays. To overcome these problems, the effect is detected offline after a change of solvent subsequent to the fractionation. It must be noted that the biological effect can be detected only in the collected fractions, which again reduces the original chromatographic resolution. For fast reacting enzyme inhibition tests, some online HPLC-EDA methods have been presented, but as already mentioned, the influence of the organic solvents has to be considered [13–15].

In contrast to HPLC, high-performance thin-layer chromatography (HPTLC) is an open separation system. The separation is carried out continuously along the HPTLC plate based on the retention of compounds on the silica surface. A compound with low mobility migrates a short distance on the HPTLC plate, whereas a compound with high mobility migrates a longer distance. Compounds not migrating at all can be detected at the point of sample application. The separation performance of HPTLC is not as high as that of HPLC, but the big advantage of the HPTLC approach is that the solvent used as the mobile phase during chromatography is completely evaporated after the separation. Therefore, solvents have no effect on bioassays that can be subsequently performed directly on the surface of the HPTLC plate to perform an effect-directed detection of compounds. Furthermore, HPTLC has a high matrix tolerance [16]. Because the solvent from sample extracts is removed after application by evaporation – and thus has no effect on the chromatography and subsequent bioassays – it is also possible to vary the applied sample volume over a wider range to increase the sensitivity of the method. However, because of the evaporation steps, it is impossible to investigate volatile compounds with HPTLC. If the parameters are selected appropriately, parallel chromatography of multiple samples allows the testing of many samples in a short period of time [17]. Problems are also reduced by using a new plate for each analytical run. Even substances that cannot be chromatographed are still detectable. As HPTLC is an open separation system, the substances are available on the plate after separation. In addition, HPTLC offers great flexibility in the selection of the detection technique, without the need for a reconfiguration of the instruments [18, 19].

However, the coupling of HPTLC with bioassays is a particular challenge, especially in the case of bioassays with long incubation times because of the diffusion of already separated compounds resulting in the broadening of signals. The only partial automation of the overall process from sample application to signal detection is a further obstacle for routine use in environmental analysis. As common for other coupling techniques [20], the naming of the individual HPTLC-EDA methods indicates the order of their use, with a hyphen between the various techniques; for example, HPTLC-acetylcholinesterase inhibition assay.

Combining EDA with planar chromatography has a long tradition. In 1946 Goddall and Levi presented an EDA method for the identification of various types of penicillin using paper chromatography [21]. Subsequently, a very wide variety of in vitro systems for effect-directed detection have been transferred to planar chromatography. The steps in HPTLC-EDA are sample preparation [11, 22–25], separation, physical detection (non-destructive), detection of biological effects, identification, and verification of the active substances (Fig. 1). Final identification of the bioactive compound is not absolutely necessary for each application and depends on the underlying question.

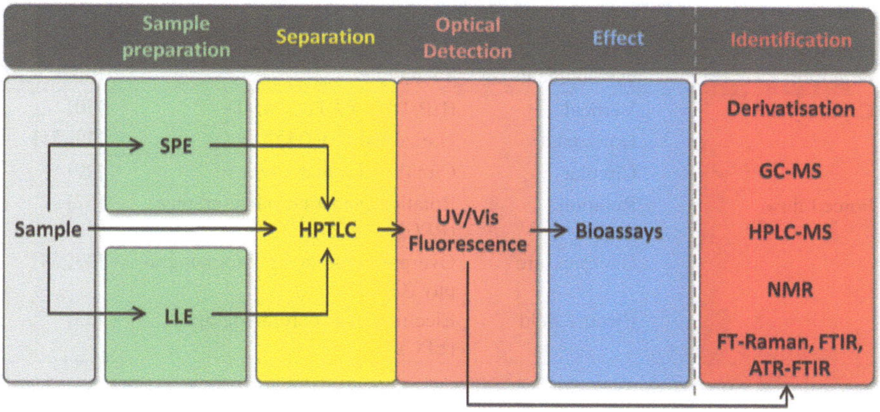

Fig. 1 Scheme of effect-directed analysis using HPTLC

Table 1 Characteristics of TLC, HPTLC, and UTLC layers (according to [26])

	TLC	HPTLC	UTLC
Plate size (cm)	20 × 20	20 × 10	3 × 3.6
Layer thickness (μm)	100–250	100–200	5–10
Particle size (μm)	8–10	6–8	Monolithic films, nanofibrous layers
Separation distance (cm)	6–15	3–7	1–3
Separation time (min)	30–200	3–20	1–5
Theor. plate height (μm)	35–75	23–25	–
LOD remission (ng)	1–5	0.5–1	0.5
LOD fluorescence (ng)	005–0.10	0.005–0.010	0.005

2 Thin-Layer Chromatography

2.1 Separation Technique

Planar chromatography encompasses all the chromatographic techniques in which separation is carried out on a planar stationary phase, including paper chromatography (PC), thin-layer chromatography (TLC) and its further refinement high-performance thin-layer chromatography (HPTLC), and ultra-thin-layer chromatography (UTLC). A comparison of the most important parameters and characteristics of TLC, HPTLC, and UTLC is given in Table 1.

As support material for the stationary phase, various materials can be used such as plastic film, aluminum foil, or glass plates. The stationary phase is usually bound to the support material by calcium sulfate, starch, or polymers [27]. The most frequently used sorbent in (HP)TLC is silica gel. On this so-called normal phase (NP), adsorption is the dominating separation mechanism. There are also various reversed phases (RP) containing *n*-alkyl moieties with different chain lengths that are bound as siloxanes to a silica gel (e.g., RP-18, *n*-alkyl moieties with 18-C

Table 2 Overview of HPTLC separation techniques

Principle of solvent migration	Mode of development	Technique	References
Capillary flow	Vertical	(HP)TLC, HPTLC/AMD	[30]
	Horizontal	Horizontal-(HP)TLC	[29, 31]
	Circular	Circular-chromatography	[29]
Forced flow	Rotation	Rotation planar chromatography (RPC)	[32]
	Overpressure	Overpressured-layer chromatography (OPLC)	[33, 34]
	Electric field	Electro-planar chromatography (EPC)	[33]

atoms). Modified silica gels with *n*-propyl chains are also available, which have a diol group or terminal $-NH_2$ or –CN group. Depending on the eluent, these phases are of NP or RP character. The different phases are classified qualitatively as follows from polar to non-polar:

$$SiO_2 > -NH_2 > -CN, -\text{Diol} > \text{RP-2} > \text{RP-8} > \text{RP-18}$$

The chromatographic development of a TLC or HPTLC plate takes place in a closed development chamber filled with a few milliliters of an organic solvent or solvent mixture as mobile phase, selected in accordance with the separation problem. For separation, the lower edge of the HPTLC plate is immersed in the mobile phase. Because of capillary forces, the solvent migrates through the stationary phase. As the migration distance increases, the speed of the eluent front decreases. In parallel, diffusion causes an increasing broadening of the compound bands over time. In most cases there is no further improvement of compound separation on HPTLC plates at a migration distance of more than about 60 mm [28].

Another type of HPTLC based on capillary flow uses a horizontal chamber for the development of the separation plate. This technique involves developing the plate from opposite sides toward the center, resulting in a twofold increase in sample throughput [29]. There are now various development techniques available which are shown in Table 2.

A further development of capillary flow HPTLC is automated multiple development (AMD). This chromatographic technique for TLC was first presented in 1973 by Perry [35] for multiple development with just one eluent (programmed multiple development, PMD). Later, Burger [36] further developed this technique for AMD to achieve automated gradient elutions. The principle of HPTLC/AMD is illustrated in Fig. 2. In HPTLC/AMD the plate is developed in multiple steps. Between each development step, the polarity of the eluent decreases and the migration distance is increased continuously. Gradients of 10–40 individual runs are typical, with increases in migration distance of 1–3 mm. For the automatic composition of the mobile phase, a typical AMD can access solvents from five supply bottles.

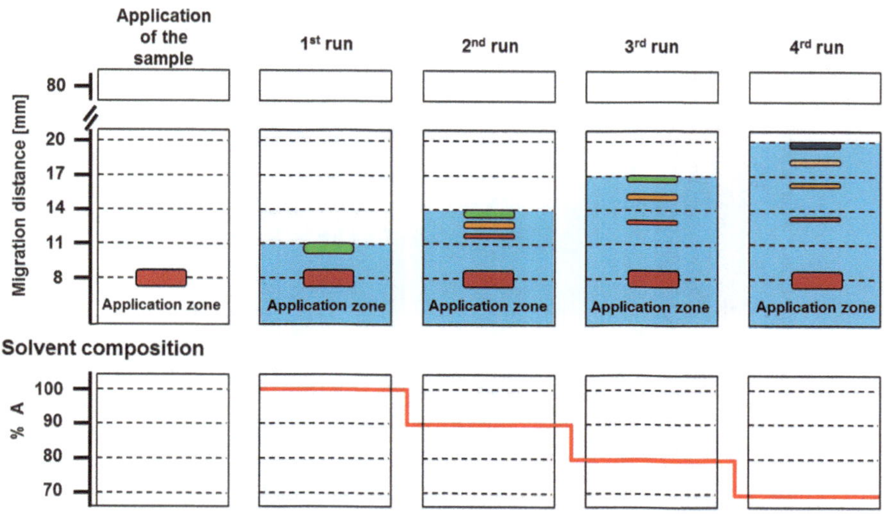

Fig. 2 Schematic illustration of gradient elution in automated multiple development (modified according to [37]). The percentage of the more polar solvent A is reduced after each separation step by the successive increase of the percentage of a less polar solvent

After sample application and the evaporation of residual solvent by the application of a vacuum, the first development run in AMD starts with the addition of the most polar eluent to the separation chamber. When the eluent has passed a prescribed migration distance, it is automatically removed from the chamber by a pump. The next separation cycle begins with drying, followed by the filling of the chamber with a less polar eluent, and the development of the HPTLC plate with a specified longer migration distance. This procedure is repeated until the entire prescribed development program has been completed. The procedure described above is carried out fully automatically in the AMD development unit.

In this development technique, in each step, the lower edge of a sample band is reached by the eluent first, so that the molecules located at the lower end start migrating before the molecules in the upper part of the band. This effect results in a continuous focusing of the compounds to be separated, leading to a suppression of band broadening caused by diffusion. Therefore, despite the multiple developments, the separation zones in AMD are very narrow. By multiple developments with the same solvent and the same migration distance, the sample can be focused in a narrow zone prior to compound separation. Figure 3a shows an example of an HPTLC/AMD gradient with a threefold focusing of the sample.

In comparison to classic HPTLC, the focusing effect of the AMD results in a higher separation capacity and lower limits of detection. In addition, the automated development results in a highly reproducible chromatographic separation. Figure 3b shows the stability of the Rf values for nine reference compounds over a time period of 6 months. The gradient shown in Fig. 3a was used for the chromatography. The substances were applied as a mix to the HPTLC plate.

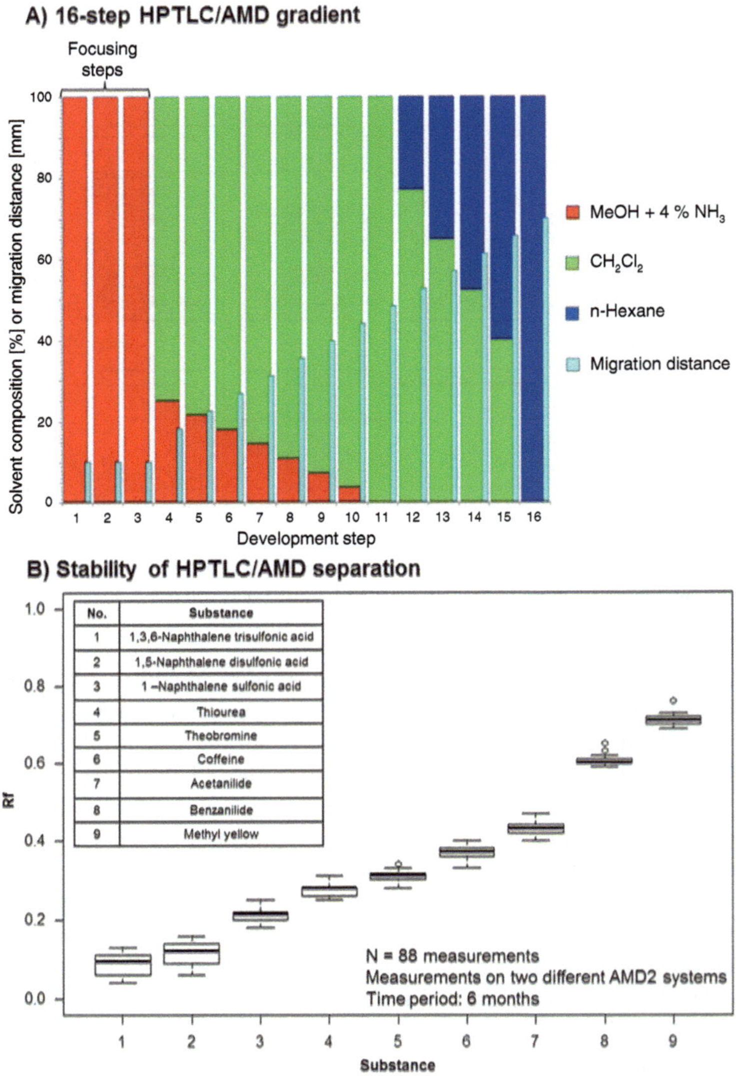

Fig. 3 (**a**) Diagram of a 16-step HPTLC/AMD gradient. (**b**) Stability of the HPTLC/AMD separation using this gradient for nine reference substances, shown as box plots. The amount applied for all substances was 300 ng

Even though AMD gradient elution is still not widely used, its potential has been demonstrated for various analytical questions. For instance, there are applications for the analysis of lipids [38, 39], foodstuffs [40], phytopharmaceuticals [41], and environmental samples [42, 43]. For environmental analysis, a standard has been developed for pesticide screening in Germany [44].

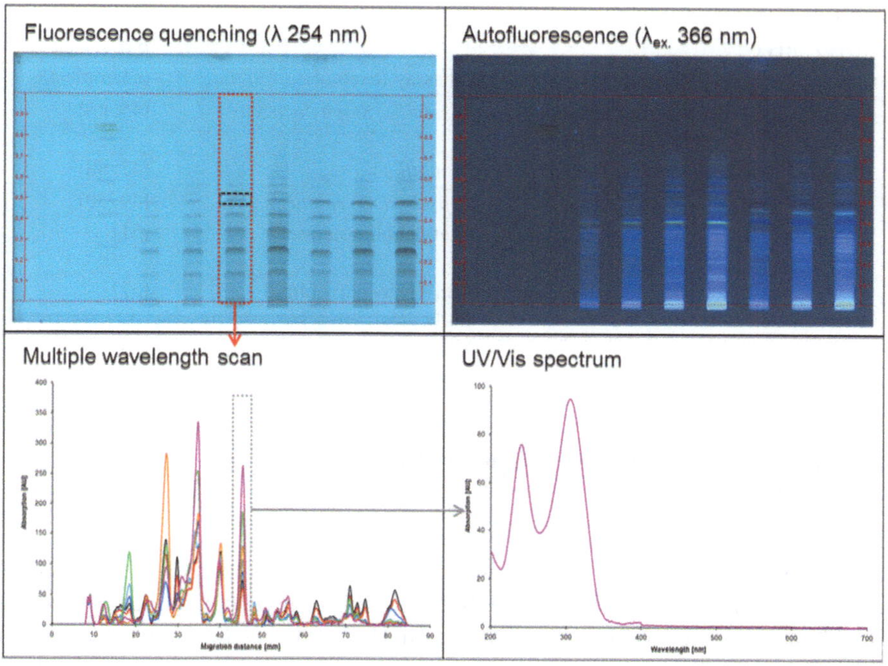

Fig. 4 Use of manifold optical detection methods

2.2 *Optical Detection*

The development of the HPTLC plate is followed by a first physicochemical detection of the separated compounds to obtain an initial overview of the samples applied. For this purpose, the plate is examined by using white light and UV light of wavelength 366 nm. At 366 nm, fluorescent substances are visible. If an HPTLC plate with fluorescence indicator is used, compounds that quench fluorescence at 254 nm (e.g., aromatic compounds) are detected by an additional inspection of the plate at 254 nm. Pictures of the plate at different wavelengths are captured for documentation. For a detailed examination, all UV/Vis absorbing substances and their positions can be recorded by means of a multi-wavelength scan. Fluorescence can also be detected in this way. It is also possible to record a UV/Vis remission spectrum of the individual substances. Examples of these various optical detection methods are shown in Fig. 4.

Table 3 Overview of the different HPTLC-EDA techniques

HPTLC-EDA technique	Subgroup	Bioassay (examples)	References (examples)
Bioautography	Diffusion/ Contact	*B. subtilis*, *Staphylococcus*	[45, 46]
	Overlay	*B. subtilis*	[47, 48]
	Direct	*A. fischeri*, *B. subtilis*	[49, 50]
Biochemical detection	Enzyme inhibition	Acetylcholinesterase	[51]
	Immunostaining	Anti-glycosphingolipid antibodies	[52]
Specific chemical characteristic	–	2,2-Diphenyl-1-picrylhydrazyl radical (DPPH)	[53]

3 Overview of Bioassay Application Techniques in HPTLC-EDA

In HPTLC used for effect-directed detection, the terms bioautography and biochemical detection have to be differentiated. Bioautography is used to describe the coupling of a bioassay using whole cells with the HPTLC, whereas biochemical detection describes methods based, for example, on the detection of enzymatic activity. There are several examples of a successful coupling of a biological or biochemical assay directly to HPTLC, such as the detection of cell death, specific effects such as estrogenic activity or genotoxicity, and enzyme inhibition or binding of antibodies. In addition to these two groups, the detection of specific chemical characteristics is often included in EDA methods. However, this is not absolutely correct in a narrow sense, because in this latter case the biological effect is not detected directly and only 'extrapolated' from a given chemical characteristic. Table 3 gives some examples of HPTLC-EDA that are explained in more details in the following sections.

3.1 Bioautography

In bioautography, detection of the biological effect is carried out by using microorganisms. Based on the method used for applying the organisms to the surface of the (HP)TLC-plate, that is, to expose the microorganisms to the separated substances, bioautography is differentiated into agar diffusion biography, agar overlay bioautography, or direct bioautography [54, 55].

In the case of agar diffusion bioautography, the stationary phase is placed on the surface of an agar medium inoculated with the selected microorganism. After a certain incubation time, the substances are diffused from the (HP)TLC plate into the agar medium. After removal of the (HP)TLC plate and a further incubation step, the biological effect is detected [21, 46, 56]. In agar overlay bioautography, the (HP)TLC plate is either immersed in an inoculated agar medium or covered with an inoculated agar layer. The agar layer then hardens on top of the stationary phase. During the incubation phase and the subsequent detection of the effect, the agar layer remains on the plate [47]. A disadvantage of the techniques described above is the possibility of false negative test results in cases where hydrophobic compounds diffuse into the agar only to a limited extent. The resulting low concentrations in the agar containing the microorganisms may be insufficient to trigger the biological effect under investigation.

For direct bioautography (DB), the (HP)TLC plate is immersed in a suspension of the used microorganism in a specific nutrient solution [57]. The subsequent incubation is carried out in a moist atmosphere. For some bacteria and fungi the growth conditions for TLC-DB have been optimized [58–62]. In DB it is crucial to ensure that the selected microorganisms are viable in direct contact with the stationary phase.

3.2 Biochemical Detection

In addition to bioautographic methods, in HPTLC-EDA, biochemical detection techniques are also used to detect effects. For the detection of enzyme-inhibiting substances after chromatography, an enzyme solution is applied to the plate. After an incubation period, visualization of the inhibition is achieved using a substrate that is converted by the selected enzyme, usually into a colored product. The enzyme is usually applied to the surface of the plate in combination with a buffer providing optimal conditions for the given enzyme, for example, optimal pH value, supply of required cofactors. To stop the enzymatic activity after a specific incubation time, the (HP)TLC plate is often heated briefly on a heating plate [63].

Magnani et al. [64] introduced a TLC immunostaining method based on the principle of enzyme-linked immunosorbent assays (ELISA). In the first step, antibodies bind to their matching counterparts. Subsequently, secondary antibodies, marked with an enzyme, bind to the already bounded primary antibodies. Finally, a substrate is applied, which is converted by the enzyme to a colorant [65]. It should be noted that TLC immunostaining always includes an incubation phase and a washing step between the steps described. In these steps, the silica gel can easily become detached from the support material. Therefore, the silica gel is fixated prior to immunostaining, usually with polyisobutylmethacrylate. Several TLC immunostaining methods have been reported for the detection of various steroidal alkaloid glycosides [66], glycosphingolipids [67, 68], gangliosides [69, 70], or antiphospholipid antibodies [71–74]. A detailed description of HPTLC immunostaining is found in Meisen et al. [65] and Conti et al. [75].

4 Endpoints Used in HPTLC-EDA

4.1 Bioluminescence Inhibition of Aliivibrio fischeri

For the (HP)TLC bioluminescence inhibition test, the plate is dried after separation and immersed in a suspension of the luminescent bacterium *A. fischeri* (direct bioautography) [49, 76]. The biochemical reactions resulting in the bacterial bioluminescence are directly linked to its metabolism. Therefore, metabolic disorders can be detected very quickly by changes in the intensity of the bioluminescence.

The inhibition of bioluminescence is detected in a dark chamber with a sensitive CCD camera. The (HP)TLC plate coated with the luminescent bacteria is placed on a slide, which is then covered with a glass plate to prevent drying (Fig. 5a). The slide also contains a liquid reservoir to maintain humidity. Because the plate is coated only with a very thin film of luminescent bacteria, prolonged exposure times of 45–55 s are required to detect the weak bioluminescence signal. Figure 5b shows the inhibition zones as dark bands against a light background.

In analogy to the classic bioluminescence inhibition test in cuvettes [77], the inhibitory effect can be quantified. This is done by using software for image analysis. The areas adjacent to a detected inhibition zone serve as a reference for the quantification of the observed inhibition. The calculation of the inhibition values at (Formula 1) for specific locations creates a so-called inhibition value chromatogram. With these inhibition values, a gamma (Γ)-value chromatogram can be calculated according to Formula (2) [78], resulting in sharper bands, particularly for bands with high inhibition values compared to the inhibition value chromatogram.

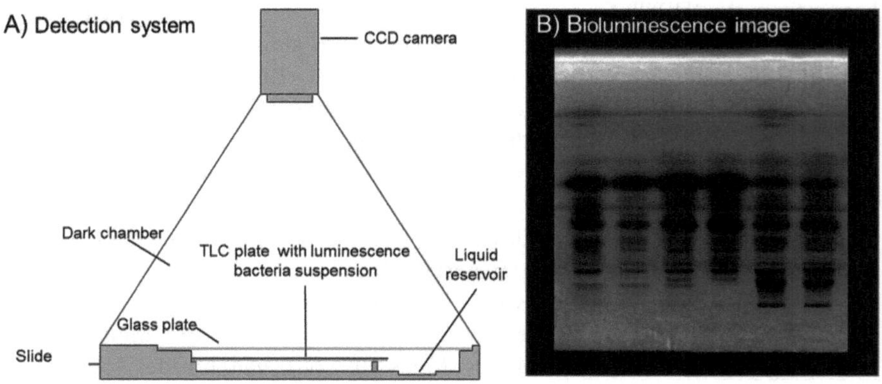

Fig. 5 Detection of bioluminescence inhibition. (**a**) Schematic illustration of a dark chamber with a CCD camera and a slide for the TLC plate. (**b**) Example of the detection of bioluminescence on a TLC plate showing multiple inhibition bands (*black*)

$$\mathrm{In}h_t = \frac{i_{0,t-}i_t}{i_0} \quad (1)$$

$$\Gamma_t = \frac{\mathrm{In}h_t}{1 - \mathrm{In}h_t} \quad (2)$$

where $\mathrm{In}h_t$ = inhibition, $i_{0,\mathrm{t}}$ = luminescence light intensity of the reference (areas adjacent to the band to be quantified) after the incubation time t, and i_t = the luminescence light intensity of the band after the incubation time t.

To ensure that the observed effects can be compared within a sample and between samples, the reciprocal iso-inhibition volume (RIV) can be used. Here, the volume of an applied sample extract required to achieve a 50% inhibition of the bioluminescence is calculated as a reference point. Because the inhibition of bioluminescence by a compound is a function of its particular but unknown concentration and the volume is inversely proportional to the concentration when the amount of the compound is constant, the inverse of the calculated volume which

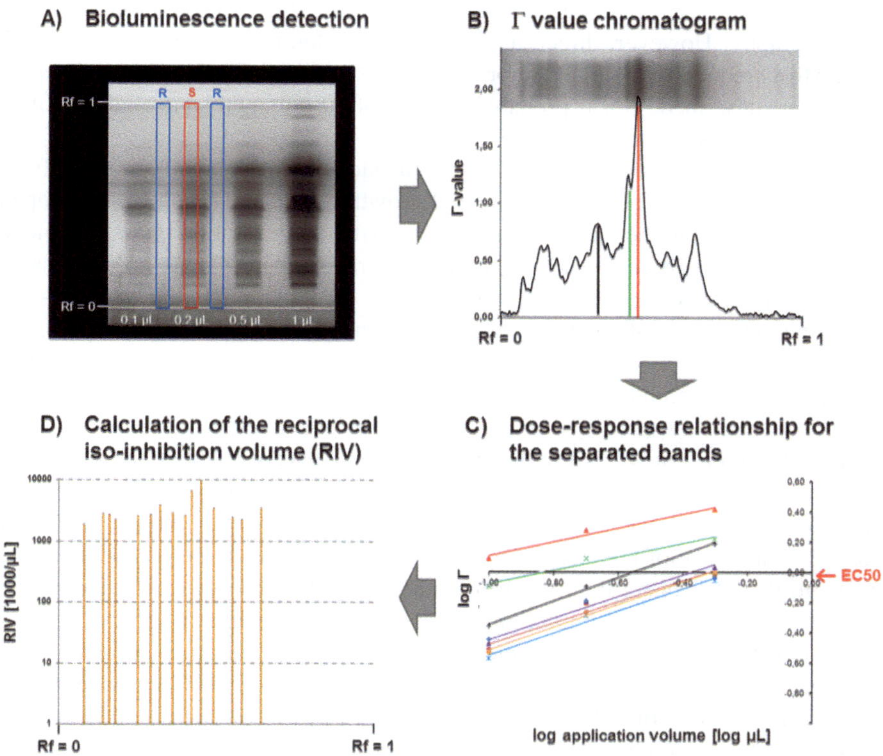

Fig. 6 Scheme of the process of evaluating the bioluminescence inhibition to calculate the reciprocal iso-inhibition volume (RIV). (**a**) Picture of the bioluminescence, R: areas for the reference, S: sample. (**b**) (Γ)-value chromatogram for the sample S. (**c**) Dose–response relationship in log-space. Log (Γ) = 0 indicates the 50% effect-level from which the EC50 is derived. (**d**) Toxicity of the detected bands expressed as iso-inhibition volume (RIV)

causes 50% inhibition of the bioluminescence (RIV) is used as a measure of the toxicity [79]. When the RIV is calculated for each inhibition band of a sample, it is possible to compare different samples via the RIV pattern (Fig. 6).

The HPTLC bioluminescence inhibition test is used in various areas, such as the testing of natural substances [80–82], personal care products [83, 84], or eluates of rubber products [48, 85]. The HPTLC bioluminescence inhibition test has also been applied successfully to the analysis of surface water [86], wastewater [87], and landfill leachates [85]. Furthermore, this method can also be used to monitor the inhibition of metabolites formed in ground water [88] or secondary products generated by ozonation of water samples [85, 89, 90].

4.2 Antibacterial and Fungicidal Effects

To detect antibacterial substances, various Gram-negative and Gram-positive bacteria are used. The detection can be done with all three overlay techniques of bioautography. However, direct autobiography predominates in the recent literature. Most frequently the applications with the Gram-positive bacterium *Bacillus subtilis* are described. This is probably because of the commercial availability of an (HP)TLC-DB test [91].

For direct bioautography with *B. subtilis*, a batch of the bacteria is prepared from a spore suspension. When the exponential growth phase is reached, the developed plate is immersed in the bacterial suspension. After an incubation phase in a moist chamber, the dehydrogenase activity of the bacteria is tested using tetrazolium salts. Usually a reagent containing the yellow 3-[4,5-dimethylthiazol-2-yl]-2,5-diphenyltetrazolium salt (MTT) is sprayed on the plate. After a few minutes, the inhibition zones appear as yellow bands against a violet background (Fig. 7). The contrast between inhibited and uninhibited zones improves with increasing reaction time.

The inhibition zones can also be analyzed quantitatively based on the size of the inhibition zone [93]. It should be noted that, because of diffusion, the inhibition zone grows larger as the amount of substance applied rises, even though the end of the sigmoid course of the dose-effect relationship has already been reached. On a wettable RP-18W plate, the zone broadening is hardly observable compared to a silica gel plate because of limited diffusion of polar compound [94].

Besides the detection of antibacterial substances from various plant extracts, it has been shown that it is possible to detect antibiotics in foodstuff such as milk [93] but also in enriched run-off samples from sewage treatment plants and in surface water [95]. Bacteria other than *B. subtilis* can also be used for this assay. There are examples reported using *Escherichia coli* [61, 96], *Staphylococcus aureus,* or the bacterium *Pseudomonas syringae*. An overview of bioautography applications with various bacterial strains to detect antibiotic effects can be found in Choma and Jesionek [55].

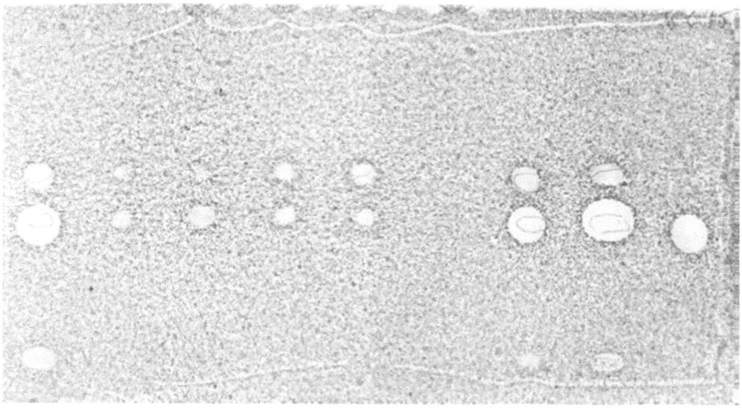

Fig. 7 TLC-DB of eluates from milk spiked at 0.05 ppm level. The amounts of ciprofloxacin and enrofloxacin in the eluates were 0.5 ppm each. *Upper spots*, enrofloxacin; *lower spots*, ciprofloxacin. From *left* to *right*: 1 ppm standard, 0.5 ppm standard, three eluates from three cartridges, blank, 1 ppm standard, two standards at 1 ppm (HPLC mobile phase and water instead of methanol). The volume spotted was 50 μL. Test bacteria: *B. subtilis*. With permission from [92]

The detection of substances with fungicidal effect after TLC separation was first described by Homans and Fusch [97]. The authors used the spore-forming fungi *Aspergillus*, *Penicillium*, and *Cladosporium*. Subsequently, other fungi have been used for the detection of fungicidal effects. The effect is also detected with MTT. Most studies were focused on fungicidal effects of plant components on human and plant pathogenic fungi species. Some applications in environmental analysis have also been described [95, 98]. For instance, it was possible to detect the fungicide metalaxyl in soil using HPTLC-EDA [98].

4.3 Estrogenic Effect

For the detection of potential estrogenic compounds in a sample, the yeast estrogen screen (YES) using a genetically modified yeast (*Saccharomyces cerevisiae*) was transferred to the HPTLC plate (HPTLC-YES, in the literature also referred to as planar YES or pYES). The YES is a reporter gene assay, in which the reporter gene *lacZ* encoding the enzyme β-galactosidase is controlled by an estrogen-dependent promoter. Therefore the expression of the *lacZ* gene depends on the presence of agonists of the estrogen receptor and the activity of the β-galactosidase is a measure for the estrogenic potential of a sample or substance. For the modified test, the developed HPTLC plate is immersed overnight in a culture of the genetically engineered yeast cells, followed by an incubation at 30°C in a humid atmosphere. To detect the expressed β-galactosidase the substrate 4-methylumbelliferyl-D-galactopyranoside (MUG) is sprayed as an artificial substrate on the surface of

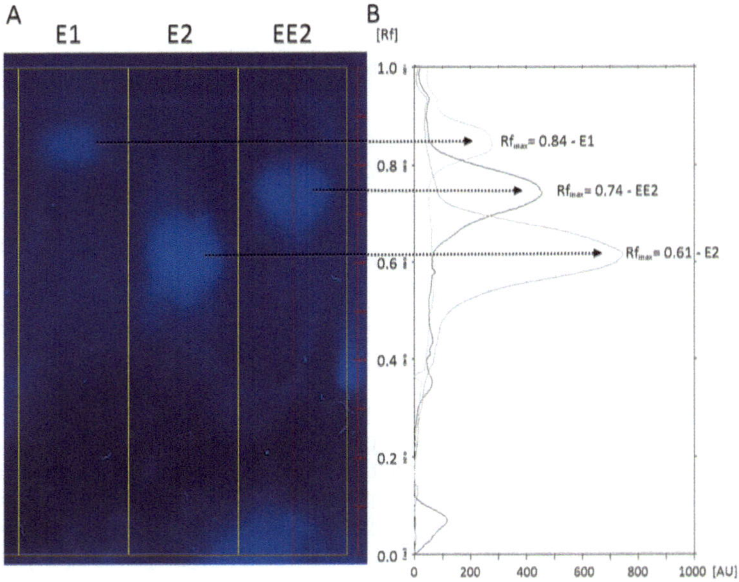

Fig. 8 Detection of estrone, 17β-estradiol, and 17α-ethinylestradiol by the pYES. 17-α-Ethinylestradiol (EE2), 17β-estradiol (E2), and estrone (E1) were applied on a TLC plate, 100 pg each. After the pYES with chromatographic separation the signal detection was done. (**a**) By fluorescence imaging (366 nm) and (**b**) fluorescence scan (320 nm). With permission from [99]

the TLC-plate. After a second incubation at 37°C for some minutes, the cleaved MUG is detectable by fluorescence (Fig. 8).

In HPTLC-YES, two different genetically altered strains of *S. cerevisiae* have been used so far. One is the strain BJ3505 modified by McDonnell et al. [100]; the other strain was constructed by Routledge and Sumpter [101].

Müller et al. [102] were the first to transfer the YES test successfully to the HPTLC plate. However, the required incubation of 30 h resulted in a massive band broadening. With the approach of Schönborn and Grimmer [103], it is possible to reduce the required incubation phase to a total time of 5 h and simultaneously lower the limit of detection to 0.5 pg/band 17β-estradiol. Spira et al. [99] shortened the incubation phases further to 3.5 h. The shorter incubation phases noticeably suppress the diffusion effect. However, although unwanted in general, the diffusion of compounds allows the detection of endocrine effects, even if the compound of interest is cytotoxic in higher concentrations, because the diffusion causes a concentration gradient surrounding the center of the compound. In the area of high concentrations, cytotoxic effect may occur and a dark area is formed within a fluorescing zone, indicating the estrogenic potential of this compound at lower concentrations [104]. As already described for the anti-bacterial effects, compound diffusion and thus signal broadening is reduces substantially by using RP-18W plates [94]. This allows a clearer association of the observed effect with a substance. It was observed sometimes that the quality of these reverse-phase plates was insufficient for the HPTLC-YES. Therefore, before performing the HPTLC-YES, the wettability of an RP-18W plate batch has to be tested.

HPTLC-YES with MUG is a very sensitive method; for example, 17-α-ethinylestradiol can be detected down to 0.5 pg on the silica gel plate [99]. Via a dose-effect relationship, it is possible to quantify the estrogenic effect [94, 99, 103, 104]. The strength of the effect can be given as a concentration equivalent to the 50% effect dose (ED50) of 17β-estradiol [104].

With this HPTLC-EDA method, estrogenic effects have already been detected in sediment pore water [99, 104], in water discharged following the first treatment step in a municipal sewage treatment plant [103], and in the water of a fish aquaculture system [103]. Likewise, an estrogenic effect has also been detected in beer [94] and in a sunscreen cream [104]. A comparison of the influent and effluent water of a sewage treatment plant has shown the possible formation of possible transformation products with estrogenic activity [104].

4.4 Genotoxic Effect

Bacterial tests available for the detection of genotoxic substances include the Ames test [105] and the SOS-*umu* test [106]. Both assays have already been combined with TLC. For instance, Bjørseth, Eidsa et al. and Houk and Claxton have applied the agar used in the Ames test and the strains *Salmonella typhimurium* TA100/TA98 directly to the developed TLC plate [107, 108]. After a 3-day incubation period, the number of colonies formed around the substance zone was detected. Because the colonies did not form directly on the substance zone, no exact allocation of the mutagens was possible. In addition to this disadvantage, Houk and Claxton describe that, for chromatographic reasons, it was only possible to apply a small amount of sample. In this situation, mutagens in low concentration remain undetected.

An initial proposal to transfer the SOS-*umu* test to an HPTLC plate was given by Baumann et al. [109]. They used the genetically modified strain *S. typhimurium* TA1535/pSK 1002. In this strain, the transcription of the *umuC* gene is linked to the transcription of the *lacZ*-gene encoding the enzyme β-galactosidase. In the presence of genotoxic compounds the transcription of these two genes is triggered by the SOS response of the bacterium. The activity of the β-galactosidase was detected by Baumann et al. with the substrate *o*-nitrophenyl-β-D-galactopyranoside (ONPG) and the resulting formation of the yellow nitrophenol. The method was performed successfully with the substance 4-nitrochinoline N-oxide (4-NQO) according to DIN 38415-3 [110], with some modifications. Instead of ONPG, the substance 5-bromo-4-chloro-3-indolyl-β-D-galactopyranoside (X-gal) was used as substrate. X-gal leads to the formation of a blue carbonyl dye [111]. The method still has to be developed further to include a metabolization system similar to the frequently used S9-mix for indirectly acting genotoxic substances.

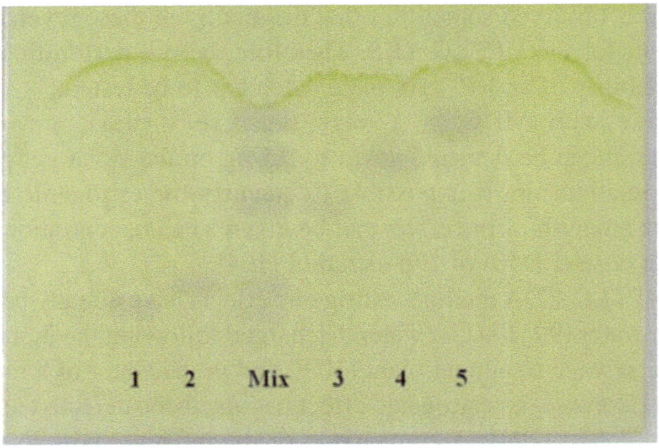

Fig. 9 Separated triazine herbicides after HILL-reaction using 2,6-dichloro-phenolindophenol staining. (*1*) Atraton (2 ng per zone), (*2*) terbumeton (1 ng per zone), (Mix) mixture off all herbicides, (*3*) simazine (2 ng per zone), (*4*) atrazine (1 ng) per zone), and (*5*) terbuthylazine (1 ng per zone) on silica gel with cyclohexane-methyl-*tert*-butyl ether (1 + 1, v/v) as mobile phase. With permission from [114]

4.5 Chloroplastida Inhibition

The detection of substances that inhibit photosynthesis is done by detecting the inhibition of the Hill reaction. In this reaction, O_2 is formed from H_2O in the thylakoid membranes of the chloroplasts. In addition to H_2O, the Hill reaction also requires a natural electron acceptor which can be replaced by an artificial electron acceptor (Hill reagent), such as the blue 2,6-dichloro-phenolindophenol (DCPIP). Chloroplasts that carry out photosynthesis reduce DCPIP to a colorless aminodiphenol. When a photosynthesis inhibitor is present, the color change does not take place. Inhibition of photosynthesis is the mode of action of many broad-band herbicides.

For the HPTLC test, the plate is immersed in a buffered suspension of chloroplasts and DCPIP [112, 113]. The chloroplasts used for this purpose can be obtained from spinach leaves or from the leaves of garden beans [112]. This is followed by an exposure time of a few minutes under white light. Zones in which the Hill reaction is inhibited appear bluish against a green background (Fig. 9). For contrast enhancement, the plate is immersed in a PEG-600 solution [114]. The strength of inhibition is determined through densitometric absorption measurement of the blue zones. Because the zones fade quickly, the detection must be carried out immediately [114].

In addition to photosynthesis inhibition, it is possible to detect algicidal effects. This can be done by applying a suspension of the fresh water alga *Pseudokirchneriella subcapitata* to the developed HPTLC plate. After applying the algal suspension, the plate is incubated in a moist, transparent chamber at room temperature. Subsequently the plate is sprayed with an MTT solution and incubated

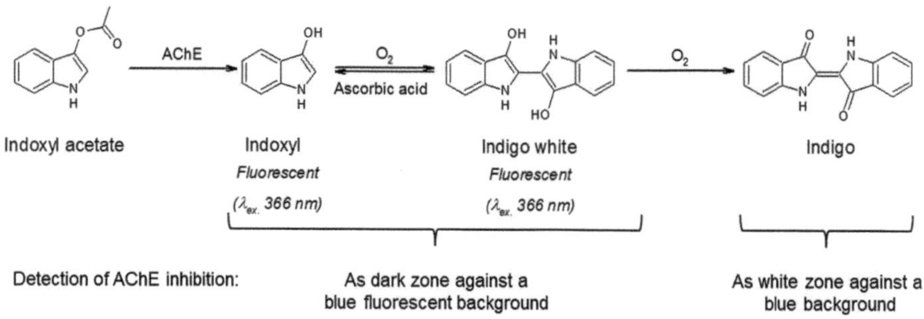

Fig. 10 Reaction of the hydrolysis of indoxyl acetate by acetylcholinesterase to indoxyl and subsequent oxidation to indigo

for another 1–2 days. Areas with a toxic effect for algae now appear as bright zones against a violet background [109].

4.6 Neurotoxic Effect

Based on the central role of acetylcholinesterase (AChE) in the transmission of signals in the synaptic gap, substances that inhibit this enzyme are considered potentially neurotoxic. The most important AChE-inhibiting substances include organophosphates and carbamates, which are often used as insecticides. In the 1960s and 1970s the first attempts to detect organophosphates and carbamates biochemically by the inhibition of bovine liver esterases after paper chromatography or TLC were described [115–124].

For the detection of the AChE inhibition the HPTLC plate is immersed in an acetylcholinesterase solution. Adding 0.1% bovine serum albumin (BSA) to the solution, especially in the case of activated silica gel and diol phases, suppresses the influence of the stationary phase on the inhibition of AChE activity [125]. Various substrates have been described in the literature to detect the AChE-inhibition after an incubation phase. One is the widespread Ellman reagent, which is converted to a yellow dye [126]. With this substrate, however, it should be noted that, in the presence of certain amines and aldehydes, false positive results can be obtained [127]. Alternatively, indoxyl acetate can be used. This substrate is converted by AChE to the fluorescent indoxyl before the blue colorant indigo develops [128, 129] (Fig. 10). The AChE inhibition is detectable as dark zones against a blue fluorescing background or as white zones against a blue background (Fig. 11).

Other substrates such as *N*-(4-(7-diethylamino-4-methylcoumarin-3yl)phenyl) maleimide (CPM maleimide), 1-naphthylacetate or 2-naphthylacetate were also used [129–131]. The first quantitative analysis of the inhibition of acetylcholinesterase by organophosphates was performed successfully by Štefanac et al. [132] using a densitometer. It is also possible to calculate inhibition factors for the

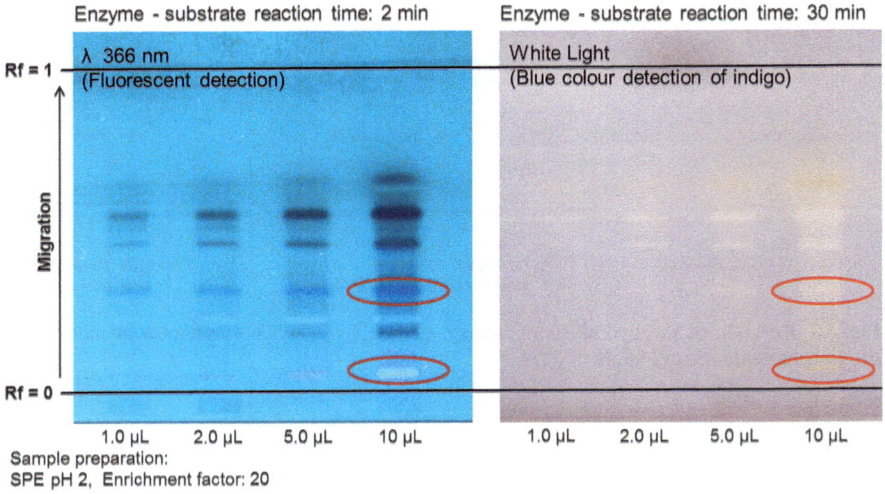

Fig. 11 Application of the HPTLC-AChE-inhibition test on water sample with the substrate indoxyl acetate. The substrate was applied after an incubation time of 5 min. *Left*: Detection of inhibition as dark zones against a fluorescent background. *Right*: Detection of inhibition as white zones against a blue background from the formed indigo. Marked in *red* are zones with a native fluorescence; thus no inhibition in these zones can be detected. However, by the detection of the *blue color*, an inhibition can be detected in these zones

detected substances. These inhibition factors correlate well with the inhibition constants obtained by the classic microtiter plate test [51].

Organothiophosphates have to be oxidized into active oxones prior to the detection of their biological effect. For this purpose, bromine is frequently used [120, 133]. Oxidation with an aqueous bromine solution was reported to be more effective than exposing the TLC plate to bromine vapor [122]. Using a multi-enzyme inhibition test, consisting of cutinase from *Fusarium solani pizi* and esterases from *B. subtilis* and from rabbit liver, it is possible to detect organothiophosphates without prior oxidation in addition to organophosphates and carbamates [134]. The limits of detection can be improved substantially by oxidation with bromine [63].

In addition to the detection of insecticides in foods and water, the test is used to detect AChE-inhibiting substances in plant extracts. These substances can possibly be used to ameliorate the symptoms of Alzheimer's disease. AChE inhibitors already being used as therapeutic agents including tacrine and edrophonium. With life expectancy increasing, it can be assumed that such compounds are increasingly be found in wastewater in the future. They can then reach the aquatic environment via sewage treatment plants.

Table 4 Selected derivatization reagents for HPTLC

Substance class	Reagent	Example (LOD)	References
Non-specific	Sulfuric acid	–	[135]
Non-specific	Iodine starch	–	[135]
Primary amines	Bratton–Marshall	Amitrol (1 ng)	[135–137]
Primary and secondary amines, phenols, thiols	NBD chloride	Proline (5 ng)	[135]
Carbonyl	2,4-Dinitrophenyl-hydrazine (DNPH)	Dehydroascorbic acid (10 ng)	[135]
Aldehyde	Purpald	Palmitaldehyde (1 μg)	[138]

5 Identification

5.1 Derivatization

Analytes that do not absorb UV/Vis light must be derivatized with a suitable reagent to allow a photometric detection. The sensitivity and selectivity of the analytical method can also be increased by adding chromophores or fluorophores to the molecule [135]. In the characterization of unknown compounds, derivatization can play an important role. By using specific reactions, it is possible to detect functional groups in molecules. Numerous reagents for qualitative and quantitative HPTLC analysis have been described (Table 4) [135, 139–142].

These derivatizations are done either before or after separation. By means of a pre-chromatographic derivatization, similar substances can be better separated chromatographically; volatile compounds are converted to less volatile derivatives [135]. The identification of unknown substances using pre-chromatographic derivatization is hindered by the altered migration behavior of the derivatives during the chromatography. In contrast, the separation of the substances on the HPTLC plate is unaffected by a post-chromatographic derivatization. In this process the reaction products can be allocated directly to their starting compounds based on the same migration distance [88, 143]. Another advantage compared to pre-chromatographic derivatization is that it is possible to take images of the UV/Vis spectra of the separated substances before and after derivatization with the respective reagents. This additional information can help to identify unknown compounds.

5.2 TLC-MS

The numerous techniques for transferring the analytes from the TLC plate to MS can be divided into desorption methods and elution methods [144]. A widely used representative of the desorption technique is matrix-assisted laser desorption/ionization (MALDI) [65, 145, 146]. In this method, an organic acid (e.g.,

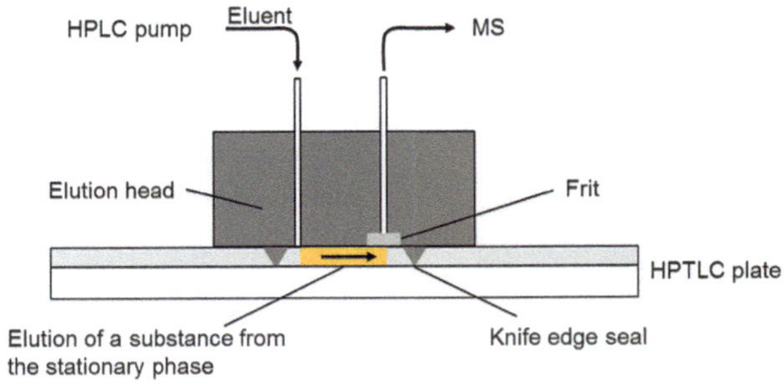

Fig. 12 Functioning of the elution method based on the technique of Luftmann [152]

2,5-dihydroxybenzoic acid or sinapinic acid) is applied to the plate homogeneously as a matrix [147]. When the solvent evaporates, the analytes are integrated in the crystal lattice of this matrix. In the ion source of the MS, the sample is exposed to an intensive pulse of a laser beam for a few nanoseconds. In this process, the sample molecules are ionized and migrate into the gas phase. Another desorption technique is "direct analysis in real time" (DART) [148–150]. In this process, the zone to be analyzed is exposed to a heated stream of gas consisting of excited helium. In this way, the analytes are desorbed and protonated. It has recently been shown that DART facilitates a quantitative detection of compounds directly from the HPTLC plate [151]. With both of the desorption techniques presented, it is possible to image the developed HPTLC plate completely with mass spectra.

The most widespread elution method for a subsequent MS analysis is the technique presented by Luftmann in 2004 using an elution head [152]. To transfer the sample molecules into the MS, the elution head is pressed onto the substance band so that a knife edge seal separates the zone from the rest of the plate. In the extraction mode of the interface, a solvent mixture is pumped through the elution head pressed against the HPTLC plate. A frit is used to hold back particles that might be detached from the stationary phase during the extraction process. Subsequently, the extract is transferred directly to the ion source of an MS (Fig. 12). This method can utilize the same ion sources as HPLC-MS coupling. Alternatively, it is possible to collect the extract in a glass vial. Collecting the extract has the advantage that another separation can be done before MS analysis, for example, by HPLC [143] and more MS experiments can be realized from one extract. This combination permits 2D chromatography on a normal phase (HPTLC) and a reverse phase (HPLC), which increases the separation capacity [153]. This extract can also be tested with spectroscopic methods, such as NMR [154].

Today the TLC-MS interface has become a well-established tool for the identification of unknown compounds [80, 88, 155]. In addition to qualitative analysis, quantification of compounds is also possible with the TLC-MS interface [156]. Because of the lack of automation, the quality of the calibration still depends mainly on the user. This variation can be improved substantially through the use of internal standards [157]. However, there have already been initial successful attempts to automate the elution completely [158].

In EDA, when using TLC-ESI-MS coupling, it is recommended that a second plate should be developed under the same conditions and that the substances of interest be eluted from this second plate. It is of course possible to use the plate on which the bioassay was performed, but, because of the high salt load from the nutrient medium, particularly with a luminescent bacteria assay, it is necessary to clean the MS more often. In addition, new metabolites can be generated by the applied microorganism on the plate, which makes it more difficult to match an effect with a substance. Irreversible binding of the molecule to cellular structures is also conceivable. In MS analysis of deuterated bisphenol A directly from the HPTLC-YES plate after the performance of the bioassay, it was found that the recovery was only 9%, whereas the recovery without a prior performance of the bioassay was around 100% [104].

5.2.1 Combining Derivatization with Mass Spectrometry

The combination of information from the derivatization reactions and from HPLC coupled to a high resolution MS often advances the identification of substances decisively. In some cases, for example, the quantity of a specific functional grouping of a molecule is to be determined. To answer such questions it is possible to elute the reaction products after a derivatization from the HPTLC plate for subsequent analysis by mass spectrometry. Based on the knowledge about the derivatization reaction, it is possible to calculate the expected masses for the respective number of functional groups. In this way, it was, for example, possible to demonstrate the ring cleavage for 4- and 5-methyl-benzotriazole during ozonation and the resulting formation of dialdehydes [143].

A particular type of derivatization is the H/D exchange. In this process, all exchangeable protons are replaced by deuterium. This occurs quickly when the analyte is dissolved in D_2O or CH_3OD. In an ESI mass spectrum, for each hydrogen atom that is replaced by deuterium a mass shift of 1.0063 u can be observed. Because of this mass shift, it is possible to restrict the number of possible proposed structures for a given sum formula. HPTLC is particularly well suited for this approach, because there is a change in solvent between the separation and the transfer of the substances to an MS. Only a small amount of D_2O or CH_3OD is required during the extraction of the compound using the TLC-MS interface described above to ensure a nearly complete exchange. In practice the sample is applied twice, and after HPTLC separation the extraction of zones of interest is carried out with a mixture of H_2O/acetonitrile and 5 mmol ammonium acetate and with D_2O/acetonitrile and 5 mmol ammonium acetate using the second sample application (Fig. 13) [159].

If no TLC-MS interface is available, a plate of aluminum foil, cut to size, can be placed at the input port of an ESI-MS. The D_2O is dropped onto the cut-out of the TLC plate. Because of the applied potential difference, a sufficient amount of the analyte enters the mass spectrometer for the determination of the mass shift [160].

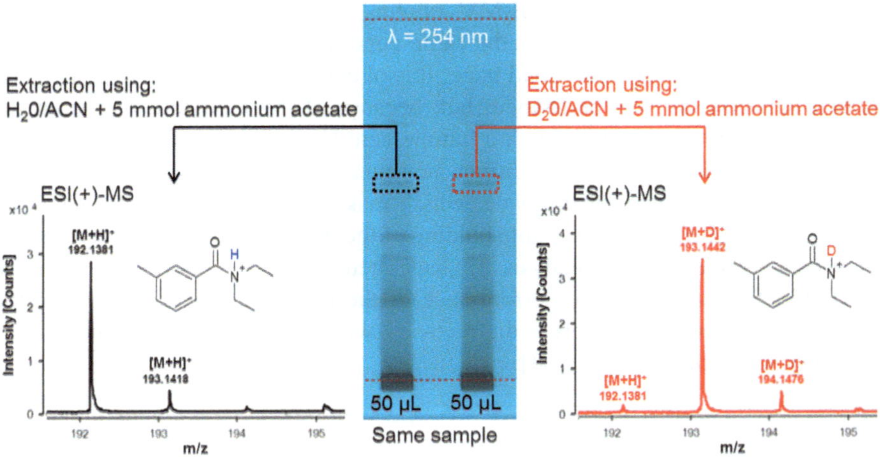

Fig. 13 Principle of combining TLC-MS with the H/D exchange for the identification of unknown organic substances. The example shows the mass shift of 1.0061 u for the quasi molecular ions of crotamiton in an ESI(+) mass spectrum

Table 5 Summary of HPTLC-EDA applications with various bioassays for examination of environmental samples

Sample type	Endpoint	Test system	References
Wastewater	Bioluminescence inhibition	*A. fischeri*	[87, 95, 165, 166]
	Endocrine effect	YES	[103, 104]
Surface water	Bioluminescence inhibition	*A. fischeri*	[85, 95]
	Antibacterial effect	*B. subtilis*	[95]
	Fungicidal effect	*Penicillium spec.*	[95]
	Neurotoxicity	Acetylcholinesterase	[95]
Landfill leachate	Bioluminescence inhibition	*A. fischeri*	[85]
Sediments	Endocrine effect	YES	[99, 104]
	Photosynthesis inhibitor	Bean leaf chloroplasts	[113]
	Fungicidal effect	*Phytophthora boehmeriae*	[98]

6 Applications in Environmental Analysis

Because of its advantages in the testing of complex matrices, HPTLC is a widely used analytical technique for the examination of plant extracts [161, 162]. Many applications of HPTLC-EDA in this area are reported [163, 164]. However, several applications have also been described for the analysis of various environmental samples (Table 5). Selected examples are presented to elucidate the increasing

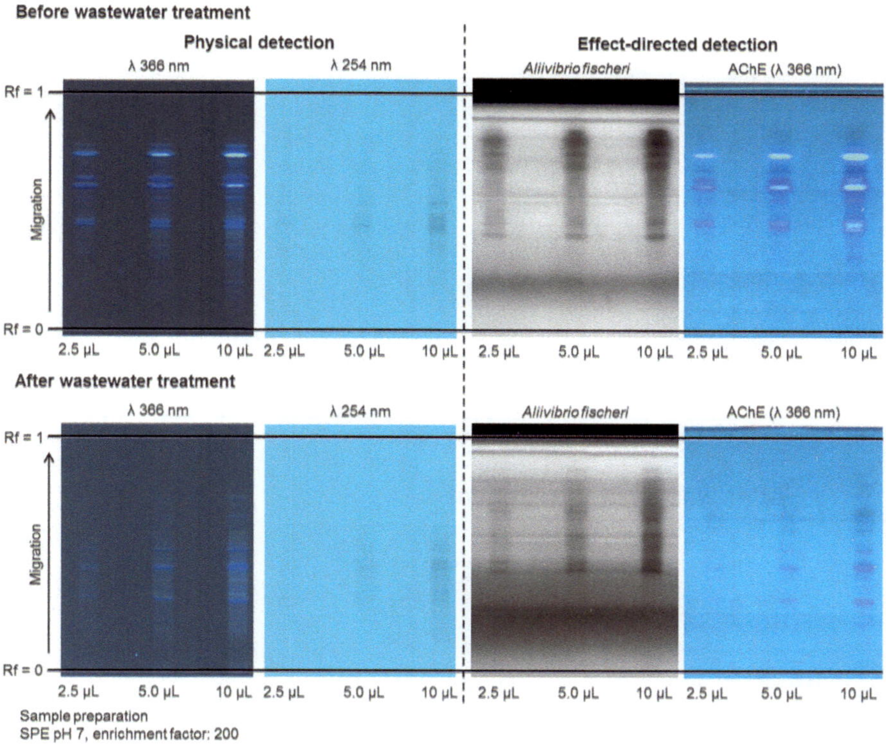

Fig. 14 Investigation of water samples from a wastewater treatment plant by using HPTLC-EDA with bioluminescence inhibition test (*A. fischeri*) and acetylcholinesterase inhibition test

significance of HPTLC-EDA as an analytical technique complementary to the classic physicochemical approaches in environmental analysis.

6.1 Process Characterization of Wastewater Treatment

The leachates of landfills are usually treated in wastewater treatment facilities. As a supplementary testing program to assess the efficiency of this treatment process, HPTLC-EDA methods using *A. fischeri* and acetylcholinesterase were used. Samples were taken before and after wastewater treatment; 200 mL of the samples were enriched by solid phase extraction (SPE) at pH values of 2, 7, and 9. After careful evaporation of the solvent under a stream of nitrogen at 40°C to near dryness, the residue was reconstituted in 1 mL of methanol. Figure 14 summarizes the results for the extraction at pH 7 for all detection techniques used. In each case an application volume of 2.5, 5.0, and 10 µL extract was used for the HPTLC separation. The left part of the illustration shows the results obtained by physical detection; the right part of the figure shows the results of the applied bioassays. The bioluminescence

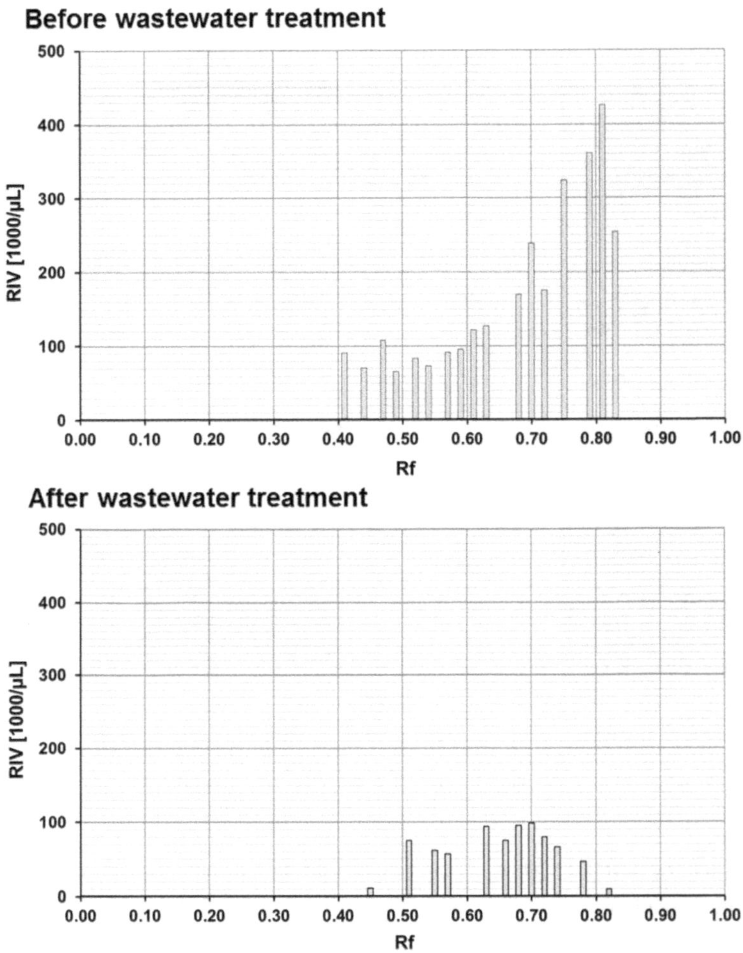

Fig. 15 Comparison of RIV values for bioluminescence inhibition from *A. fischeri* before and after wastewater treatment for the SPE pH 7 extract

inhibition test with *A. fischeri* shows clearly that especially substances with a high Rf, that is, rather nonpolar substances, are removed through wastewater treatment (Fig. 14, lower part). Using the AChE inhibition test, some potential neurotoxic substances are detected before wastewater treatment, which are removed almost completely by the treatment. Likewise, physical detection shows fewer substances after wastewater treatment than before.

For a semi-quantitative assessment of the elimination efficiency of the wastewater treatment plant, the data obtained were evaluated according to the RIV concept explained above. The results in Fig. 15 show that mainly toxic effects of substances with an $Rf > 0.70$ were reduced by the treatment. In this range, the RIV-values decrease by a factor of 2–6.

6.2 *Investigation of a Possible Impact of Leachates from Artificial Turf on Ground Water*

Environmentally relevant substances can leach from sports flooring such as artificial turf and might infiltrate into ground water if there is no or an inadequate drainage system of the sports facility. Artificial turf surfaces are increasingly being used in football and other types of sports. The elastic layer of the artificial turf consists mostly of polyurethane-bonded rubber granulates. For this layer, besides new synthetic rubber granulates (ethylene propylene diene monomer, EPDM), recycled rubber, for instance from old tires, is also used. In addition to meeting the requirements for the individual types of sports and the protective function for the athletes, the surface must also meet certain environmental requirements [167]. In the present case, artificial turf was used for construction in the catchment area of a ground water well. By analyzing samples taken from the elastic layer of the artificial turf it was investigated as to whether toxic compounds can eventually be leached out of the material and can migrate to the connected ground water resource. For this "non-target question" the HPTLC bioluminescence test with *A. fischeri* was selected because of its broad compound specificity.

To simulate the leaching by rain, 40 g of the elastic layer from the freshly installed artificial turf was eluted for 24 h with 80 g ultra-pure laboratory water in a batch test according to DIN EN 12457-01 [168], followed by an extraction of the aqueous eluate with 6 mL *tert*-butyl methyl ether (TBME) and 12 g NaCl. Then 20, 40, and 60 µL of the extract were applied to an HPTLC plate. After HPTLC/AMD separation using the gradient shown in Fig. 3 and immersion of the plate into a luminescent bacteria suspension, the luminescent activity of the bacteria was determined. Several inhibition bands were detected, indicating the elution of bacteriotoxic compounds from the sample (Fig. 16a).

With the UV/Vis remission spectra measured directly on the plate and the measured respective Rf-values, a query was examined in an in-house UV/Vis

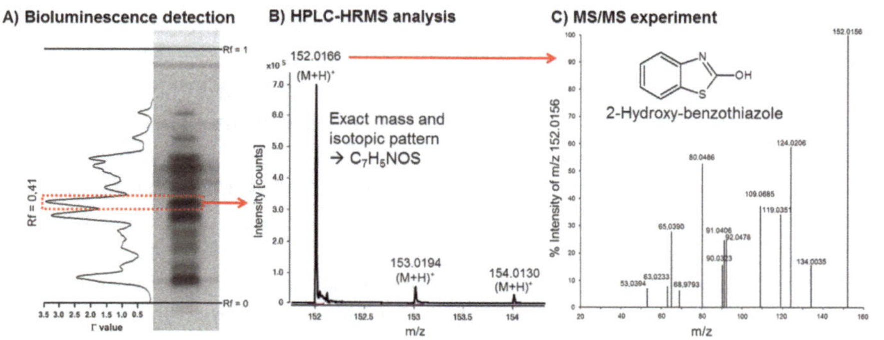

Fig. 16 Process of the HPTLC-EDA investigation for water elutable substances from an elastic layer of an artificial turf, resulting in the identification of 2-hydroxy-benzothiazole for the zone at Rf = 0.41 (outlined in *red* in the bioluminescence detection)

spectrum library. To identify individual bacteriotoxic substances, a second HPTLC plate was developed using the same separation conditions. From this second HPTLC plate, the individual substance bands were first extracted in a vial using the described elution TLC-MS interface and directed to an HPLC coupled to a quadrupole time-of-flight mass spectrometer (QTOF/MS) system. Using the exact mass and the isotope patterns of the detected signals (Fig. 16b), sum formulas and, if possible, structure predictions were generated and verified via the MS/MS spectrum where possible (Fig. 16c). By combining the measured UV/Vis spectrum and MS data with a prioritization based on the detected effect, it was possible to demonstrate the presence of relevant environmental contaminants in the elutes from the artificial turf. As one example, the marked band (Rf = 0.41) in Fig. 16a was identified as 2-hydroxybenzothiazole. Furthermore, it was also possible to identify 2-aminobenzothiazole (Rf = 0.36) and benzothiazole (Rf = 0.58). To detect these substances in the nearby ground water catchment, a direct injection HPLC-MS/MS analysis with a triple quadrupole mass spectrometer was performed. In the ground water samples examined, these benzothiazoles were not detected. Nevertheless, a further regular monitoring of the ground water is reasonable, because the artificial turf was installed only a short time before the investigation described above was performed, and thus the contaminants detected may not yet have reached the ground water.

7 Conclusion

Because of the increased awareness of the need for EDA in environmental analysis, the interest in HPTLC coupled with in vitro assays is rising continuously. HPTLC-EDA represents an orthogonal analytical technique in the common physicochemical analytical methods. The major advantage of HPTLC in comparison to the often described HPLC approach is that the separated substances are present free of solvent after separation, which is precisely what allows the direct coupling with various biological systems. EDA is supported by identification strategies using specific derivatization reagents and spectroscopic techniques. Based on the effect determined, it is possible to prioritize the substances to be identified. In this way, along with information on the effect, HPTLC-EDA can also provide structural information on active compounds.

It should be mentioned that the HPTLC-EDA methods are screening methods that are used to provide an initial estimate of biological effects induced by a sample. Because of the separation, the possibility of false-positive and false-negative results are reduced. In addition, there is often a distinct increase in sensitivity in comparison to tests performed in cuvettes or microtiter plates. An important challenge for the future is the evaluation, assessment, and communication of the analytical results obtained. The correlation between the results of HPTLC-EDA and the results from tests to determine chronic effects should be a topic of future studies.

HPTLC-EDA is an analytical method with a high degree of freedom. This provides for high flexibility regarding the selection of bioassays and the identification techniques. However, in comparison to other physicochemical analytical techniques, for example, HPLC-MS, it requires a higher degree of technical skill and experience. It is therefore very important for expert groups to publish guidelines to lower the hurdles for new users of this technique. The possibilities and limitations of the individual HPTLC-EDA tests can be revealed only by the experience of many users in many different areas. The standardization of HPTLC-EDA is important for the recognition and use of this approach by official agencies.

References

1. Farré M, Pérez S, Kantiani L, Barceló D (2008) Fate and toxicity of emerging pollutants, their metabolites and transformation products in the aquatic environment. TrAC Trends Anal Chem 27(11):991–1007
2. Schwarzenbach RP, Escher BI, Fenner K, Hofstetter TB, Johnson CA, von Gunten U, Wehrli B (2006) The challenge of micropollutants in aquatic systems. Science 313 (5790):1072–1077. doi:10.1126/science.1127291
3. Kümmerer K (2009) The presence of pharmaceuticals in the environment due to human use - present knowledge and future challenges. J Environ Manag 90:2354–2366. doi:10.1016/j.jenvman.2009.01.023
4. Pal A, Gin KY-H, Lin AY-C, Reinhard M (2010) Impacts of emerging organic contaminants on freshwater resources: review of recent occurrences, sources, fate and effects. Sci Total Environ 408(24):6062–6069. doi:10.1016/j.scitotenv.2010.09.026
5. Escher BI, Fenner K (2011) Recent advances in environmental risk assessment of transformation products. Environ Sci Technol 45(9):3835–3847. doi:10.1021/es1030799
6. Farré M, Barceló D (2003) Toxicity testing of wastewater and sewage sludge by biosensors, bioassays and chemical analysis. Trends Anal Chem 22(5):299–310
7. Kokkali V, van Delft W (2014) Overview of commercially available bioassays for assessing chemical toxicity in aqueous samples. TrAC Trends Anal Chem 61:133–155. doi:10.1016/j.trac.2014.08.001
8. Tuikka AI, Schmitt C, Höss S, Bandow N, von der Ohe PC, de Zwart D, de Deckere E, Streck G, Mothes S, van Hattum B, Kocan A, Brix R, Brack W, Barceló D, Sormunen AJ, Kukkonen JVK (2011) Toxicity assessment of sediments from three European river basins using a sediment contact test battery. Ecotoxicol Environ Saf 74(1):123–131. doi:10.1016/j.ecoenv.2010.08.038
9. Bletsou AA, Jeon J, Hollender J, Archontaki E, Thomaidis NS (2015) Targeted and non-targeted liquid chromatography-mass spectrometric workflows for identification of transformation products of emerging pollutants in the aquatic environment. Trends Anal Chem 66:32–44. doi:10.1016/j.trac.2014.11.009
10. Leendert V, Van Langenhove H, Demeestere K (2015) Trends in liquid chromatography coupled to high-resolution mass spectrometry for multi-residue analysis of organic micropollutants in aquatic environments. TrAC Trends Anal Chem 67:192–208. doi:10.1016/j.trac.2015.01.010
11. Brack W (2003) Effect-directed analysis: a promising tool for the identification of organic toxicants in complex mixtures? Anal Bioanal Chem 377(3):397–407
12. Weiss J, Simon E, Stroomberg G, de Boer R, de Boer J, van der Linden S, Leonards P, Lamoree M (2011) Identification strategy for unknown pollutants using high-resolution mass

spectrometry: androgen-disrupting compounds identified through effect-directed analysis. Anal Bioanal Chem 400(9):3141–3149. doi:10.1007/s00216-011-4939-x
13. Rhee IK, Appels N, Hofte B, Karabatak B, Erkelens C, Stark LM, Flippin LA, Verpoorte R (2004) Isolation of the acetylcholinesterase inhibitor ungeremine from *Nerine bowdenii* by preparative HPLC coupled on-line to a flow assay system. Biol Pharm Bull 27 (11):1804–1809. doi:10.1248/bpb.27.1804
14. Fabel S, Niessner R, Weller MG (2005) Effect-directed analysis by high-performance liquid chromatography with gas-segmented enzyme inhibition. J Chromatogr A 1099 (1-2):103–110. doi:10.1016/j.chroma.2005.08.081
15. Reinen J, Kool J, Vermeulen NE (2008) Reversed-phase liquid chromatography coupled on-line to estrogen receptor bioaffinity detection based on fluorescence polarization. Anal Bioanal Chem 390(8):1987–1998. doi:10.1007/s00216-008-1833-2
16. Morlock GE, Prabha S (2007) Analysis and stability of sucralose in a milk-based confection by a simple planar chromatographic method. J Agric Food Chem 55(18):7217–7223. doi:10.1021/jf071719u
17. Morlock G, Schwack W (2008) Planar chromatography - back to the future? LC-GC Europe 21:366–371
18. Aranda M, Morlock G (2006) Simultaneous determination of riboflavin, pyridoxine, nicotinamide, caffeine and taurine in energy drinks by planar chromatography-multiple detection with confirmation by electrospray ionization mass spectrometry. J Chromatogr A 1131 (1-2):253–260
19. Weber WH, Seitz W, Schulz W, Aichinger A (2007) Evidence of the metabolites desphenyl-chloridazon and methyl-desphenyl chloridazon in surface, ground and drinking water. Camag Bibliogr Serv 98:12–15
20. Hirschfeld T (1980) The hy-phen-ated methods. Anal Chem 52(2):297A–312A. doi:10.1021/ac50052a870
21. Goodall RR, Levi AA (1946) A microchromatographic method for the detection and approximate determination of the different penicillins in a mixture. Nature 158:675–676. doi:10.1038/158675a0
22. Pailler JY, Krein A, Pfister L, Hoffmann L, Guignard C (2009) Solid phase extraction coupled to liquid chromatography-tandem mass spectrometry analysis of sulfonamides, tetracyclines, analgesics and hormones in surface water and wastewater in Luxembourg. Sci Total Environ 407(16):4736–4743. doi:10.1016/j.scitotenv.2009.04.042
23. Faber P, Bierl R (2012) Influence of different flow conditions on the occurrence and behavior of potentially hazardous organic xenobiotics in the influent and effluent of a municipal sewage treatment plant in Germany: an effect-directed approach. Environ Sci Eur 24 (1):1–13. doi:10.1186/2190-4715-24-2
24. Oellig C, Schwack W (2011) Planar solid phase extraction--a new clean-up concept in multi-residue analysis of pesticides by liquid chromatography-mass spectrometry. J Chromatogr A 1218(37):6540–6547. doi:10.1016/j.chroma.2011.06.108
25. Ramos L, Ramos JJ, Brinkman UAT (2005) Miniaturization in sample treatment for environmental analysis. Anal Bioanal Chem 381(1):119–140
26. Poole SK, Poole CF (2011) High performance stationary phases for planar chromatography. J Chromatogr A 1218(19):2648–2660. doi:10.1016/j.chroma.2010.10.072
27. Poole CF (2015) High-performance precoated stationary phases. In: Poole CF (ed) Instrumental thin-layer chromatography. Elsevier, Amsterdam, pp 31–51. doi:10.1016/B978-0-12-417223-4.00002-9
28. Poole CF, Poole SK (1995) Multidimensionality in planar chromatography. J Chromatogr A 703(1–2):573–612. doi:10.1016/0021-9673(94)01286-N
29. Dzido TH (2000) Chromatography: thin-layer (planar)/modes of development: conventional. In: Wilson ID, Adlard ER, Cooke M, Poole CF (eds) Encyclopedia of separation science. Academic, San Diego, pp 866–876

30. Poole CF (1999) Planar chromatography at the turn of the century. J Chromatogr A 856 (1-2):399–427
31. Morlock G, Schwack W (2007) The contribution of planar chromatography to food analysis. J Planar Chromatogr - Mod TLC 20(6):399–406
32. Móricz Á, Kalász H (2010) Centrifugal layer chromatography — rotation planar chromatography. J Planar Chromatogr - Mod TLC 23(6):415–419
33. Nyiredy S (2003) Progress in forced-flow planar chromatography. J Chromatogr A 1000 (1–2):985–999. doi:10.1016/s0021-9673(03)00308-x
34. Tyihák E, Mincsovics E, Móricz ÁM (2012) Overpressured layer chromatography: from the pressurized ultramicro chamber to BioArena system. J Chromatogr A 1232:3–18. doi:10.1016/j.chroma.2011.11.049
35. Perry JA, Haag KW, Glunz LJ (1973) Programmed multiple development in thin layer chromatography. J Chromatogr Sci 11(9):447–453. doi:10.1093/chromsci/11.9.447
36. Burger K (1984) DC-PMD, Dünnschicht-Chromatographie mit Gradienten-Elution im Vergleich zur Säulenflüssigkeits-Chromatographie. Fresenius J Anal Chem 318 (3):228–233. doi:10.1007/bf00528586
37. Burger K (1995) Thin layer chromatography with automated multiple development (AMD-TLC). Chem Plant Protect 12:181–195
38. Fuchs B, Süß R, Teuber K, Eibisch M, Schiller J (2011) Lipid analysis by thin-layer chromatography--a review of the current state. J Chromatogr A 1218(19):2754–2774. doi:10.1016/j.chroma.2010.11.066
39. Cebolla VL, Jarne C, Domingo P, Domínguez A, Delgado-Camón A, Garriga R, Galbán J, Membrado L, Gálvez EM, Cossío FP (2011) Fluorescence detection by intensity changes for high-performance thin-layer chromatography separation of lipids using automated multiple development. J Chromatogr A 1218(19):2668–2675. doi:10.1016/j.chroma.2010.11.033
40. Sherma J (2000) Thin-layer chromatography in food and agricultural analysis. J Chromatogr A 880(1–2):129–147. doi:10.1016/s0021-9673(99)01132-2
41. Galand N, Pothier J, Viel C (2002) Plant drug analysis by planar chromatography. J Chromatogr Sci 40(10):585–597. doi:10.1093/chromsci/40.10.585
42. De La Vigne U, Jänchen DE, Weber WH (1991) Application of high-performance thin-layer chromatography and automated multiple development for the identification and determination of pesticides in water. J Chromatogr A 553:489–496
43. Burger K, Jork H, Köhler J (1996) Application of AMD to the determination of crop protection agents in drinking water. Part IV: Fundamentals of a confirmatory test. Acta Hydrochim Hydrobiol 24(1):6–15
44. DIN V 38407-11:1995-01 (Prestandard) (1995) German standard methods for the examination of water, waste water and sludge - jointly determinable substances (group F) - Part 11: Determination of selected organic plant protecting agents by Automated Multiple Development (AMD)-technique (F 11) (in German)
45. Goodall RR, Levi AA (1947) A micro-chromatographic method for the detection and approximate determination of the different penicillins in a mixture. Analyst 72 (856):277–290. doi:10.1039/an9477200277
46. Meyers E, Smith DA (1964) Bioautography of antibiotic spread-layer chromatograms. J Chromatogr A 14:129–132. doi:10.1016/S0021-9673(00)86603-0
47. Rahalison L, Hamburger M, Hostettmann K, Monod M, Frenk E (1991) A bioautographic agar overlay method for detection of antifungal compounds from higher plants. Phytochem Anal 2:199–203
48. Baumann U, Indermaur H, Pletscher E (2000) Untersuchungskonzept zur Charakterisierung toxischer Stoffe in belastetem Wasser. Umweltwiss Schadst Forsch 12(2):63–67. doi:10.1007/bf03038159
49. Eberz G, Rast H, Burger K, Kreiss W, Weisemann C (1996) Bioactivity screening by chromatography-bioluminescence coupling. Chromatographia 43(1):5–9

50. Choma I, Choma A, Staszczuk K (2002) Direct bioautography-thin-layer chromatography of flumequine and doxycycline in milk. J Planar Chromatogr - Mod TLC 15(3):187–191
51. Akkad R, Schwack W (2010) Multi-enzyme inhibition assay for the detection of insecticidal organophosphates and carbamates by high-performance thin-layer chromatography applied to determine enzyme inhibition factors and residues in juice and water samples. J Chromatogr B 878(17-18):1337–1345
52. Schweppe CH, Hoffmann P, Nofer J-R, Pohlentz G, Mormann M, Karch H, Friedrich AW, Müthing J (2010) Neutral glycosphingolipids in human blood: a precise mass spectrometry analysis with special reference to lipoprotein-associated Shiga toxin receptors. J Lipid Res 51 (8):2282–2294. doi:10.1194/jlr.M006759
53. Cieśla Ł, Kryszeń J, Stochmal A, Oleszek W, Waksmundzka-Hajnos M (2012) Approach to develop a standardized TLC-DPPH test for assessing free radical scavenging properties of selected phenolic compounds. J Pharm Biomed Anal 70:126–135. doi:10.1016/j.jpba.2012.06.007
54. Rios JL, Recio MC, Villar A (1988) Screening methods for natural products with antimicrobial activity: a review of the literature. J Ethnopharmacol 23(2–3):127–149. doi:10.1016/0378-8741(88)90001-3
55. Choma I, Jesionek W (2015) Effects-directed biological detection: bioautography. In: Poole CF (ed) Instrumental thin-layer chromatography. Elsevier, Amsterdam, pp 279–312. doi:10.1016/B978-0-12-417223-4.00011-X
56. Wagman GH, Bailey JV (1969) Use of silicic acid—glass fiber sheets for bioautography of antimicrobial substances. J Chromatogr A 41:263–264. doi:10.1016/0021-9673(64)80132-1
57. Betina V (1973) Bioautography in paper and thin-layer chromatography and its scope in the antibiotic field. J Chromatogr A 78:41–51. doi:10.1016/S0021-9673(01)99035-1
58. Nagy S, Kocsis B, Kőszegi T, Botz L (2002) Optimization of conditions for culture of the test bacteria used for direct bioautographic detection. 1. The Gram-positive test bacterium Bacillus subtilis. J Planar Chromatogr - Mod TLC 15(2):132–137
59. Nagy S, Kőszegi T, Botz L, Kocsis B (2003) Optimization of conditions for culture of test bacteria used for direct bioautographic TLC detection. 2. Gram-negative test bacterium: Escherichia coli. J Planar Chromatogr - Mod TLC 16(2):121–126
60. Nagy S, Kocsis B, Kőszegi T, Botz L (2007) Optimization of growth conditions for test fungus cultures used in direct bioautographic TLC detection. 3. Test fungus: Candida albicans. J Planar Chromatogr - Mod TLC 20(5):385–389
61. Grzelak EM, Majer-Dziedzic B, Choma IM (2011) Development of a novel direct bioautography-thin-layer chromatography test: optimization of growth conditions for Gram-negative bacteria, Escherichia coli. J AOAC Int 94(5):1567–1572. doi:10.5740/jaoac.10-385
62. Grzelak EM, Majer-Dziedzic B, Choma IM, Pilorz KM (2013) Development of a novel direct bioautography-thin-layer chromatography test: optimization of growth conditions for Gram-positive bacteria, *Bacillus subtilis*. J AOAC Int 96(2):386–391. doi:10.5740/jaoacint.11-466
63. Akkad R, Schwack W (2011) Effect of bromine oxidation on high-performance thin-layer chromatography multi-enzyme inhibition assay detection of organophosphates and carbamate insecticides. J Chromatogr A 1218(19):2775–2784. doi:10.1016/j.chroma.2011.02.029
64. Magnani JL, Smith DF, Ginsburg V (1980) Detection of gangliosides that bind cholera toxin: direct binding of 125I-labeled toxin to thin-layer chromatograms. Anal Biochem 109 (2):399–402. doi:10.1016/0003-2697(80)90667-3
65. Meisen I, Mormann M, Müthing J (2011) Thin-layer chromatography, overlay technique and mass spectrometry: a versatile triad advancing glycosphingolipidomics. Biochim Biophys Acta 1811(11):875–896. doi:10.1016/j.bbalip.2011.04.006
66. Tanaka H, Putalun W, Tsuzaki C, Shoyama Y (1997) A simple determination of steroidal alkaloid glycosides by thin-layer chromatography immunostaining using monoclonal antibody against solamargine. FEBS Lett 404(2–3):279–282. doi:10.1016/S0014-5793(97)00106-3

67. Muthing J, Cacic M (1997) Glycosphingolipid expression in human skeletal and heart muscle assessed by immunostaining thin-layer chromatography. Glycoconj J 14(1):19–28. doi:10.1023/a:1018552729572
68. Kushi Y, Ogura K, Rokukawa C, Handa S (1990) Blood group A-active glycosphingolipids analysis by the combination of TLC-immunostaining assay and TLC/SIMS mass spectrometry. J Biochem 107(5):685–688
69. Portoukalian J, Bouchon B (1986) Hydrolysis of all gangliosides, including GM1 and GM2, on thin-layer plates by Vibrio cholerae neuraminidase. J Chromatogr B Biomed Sci Appl 380:386–392. doi:10.1016/S0378-4347(00)83668-3
70. Yu RK, Ariga T (2000) Ganglioside analysis by high-performance thin-layer chromatography. In: Alfred HM, Yusuf AH (eds) Methods in enzymology, vol 312. Academic, London, pp 115–134. doi:10.1016/S0076-6879(00)12903-9
71. Sorice M, Griggi T, Circella A, Garofalo T, d'Agostino F, Pittoni V, Pontieri GM, Lenti L, Valesini G (1994) Detection of antiphospholipid antibodies by immunostaining on thin layer chromatography plates. J Immunol Methods 173(1):49–54. doi:10.1016/0022-1759(94)90282-8
72. Sorice M, Circella A, Misasi R, Pittoni V, Garofalo T, Cirelli A, Pavan A, Pontieri GM, Valesini G (2000) Cardiolipin on the surface of apoptotic cells as a possible trigger for antiphospholipid antibodies. Clin Exp Immunol 122(2):277–284. doi:10.1046/j.1365-2249.2000.01353.x
73. Alessandri C, Sorice M, Bombardieri M, Conigliaro P, Longo A, Garofalo T, Manganelli V, Conti F, Esposti M, Valesini G (2006) Antiphospholipid reactivity against cardiolipin metabolites occurring during endothelial cell apoptosis. Arthritis Res Ther 8(6):R180
74. Conti F, Alessandri C, Sorice M, Capozzi A, Longo A, Garofalo T, Misasi R, Bompane D, Hughes GRV, Khamashta MA, Valesini G (2012) Thin-layer chromatography immunostaining in detecting anti-phospholipid antibodies in seronegative anti-phospholipid syndrome. Arthritis Res Ther 167(3):429–437. doi:10.1111/j.1365-2249.2011.04532.x
75. Conti F, Alessandri C, Spinelli F, Capozzi A, Martinelli F, Recalchi S, Misasi R, Valesini G, Sorice M (2014) TLC immunostaining for detection of "antiphospholipid" antibodies. In: Eggleton P, Ward FJ (eds) Systemic lupus erythematosus, vol 1134. Methods in molecular biology. Springer, New York, pp 95–101. doi:10.1007/978-1-4939-0326-9_8
76. Weins C, Jork H (1996) Toxicological evaluation of harmful substances by in situ enzymatic and biological detection in high-performance thin-layer chromatography. J Chromatogr A 750(1-2):403–407. doi:10.1016/0021-9673(96)00601-2
77. DIN EN ISO 11348-3:2009-05 (2009) Water quality - determination of the inhibitory effect of water samples on the light emission of Vibrio fischeri (luminescent bacteria test) - Part 3: Method using freeze-dried bacteria (German version of ISO 11348-3:2007)
78. Schulz W, Seitz W, Weber WH (2008) HPTLC/AMD with Vibrio fischeri as an example for bioactivity-based detection – a new dimension in analytics. Paper presented at the HPTLC - International Symposium for Thin Layer Chromatography, Helsinki
79. Schulz W, Weiss SC, Weber WH (2015) The RIV concept: a novel procedure for the evaluation in effect-directed analysis with TLC - using the TLC-bioluminescence inhibition assay as an example- (Submitted)
80. Klöppel A, Grasse W, Brümmer F, Morlock G (2008) HPTLC coupled with bioluminescence and mass spectrometry for bioactivity-based analysis of secondary metabolites in marine sponges. J Planar Chromatogr - Mod TLC 21(6):431–436
81. Verbitski SM, Gourdin GT, Ikenouye LM, McChesnex JD, Hildreth J (2008) Detection of Actaea racemosa adulteration by thin-layer chromatography and combined thin-layer chromatography-bioluminescence. J AOAC Int 91(2):268–275
82. Okusa P, Stévigny C, Devleeschouwer M, Duez P (2010) Optimization of the culture medium used for direct TLC-bioautography. Application to the detection of antimicrobial compounds from Cordia gilletii De Wild (Boraginaceae). J Planar Chromatogr - Mod TLC 23(4):245–249

83. Baumgartner V, Hohl C, Hauri U (2009) Bioactivity-based analysis of sunscreens using the luminescent bacteria Vibrio fischeri. J Planar Chromatogr - Mod TLC 22(1):19–23
84. Baumgartner V, Hohl C, Schwack W (2013) Screening for antimicrobials in mouthwashes using HPTLC-bioluminescence detection. Chromatographia 76(19-20):1315–1325. doi:10.1007/s10337-012-2375-5
85. Schulz W, Seitz W, Weiss SC, Weber WH, Böhm M, Flottmann D (2008) Use of Vibrio fischeri for screening for bioactivity in water analysis. J Planar Chromatogr - Mod TLC 21 (6):427–430
86. Weber WH, Seitz W, Aichinger A, Albert R (2005) Luminographic detection of toxicity with Vibrio fischeri. Camag Bibliogr Serv 94:2–4
87. Reemtsma T, Putschew A, Jekel M (1999) Industrial wastewater analysis: a toxicity-directed approach. Waste Manag 19(2):181–188
88. Müller A, Weiss SC, Schulz W, Seitz W, Albert R, Ruck WKL, Weber WH (2010) Combination of different liquid chromatography/mass spectrometry technologies for the identification of transformation products of rhodamine B in groundwater. Rapid Commun Mass Spectrom 24(5):659–666
89. Prasse C, Wagner M, Schulz R, Ternes TA (2012) Oxidation of the antiviral drug acyclovir and its biodegradation product carboxy-acyclovir with ozone: kinetics and identification of oxidation products. Environ Sci Technol 46(4):2169–2178. doi:10.1021/es203712z
90. Luft A, Wagner M, Ternes TA (2014) Transformation of biocides irgarol and terbutryn in the biological wastewater treatment. Environ Sci Technol 48(1):244–254. doi:10.1021/es403531d
91. Choma IM, Grzelak EM (2011) Bioautography detection in thin-layer chromatography. J Chromatogr A 1218(19):2684–2691. doi:10.1016/j.chroma.2010.12.069
92. Choma I (2006) Screening of enrofloxacin and ciprofloxacin residues in milk by HPLC and by TLC with direct bioautography. J Planar Chromatogr - Mod TLC 19(108):104–108. doi:10.1556/JPC.19.2006.2.3
93. Choma IM, Kowalski C, Lodkowski R, Burmańczuk A, Komaniecka I (2008) TLC-DB as an alternative to the HPLC method in the determination of cefacetril residues in cow's milk. J Liq Chromatogr Relat Technol 31(13):1903–1912. doi:10.1080/10826070802194815
94. Klingelhöfer I, Morlock GE (2014) Sharp-bounded zones link to the effect in planar chromatography-bioassay-mass spectrometry. J Chromatogr A 1360:288–295. doi:10.1016/j.chroma.2014.07.083
95. Weins C (2008) Overview of bioactivity-based analysis by HPTLC. Bridging the gap between cause and effect — HPTLC detection of bioactive compounds in the environment and in food. J Planar Chromatogr - Mod TLC 21(6):405–410
96. Jesionek W, Grzelak E, Majer-Dziedzic B, Choma I (2013) Thin-layer chromatography — direct bioautography for the screening of antimicrobial properties of plant extracts. J Planar Chromatogr - Mod TLC 26(2):109–113. doi:10.1556/jpc.26.2013.2.1
97. Homans AL, Fuchs A (1970) Direct bioautography on thin-layer chromatograms as a method for detecting fungitoxic substances. J Chromatogr A 51:327–329. doi:10.1016/S0021-9673(01)96877-3
98. Bailey AM, Coffey MD (1984) A sensitive bioassay for quantification of metalaxyl in soils. Phytopathology 74(6):667–669. doi:10.1094/Phyto-74-667
99. Spira D, Reifferscheid G, Buchinger S (2013) Combination of high-performance thin-layer chromatography with a specific bioassay - a tool for effect-directed analysis. J Planar Chromatogr - Mod TLC 26(5):395–401. doi:10.1556/jpc.26.2013.5.2
100. McDonnell DP, Nawaz Z, Densmore C, Weigel NL, Pham TA, Clark JH, O'Malley BW (1991) High level expression of biologically active estrogen receptor in Saccharomyces cerevisiae. J Steroid Biochem Mol Biol 39(3):291–297. doi:10.1016/0960-0760(91)90038-7
101. Routledge EJ, Sumpter JP (1996) Estrogenic activity of surfactants and some of their degradation products assessed using a recombinant yeast screen. Environ Toxicol Chem 15 (3):241–248

102. Müller MB, Dausend C, Weins C, Frimmel FH (2004) A new bioautographic screening method for the detection of estrogenic compounds. Chromatographia 60(3):207–211
103. Schönborn A, Grimmer A (2013) Coupling sample preparation with effect-directed analysis of estrogenic activity - proposal for a new rapid screening concept for water samples. J Planar Chromatogr - Mod TLC 26(5):402–408. doi:10.1556/jpc.26.2013.5.3
104. Buchinger S, Spira D, Bröder K, Schlüsener M, Ternes T, Reifferscheid G (2013) Direct coupling of thin-layer chromatography with a bioassay for the detection of estrogenic compounds: applications for effect-directed analysis. Anal Chem 85(15):7248–7256. doi:10.1021/ac4010925
105. Ames BN, Durston WE, Yamasaki E, Lee FD (1973) Carcinogens are mutagens: a simple test system combining liver homogenates for activation and bacteria for detection. Proc Natl Acad Sci 70(8):2281–2285
106. Oda Y, Nakamura S-I, Oki I, Kato T, Shinagawa H (1985) Evaluation of the new system (umu-test) for the detection of environmental mutagens and carcinogens. Mutat Res/Environ Mutagen Relat Subj 147(5):219–229. doi:10.1016/0165-1161(85)90062-7
107. Bjørseth A, Eidsa G, Gether J, Landmark L, Møller M (1982) Detection of mutagens in complex samples by the Salmonella assay applied directly on thin-layer chromatography plates. Science 215(4528):87–89. doi:10.1126/science.7031897
108. Houk VS, Claxton LD (1986) Screening complex hazardous wastes for mutagenic activity using a modified version of the TLC/Salmonella assay. Mut Res/Genet Toxicol 169 (3):81–92. doi:10.1016/0165-1218(86)90087-x
109. Baumann U, Brunner C, Pletscher E, Tobler N (2003) Biologische Detektionsverfahren in der Dünnschichtchromatographie. Umweltwiss Schadst Forsch 15(3):163–167. doi:10.1065/uwsf2001.12.080
110. DIN 38415-3:1996-12 (1996) German standard methods for the examination of water, waste water and sludge - sub-animal testing (group T) - Part 3: Determination of the genotype potential of water and waste water components with the umu-test (T 3) (in German)
111. Egetenmeyer N, Weiss SC (2015) Investigations for the detection of genotoxic substances on the TLC plate (Submitted)
112. Kováč J, Henselová M (1977) Detection of triazine herbicides in soil by a Hill-reaction inhibition technique after thin-layer chromatography. J Chromatogr A 133(2):420–422. doi:10.1016/S0021-9673(00)83509-8
113. Lawrence JF (1980) Simple, sensitive and selective thin layer chromatographic technique for detecting some photosynthesis inhibiting herbicides. J Assoc Off Anal Chem 63(4):758–761
114. Broszat M, Ernst H, Spangenberg B (2011) Quantification of triazine herbicides using chloroplasts in conjunction with thin-layer chromatography. Environ Biotechnol 7(2):47–52
115. McKinley WP, Read SI (1962) Esterase inhibition technique for the detection of organophosphorus pesticides. J Assoc Off Agric Chem 45:467–474
116. McKinley WP, Johal PS (1963) Esterase inhibition technique for detection of organophosphorus pesticides - II. Simplified version for routine checking. J Assoc Off Agric Chem 46:840–842
117. Ortloff R, Franz P (1965) Zwei neue Methoden der biochemischen Lokalisierung von phosphorhaltigen Insektiziden auf Dünnschicht-chromatogrammen. Zeitschrift für Chemie 5(10):388–389. doi:10.1002/zfch.19650051020
118. Menn JJ, McBain JB, Dennis MJ (1964) Detection of naturally occurring cholinesterase inhibitors in several crops by paper chromatography. Nature 202(4933):697–698
119. Bunyan PJ (1964) The detection of organo-phosphorus pesticides on thin-layer chromatograms. Analyst 89(1062):615–618. doi:10.1039/an9648900615
120. El-Refai A, Hopkins TL (1965) Insecticide metabolism, thin-layer chromatography and cholinesterase detection of several phosphorothiono insecticides and their oxygen analogs. J Agric Food Chem 13(5):477–481. doi:10.1021/jf60141a026
121. Menn JJ, McBain JB (1966) Detection of cholinesterase-inhibiting insecticide chemicals and pharmaceutical alkaloids on thin-layer chromatograms. Nature 209(5030):1351–1352

122. Ackermann H (1968) Dünnschichtchromatographisch-enzymatischer'Nachweis phosphororganischer Insektizide: Aktivierung schwacher Esterasehemmer. J Chromatogr A 36:309–317. doi:10.1016/s0021-9673(01)92947-4
123. Geike F (1970) Dünnschichtchromatographisch--enzymatischer Nachweis von Carbamaten: I. Nachweis insektizider Carbamate mit Rinderleber-Esterase. J Chromatogr A 53 (2):269–277. doi:10.1016/s0021-9673(01)98467-5
124. Mendoza CE, Shields JB (1970) Sensitivity of pig liver esterase in detecting twelve carbamate pesticides on thin-layer chromatograms. J Chromatogr A 50:92–102. doi:10.1016/s0021-9673(00)97920-2
125. Weins C (2006) Möglichkeiten und Grenzen der wirkungsbezogenen Analytik mit der Hochleistungs - Dünnschichtchromatographie. Dissertation, Universität Basel, Basel
126. Ellman GL, Courtney KD, Andres V Jr, Featherstone RM (1961) A new and rapid colorimetric determination of acetylcholinesterase activity. Biochem Pharmacol 7(2):88–95. doi:10.1016/0006-2952(61)90145-9
127. Rhee IK, van Rijn RM, Verpoorte R (2003) Qualitative determination of false-positive effects in the acetylcholinesterase assay using thin layer chromatography. Phytochem Anal 14 (3):127–131. doi:10.1002/pca.675
128. Guilbault GG, Kramer DN (1965) Resorufin butyrate and indoxyl acetate as fluorogenic substrates for cholinesterase. Anal Chem 37(1):120–123. doi:10.1021/ac60220a031
129. Mendoza CE, Wales PJ, McLeod HA, McKinley WP (1968) Enzymatic detection of ten organophosphorus pesticides and carbaryl on thin-layer chromatograms: an evaluation of indoxyl, substituted indoxyl and 1-naphthyl acetates as substrates of esterases. Analyst 93:34–38
130. Hamada M, Wintersteiger R (2003) Fluorescence screening of organophosphorus pesticides in water by an enzyme inhibition procedure on TLC plates. J Planar Chromatogr - Mod TLC 16(1):4–10
131. Zoun PEF, Spierenburg TJ (1989) Determination of cholinesterase-inhibiting pesticides and some of their metabolites in cases of animal poisoning using thin-layer chromatography. J Chromatogr A 462:448–453. doi:10.1016/s0021-9673(00)91376-1
132. Štefanac Z, Štengl B, Vasilić Ž (1976) Quantitative determination of organophosphorus pesticides by thin-layer densitometry. J Chromatogr A 124(1):127–133. doi:10.1016/S0021-9673(00)87852-8
133. Winterlin W, Walker G, Frank H (1968) Detection of cholinesterase-inhibiting pesticides following separation on thin-layer chromatograms. J Agric Food Chem 16(5):808–812. doi:10.1021/jf60159a028
134. Akkad R, Schwack W (2008) Multi-enzyme inhibition assay for detection of insecticidal organophosphates and carbamates by high-performance thin-layer chromatography. 1. Basics of method development. J Planar Chromatogr - Mod TLC 21(6):411–415
135. Jork H, Funk W, Fischer W, Wimmer H (1990) Volume 1a: Physical and chemical detection methods: fundamentals, reagents I. Thin-layer chromatography - reagents and detection methods. VCH, Weinheim
136. Bratton AC, Marshall EK (1939) A new coupling component for sulfanilamide determination. J Biol Chem 128(2):537–550
137. Kinast A (2006) Determination of Amitrol in water by AMD. Camag Bibliogr Serv 96:2–5
138. Rahn C, Schlenk H (1973) Detection of aldehydes with 4-amino-5-hydrazino-1,2,4-triazole-3-thiol as spray reagent. Lipids 8(11):612–616
139. Frey H-P, Zieloff K (1992) Qualitative und quantitative Dünnschichtchromatographie: Grundlagen und Praxis. VCH, Weinheim
140. Jork H, Funk W, Fischer W, Wimmer H (1994) Volume 1b: Physical and chemical detection methods: activation reactions, reagent sequences, reagents II. Thin-layer chromatography - reagents and detection methods. VCH, Weinheim

141. Merck AG (2007) Dyeing reagents for thin layer and paper chromatography. http://www.clubdeccm.com/PDF/Dyeing_Reagents_TLC.pdf. Accessed 30.06.2009
142. Spangenberg B, Poole CF, Weins C (2011) Quantitative thin-layer chromatography: a practical survey, 1st edn. Springer, Berlin/Heidelberg
143. Müller A, Weiss SC, Beißwenger J, Leukhardt HG, Schulz W, Seitz W, Ruck WKL, Weber WH (2012) Identification of ozonation by-products of 4- and 5-methyl-1H-benzotriazole during the treatment of surface water to drinking water. Water Res 46(3):679–690. doi:10.1016/j.watres.2011.11.033
144. Morlock G, Schwack W (2010) Coupling of planar chromatography to mass spectrometry. TrAC Trends Anal Chem 29(10):1157–1171. doi:10.1016/j.trac.2010.07.010
145. Wilson ID (1999) The state of the art in thin-layer chromatography-mass spectrometry: a critical appraisal. J Chromatogr A 856(1-2):429–442
146. Fuchs B, Schiller J, Süß R, Zscharnack M, Bader A, Müller P, Schürenberg M, Becker M, Suckau D (2008) Analysis of stem cell lipids by offline HPTLC-MALDI-TOF MS. Anal Bioanal Chem 392(5):849–860
147. Salo P (2007) Thin-layer chromatography with ultraviolet and mass spectrometric detection: from preparative-layer to miniaturized ultra-thin-layer technique. Dissertation, University of Helsinki, Helsinki
148. Cody RB, Laramée JA, Durst HD (2005) Versatile new ion source for the analysis of materials in open air under ambient conditions. Anal Chem 77(8):2297–2302. doi:10.1021/ac050162j
149. Morlock G, Ueda Y (2007) New coupling of planar chromatography with direct analysis in real time mass spectrometry. J Chromatogr A 1143(1-2):243–251
150. Kim H, Jee E, Ahn K, Choi H, Jang Y (2010) Identification of marker compounds in herbal drugs on TLC with DART-MS. Arch Pharm Res 33(9):1355–1359. doi:10.1007/s12272-010-0909-7
151. Häbe TT, Morlock GE (2015) Quantitative surface scanning by Direct Analysis in Real Time mass spectrometry. Rapid Commun Mass Spectrom 29(6):474–484. doi:10.1002/rcm.7127
152. Luftmann H (2004) A simple device for the extraction of TLC spots: direct coupling with an electrospray mass spectrometer. Anal Bioanal Chem 378(4):964–968
153. Stan H-J, Schwarzer F (1998) On-line coupling of liquid chromatography with thin-layer chromatography. J Chromatogr A 819(1-2):35–44
154. Gössi A, Gaugler S, Scherer U, Schlotterbeck G, Wyss S, Büttler A, Hettich T, Baron A (2013) Rapid structure confirmation and quantitation by HPTLC-NMR. Camag Bibliogr Ser 110:2–4
155. Weber WH, Seitz W, Schulz W, Wagener H-A (2007) Nachweis der Metaboliten Desphenyl-chloridazon und Methyl-desphenyl-chloridazon in Oberflächen-, Grund- und Trinkwasser. Vom Wasser 105(1):7–14
156. Jautz U, Morlock G (2006) Efficacy of planar chromatography coupled to (tandem) mass spectrometry for employment in trace analysis. J Chromatogr A 1128(1-2):244–250
157. Aranda M, Morlock G (2007) New method for caffeine quantification by planar chromatography coupled with electropray ionization mass spectrometry using stable isotope dilution analysis. Rapid Commun Mass Spectrom 21(7):1297–1303
158. Luftmann H, Aranda M, Morlock GE (2007) Automated interface for hyphenation of planar chromatography with mass spectrometry. Rapid Commun Mass Spectrom 21(23):3772–3776
159. Weiss SC, Müller A, Schulz W, Weber WH (2014) HPTLC-MS combined with H/D exchange for the identification of substances in environmental analysis. Camag Bibliogr Serv 113:5–7
160. Himmelsbach M, Waser M, Klampfl C (2014) Thin layer chromatography–spray mass spectrometry: a method for easy identification of synthesis products and UV filters from TLC aluminum foils. Anal Bioanal Chem 406(15):3647–3656. doi:10.1007/s00216-014-7639-5

161. Meier B, Spriano D (2010) Modern HPTLC - a perfect tool for quality control of herbals and their preparations. J AOAC Int 93(5):1399–1409
162. Reich E, Schibli A, Debatt A (2008) Validation of high-performance thin-layer chromatographic methods for the identification of botanicals in a cGMP environment. J AOAC Int 91 (1):13–20
163. Marston A (2011) Thin-layer chromatography with biological detection in phytochemistry. J Chromatogr A 1218(19):2676–2683. doi:10.1016/j.chroma.2010.12.068
164. Hostettmann K, Terreaux C, Marston A, Potterat O (1997) The role of planar chromatography in the rapid screening and isolation of bioactive compounds from medicinal plants. J Planar Chromatogr 10:251–257
165. Reemtsma T (2001) Prospects of toxicity - directed wastewater analysis. Analytica Chimica Acta 426:279–287
166. Morlock GE, Schuele L, Grashorn S (2011) Development of a quantitative high-performance thin-layer chromatographic method for sucralose in sewage effluent, surface water, and drinking water. J Chromatogr A 1218(19):2745–2753. doi:10.1016/j.chroma.2010.11.063
167. DIN V 18035-7 (Prestandard) (2002) Sport grounds - Part 7: Synthetic turf areas (in German)
168. DIN EN 12457-1:2003-01 (2003) Characterization of waste - leaching; compliance test for leaching of granular and sludges - Part 1: One stage batch test at a liquid to solid ration of 2 l/kg with particle size below 4 mm (without or with size reduction) (German version of EN 12457-1:2002)

Adv Biochem Eng Biotechnol (2017) 157: 225–262
DOI: 10.1007/10_2015_5014

Published online: 4 August 2016

Combining Passive Sampling with Toxicological Characterization of Complex Mixtures of Pollutants from the Aquatic Environment

Annika Jahnke, Gesine Witt, Sabine Schäfer, Nora Haase, and Beate I. Escher

Abstract The combination of polymer-based passive sampling to collect complex environmental mixtures of pollutants, the transfer of these mixtures into bioassays, and their related toxicological characterization is still in its infancy. However, this approach has considerable potential to improve environmental hazard and risk assessment for two reasons. First, the passive sampler collects a broad range of chemicals representing the fraction of compounds available for diffusion and (bio) uptake, excluding a large part of the matrix; thus, extensive sample cleanup which could discriminate certain compounds can be avoided. Second, the toxicological characterization of samples using bioassays is complementary to chemical (target) analysis within environmental monitoring because it captures all chemicals exerting the same mode of toxic action and acting jointly in mixtures, thus

A. Jahnke (✉) and N. Haase
Department of Cell Toxicology, Helmholtz Centre for Environmental Research – UFZ, Leipzig 04318, Germany
e-mail: annika.jahnke@ufz.de

G. Witt
Department of Environmental Engineering, Hamburg University of Applied Sciences, Hamburg 20099, Germany

S. Schäfer
Department Biochemistry, Ecotoxicology, German Federal Institute of Hydrology, Koblenz 56068, Germany

B.I. Escher
Department of Cell Toxicology, Helmholtz Centre for Environmental Research – UFZ, Leipzig 04318, Germany

Environmental Toxicology, Center for Applied Geoscience, Eberhard Karls University Tübingen, Tübingen 72074, Germany providing a comprehensive picture of their overall combined effects. The scientific literature describes a range of examples from the water phase where passive sampling is usually carried out in the kinetic uptake regime for most chemicals although some may already have reached equilibrium. The composition of the chemical mixture changes from the water phase to the passive sampling material because of kinetic effects and polymer/water partition coefficients which depend on the chemicals' hydrophobicity. In contrast, only a few applications in sediment and biota have been described, but amongst these some pioneering studies have demonstrated the feasibility and potential of this combined approach. This chapter gives an overview of what has been carried out in this research area, focusing on opportunities and challenges, and points out desirable future developments with a focus on the importance of choosing a suitable combination of sampling and dosing to transfer (or re-establish) the environmental mixture into the bioassay.

Keywords Aquatic environment, Environmental monitoring, Hazard and risk assessment, Hydrophobic organic chemicals, Mixture toxicity, Passive sampling

Contents

1 Introduction

Passive sampling (PS) has become increasingly popular and more broadly applied in a wide range of environmental media. PS devices are used in air, sediment and soils, water including groundwater, biota including vegetation, and humans. One of the advantages of PS over active sampling is that samplers are inexpensive and can be operated basically everywhere without the need for electricity. They can provide a measure of the freely dissolved concentrations of chemicals (C_{free}) available for (bio)uptake and diffusion, and they leave most of the matrix constituents such as lipids which could disturb chemical analysis or toxicological assessment behind.

The focus of this chapter is on the combination of PS with mixture toxicity characterization of the sampled environmental pollutants in bioassays. The combination of PS with effect assessment is particularly challenging because sampling efficiency is dependent on the physicochemical properties of the chemicals, posing special demands on their dosing into the bioassays. Calibration is possible for target analytes but not for unidentified mixtures as they occur in environmental media. Nevertheless, bioassays have been applied in combination with PS extracts in a qualitative or semi-quantitative comparative manner, where the biological effect per mass of the PS phase was compared across different sites or sample types. In a few cases, attempts have been made to relate the effects observed in the PS device back to the concentration in the sampled medium.

Low-complexity in vivo bioassays such as toxicity tests with bacteria, algae, daphnids, or fish embryos or in vitro cell-based bioassays giving indications of modes of toxic action relevant for environmental and human health [1] have frequently been exposed to extracts from total extraction, but also to solvent-based PS extracts or directly to PS devices via passive dosing (PD) [2]. We complement a literature review of existing applications of PS combined with bioanalytical assessment with a critical evaluation of limitations and challenges but also opportunities of using bioassays in conjunction with PS. This chapter focuses on the aquatic environment, i.e., water, sediment, and biota.

2 Passive Sampling Approaches

The general principle of absorptive polymer-based PS for hydrophobic organic compounds (HOCs) is that a clean sampling phase is brought into contact with the medium to be sampled that is at a higher chemical activity than the sampler. The chemicals present in the sample passively diffuse towards the polymer along the activity gradient [2, 3]. Based on the initially large difference in chemical activity between the compounds in the medium of interest and the sampler, the chemical uptake into the polymer is rapid at first (kinetic linear uptake phase) and then slows down (intermediate phase) until a thermodynamic equilibrium (i.e., equal chemical activity, equilibrium phase) between the sample and the polymer is reached. The difference in chemical activity, sampling rates (R_s) and time to equilibrium varies strongly between chemicals and media. Hence, the equilibration status of single constituents of the mixture varies if sampling is performed in the kinetic mode, with some compounds having approached equilibrium and others far from being fully equilibrated. Likewise, the mixture composition at equilibrium differs from the profile of the original sample if sampling is carried out in the water phase (i.e., including the enrichment of highly hydrophobic chemicals relative to the water).

At the end of the sampling period, the polymer is removed from the sample, its surface is thoroughly cleaned to avoid a bias caused by adsorption effects, and the sampler is solvent-extracted to prepare the extracts for chemical analysis and toxicological characterization. As an alternative, the PS polymer can be used directly for PD of the mixture of chemicals into the toxicity test [4]. See the [5] for further information. There are also adsorptive samplers for polar and ionic compounds, but in this chapter we mainly focus on absorption-based samplers for hydrophobic compounds.

The material and geometry used for PS devices can be tailored to match the physicochemical properties of the compounds of interest, the desired sampling mode (i.e., kinetic or equilibrium mode), and the medium to be sampled. Generally, PS materials include sorbents such as polymer resins, e.g., Tenax, polymer sheets, e.g., low-density polyethylene (LDPE), polyoxymethylene (POM), silicone rubber (SR) such as polydimethylsiloxane (PDMS), semi-permeable membrane devices (SPMDs: lay flat LDPE tube filled with synthetic lipid), and polar organic chemical integrative samplers (POCIS: different solid-phase extraction (SPE) sorbents exposed in membranes). Triolein-filled SPMDs were designed to represent the lipid phase of organisms, whereas POCIS mimic the respiratory exposure of organisms in the aquatic environment, i.e., uptake via the gills. Strictly speaking they do not count as polymer-based PS devices as the sampling phase consists of triolein and SPE sorbents, respectively.

PS can be applied in most types of environmental media. Figure 1 depicts relevant examples for which literature is available, but is not comprehensive. Air and water are exclusively sampled with in situ PS because the PS device has to be

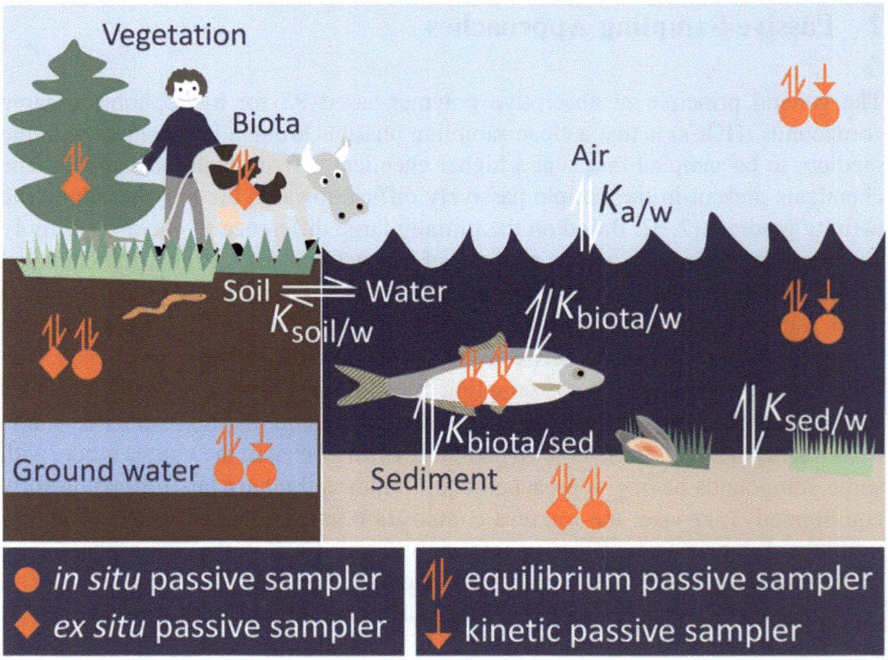

Fig. 1 Opportunities for PS devices in various environmental media, operated either in situ in the field or ex situ in the laboratory, and in either the kinetic uptake or equilibrium mode. Knowing the equilibrium partitioning concentrations in different compartments allows the determination of partition coefficients (K) between phases, e.g., sediment/water ($K_{sed/w}$), soil/water ($K_{soil/w}$), air/water ($K_{a/w}$), biota/water ($K_{biota/w}$), or biota/sediment ($K_{biota/sed}$). The discussion of methods in this chapter is restricted to PS in the aquatic environment, i.e., water, sediment, and biota. Figure adapted from Jahnke et al. [2]

equilibrated with huge volumes of the medium, rendering ex situ sampling practically impossible. In contrast, ex situ sampling is most commonly used for biota, and both modes are popular for sediment and soil; the applications depend on the research question.

PS devices can be operated in different modes: (1) in kinetic linear uptake mode with fast uptake into the sampling phase along a strong activity gradient or (2) in (near) equilibrium mode with slower and finally stagnant uptake [3]. In the kinetic mode, often used for matrices such as water with slow uptake kinetics, PS devices are considered to provide integrative, i.e., time-weighted averaged concentrations (C_{TWA}). The concentration of a chemical i in the medium to which the PS device has been exposed can be calculated (1) if the sampling rate of this chemical, $R_{s,i}$, is known.

$$C_{i,\mathrm{medium}} = \frac{C_{i,\mathrm{PS}}}{R_{s,i} \cdot t} \tag{1}$$

In the equilibrium mode, which is usually applied in media such as sediment with faster uptake kinetics, the PS device reflects the conditions at steady state and hence provides a direct measure of the concentrations when the sampler has been retrieved. If a chemical reaches equilibrium partitioning between the medium and the PS device, knowledge of $R_{s,i}$ is not required. The concentration of a chemical i in the medium to which the PS device has been exposed can be calculated (2) if the partition coefficient of this chemical between PS device and sample, $K_{\mathrm{PS/medium},i}$ is known.

$$C_{i,\mathrm{medium}} = \frac{C_{i,\mathrm{PS}}}{K_{\mathrm{PS/medium},i}} \tag{2}$$

Equilibrium sampling mode has the benefits that the precision of the measured data and the overall sensitivity are higher, and knowledge of $R_{s,i}$ is not required. On the other hand, sampling in kinetic mode delivers time-integrative information opposed to "snapshot" data, but usually requires the determination of $R_{s,i}$ and hence is subject to considerable uncertainty.

Much progress has been made regarding the calibration of PS devices in aqueous systems and consequently kinetic PS is most common for water. Moreover, R_s of HOCs in water are often quite low because of the low levels of these compounds in the water phase and the uptake being hampered by an unstirred aqueous boundary layer, often with equilibration times of months to years [6], rendering sampling in the equilibrium mode very difficult unless colloidal matter accelerates the uptake [7]. In contrast, in sediment and biota the uptake kinetics are complex and not well described, but R_s are usually considerably higher than in water, and hence most studies have used PS in equilibrium mode (Fig. 1).

Similar to the partitioning between the sample matrix and the PS device, chemicals partition between environmental media (Fig. 1) driven by the change in Gibbs free energy (ΔG) until ΔG equals zero at thermodynamic equilibrium.

Once equilibrium is established, partition coefficients (K) between phases can be calculated such as between biota and water ($K_{\text{biota/w}} = C_{\text{biota}}/C_{\text{w}}$), biota and sediment ($K_{\text{biota/sed}} = C_{\text{biota}}/C_{\text{sed}}$), soil and water ($K_{\text{soil/w}}$), and sediment and water ($K_{\text{sed/w}}$). To describe the partitioning of chemicals between water and air, the air/water partition coefficient ($K_{\text{a/w}} = C_{\text{w}}/C_{\text{air}}$) or Henry's law constant ($K_{\text{H}} = K_{\text{a/w}}{*}\,RT$, where R is the gas constant and T is the absolute temperature) is relevant. Few studies have applied the same PS material in equilibrium mode in different media to measure not only the chemical burden in one medium but also the partitioning and (dis)equilibrium of chemicals between different environmental compartments [3, 8].

The PS polymer/water partition coefficients (i.e., $K_{\text{PS/w}}$) and hence the chemical concentrations in the PS polymer are dependent on the hydrophobicity of the chemicals. Therefore, the mixture composition changes from the aqueous medium to the PS device. HOCs reach substantially higher levels in the PS device than in the aqueous phase with elevated concentrations of the more hydrophobic compounds. The PS device can hence be considered as mimicking an organism, which is also referred to as biomimetic sampling [9]. However, if bioassays are used as bioanalytical diagnostic tools, one must exert caution because the relative ratios of the chemicals in the PS device differ from the ratios in water.

In contrast to water, for sediments and biota the partition coefficient between the PS polymer and the environmental matrix is largely independent of the hydrophobicity of the chemicals because the main sorptive components for HOCs are the organic carbon (oc) in sediments and the lipids (lip) in biota, and their relationship with log octanol/water partition coefficients (K_{ow}) is quasi parallel, resulting in a largely K_{ow}-independent $K_{\text{PS/sed}}$ and $K_{\text{PS/biota}}$ [2].

3 General Principles for Toxicological Assessment of Passive Sampling Extracts

Bioassays can complement chemical analysis of PS extracts within environmental monitoring by providing information about the overall toxic potency of an environmental sample containing unidentified mixtures of contaminants. By choosing a battery of different biological endpoints (e.g., estrogenicity), it is possible to obtain integrative information on groups of chemicals with a common mode of toxic action. These compounds act together additively in a mixture and therefore bioanalytical endpoints can be regarded as sum parameters characterizing the contaminant mixture. The biological effect caused by the mixture may still be quantifiable at low levels whereas single contaminants might be too low in concentrations to be detectable by chemical analysis. An additional important aspect is the fact that many compounds are not even targeted by the analytical method [10]. Another advantage of combining PS with effect assessment is that the time-consuming enrichment of target analytes by, e.g., SPE, which is often necessary to investigate toxic effects at low concentrations in water becomes obsolete. Furthermore, a large part of disturbing matrix constituents of the sampled medium (e.g., lipids), confounding factors such as pH or salinity and inorganic compounds are left behind; thus PS provides a straightforward method for sample extraction which includes a simultaneous sample cleanup.

However, the major challenge when combining PS, solvent extraction, and dosing into bioassays is the fact that the mixture composition may change from the sampled medium to the PS device. As Fig. 2 shows, the K_{ow}-dependence of R_s means that PS of water changes the composition of the mixture in the PS device. Despite this limitation, PS of water in the kinetic mode has been applied in combination with bioassays. This procedure is appropriate for qualitative purposes and for benchmarking between samples and also if the sample is directly dosed via PD as demonstrated in a recent study [4]. In contrast, because $K_{PS/sed}$ and $K_{PS/biota}$ are largely independent of hydrophobicity, the composition of the PS extract of sediment and biota reflects their original chemical composition (Fig. 2). Thus, sampling of both sediments and biota can be realized without changing the composition of the mixture during the extraction step.

PS extracts have been dosed into different types of bioassays [2]. (1) In vivo bioassays with aquatic organisms, such as fish (embryos), daphnids or algae are mostly used to assess the effects of the chemicals – as present in the environmental medium under investigation – similar to a whole effluent toxicity test. Here, the freely dissolved aqueous concentrations, not the total or nominal concentrations, should be re-established as dose-metric. (2) Complementarily, in vitro bioassays are more often used as bioanalytical diagnostic tools, closely linked with chemical analysis, to assess the burden of chemicals that act together as mixtures triggering the mode of action targeted by the given in vitro bioassay. In vitro bioassays are mostly conducted on 96- or 384-well plates. The concentrations and relative effect

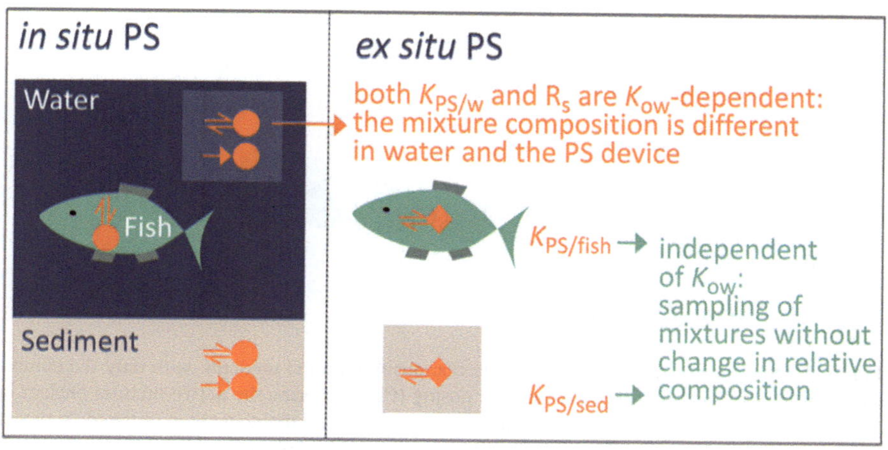

Fig. 2 Aquatic environment: dependence of PS/matrix partition coefficients on K_{ow} and implications for the combination of PS with bioassays

potencies of reference compounds are typically expressed as nominal concentrations, even if there are first attempts to model the partitioning in in vitro systems [11] and to relate observed effects to cellular or freely dissolved concentrations [12]. However, this approach has not yet been applied in conjunction with PS. Here it is important to transfer the environmental mixture in such a way that the concentration ratios remain constant and the total amount is transferred into the bioassay.

If a solvent extract of the PS device that has been exposed to an aqueous phase is dosed into the bioassay, the chemical composition in the PS device mimics the concentration in an organism (i.e., with relative enrichment of the more hydrophobic chemicals), not the external exposure concentration of the environmental sample (Fig. 3). To reflect the composition of the original aqueous sample, PS has to be combined with PD (Fig. 3a; compare [5]). Then the composition of chemicals in the bioassay medium is the same as in the aqueous phase of the original medium (bulk water or pore water of sediments). This innovative approach has recently been described by Claessens et al. [4] and is discussed in more detail

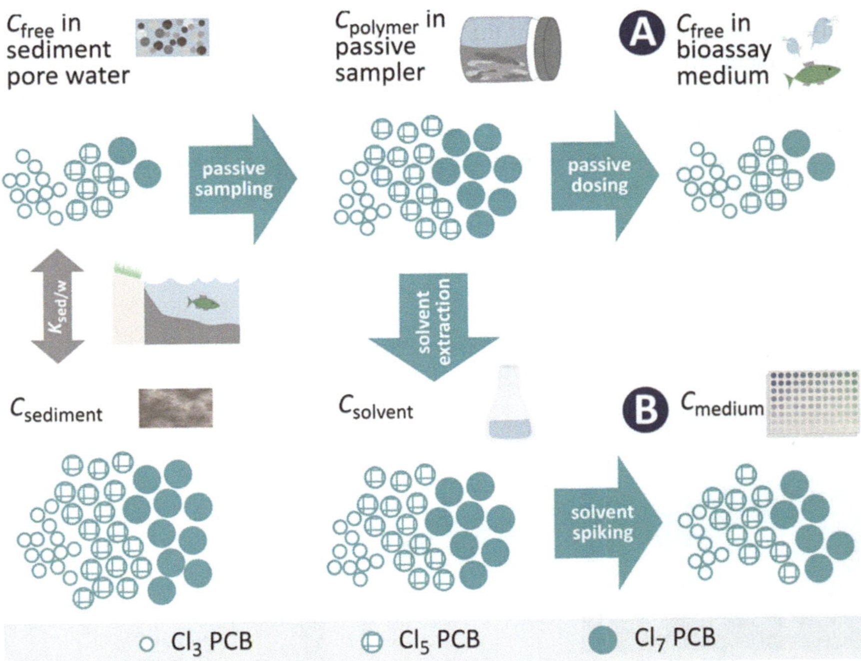

Fig. 3 Dependence of the chemical mixture composition on the medium, with only the combination of passive sampling (PS) and passive dosing (PD) mirroring the native mixture present in the sediment interstitial pore water (and likewise for bulk water) in the medium of the toxicity test (**a**). In contrast, solvent extraction of PS devices followed by solvent spiking into the bioassay results in concentration ratios similar to the ratios present in the bulk sediment (and correspondingly in biota) (**b**)

below. In this case, the PS polymer is used as a biomimetic extraction device in complex media with an aqueous phase.

In contrast, if the solvent extract of a PS device is directly dosed into an in vitro bioassay, the nominal concentrations in the microtiter plate reflect the composition in the PS material rather than the initial (aqueous) medium (Fig. 3b). This is probably the most common application of PS so far. One can argue that the nominal concentration ratios in the in vitro bioassay reflect the concentration ratios in the original medium, i.e., sediment or biota (Fig. 3b) and cellular concentrations are proportional to the nominal concentrations in the microtiter plate [2]. However, even if the concentration ratios are constant, the absolute concentrations still need to be known to allow application of PS in a quantitative way.

Results from in vivo bioassays are typically expressed as lethal concentrations causing 10% or 50% mortality (LC_{10} or LC_{50}), whereas those from in vitro bioassays are expressed as effect concentrations causing 50% of an observed effect (EC_{50}) or an induction ratio of activation of a receptor or pathway, or inhibition of an enzyme of 50% over the control, also called induction ratio 1.5 ($EC_{IR1.5}$). In the following mathematical derivations we call all types of effect concentrations EC for simplicity. Both in vivo and in vitro effects are expressed in units of relative enrichment factors (REFs) or, in the case of PS devices, per total PS or per PS mass.

It is conceptually easier to express biological effects in toxic equivalent concentrations (TEQ) in the case of in vivo toxicity assays or bioanalytical equivalent concentrations (BEQ) in the case of in vitro cell-based bioassays. TEQs and BEQs (in the following, for simplicity, "BEQ") allow the direct comparison of effects determined in a complex sample from toxicological assessments (BEQ_{bio}) with effects caused by the chemicals quantified in this sample by chemical (target) analysis (BEQ_{chem}). For considerations about advantages and challenges regarding the derivation of BEQs, we refer to Wagner et al. [13].

The BEQ_{chem} (3) is the sum of the analytically determined concentration of chemicals (C_i) multiplied by each chemical's relative effect potency (REP_i) (4) in relation to a specific potent reference compound [14]:

$$BEQ_{chem} = \sum_{i=1}^{n} REP_i \cdot C_i \tag{3}$$

$$REP_i = \frac{EC_{ref}}{EC_i} \tag{4}$$

The BEQ_{bio} is the sum of the potency-scaled concentrations of unknown chemicals that have the same mode of toxic action as the reference compound and act concentration-additively translated into concentrations of the reference compound. The BEQ_{bio} can be estimated by dividing the EC of the reference compound (EC_{ref}) by the EC of the sample (EC_{sample}, in units of relative enrichment factors, i.e. dimensionless) in the bioassay (5) and has the same unit as the EC_{ref}, e.g., mol/L or ng/L.

$$\mathrm{BEQ_{bio}} = \frac{\mathrm{EC_{ref}}}{\mathrm{EC_{sample}}} \tag{5}$$

However, these equations cannot be translated directly into the mixture effects determined with bioassays.

Using PS devices for applications in water quality testing is only suitable for combination with effect assessment if the PS/medium partition coefficients (in equilibrium mode) or sampling rates (in kinetic mode) are largely independent of K_{ow} (see below). Only then can (1) be converted to (6) and (2) to (7).

$$\mathrm{BEQ_{chem,\,medium}} = \sum_{i=1}^{n} \frac{\mathrm{BEQ}_{i,\,\mathrm{PS}}}{R_{\mathrm{s},i} \cdot t} \xrightarrow{\text{if } R_\mathrm{s} \text{ is independent of } i} = \frac{1}{R_\mathrm{s} \cdot t} \mathrm{BEQ_{chem,PS}} \tag{6}$$

$$\mathrm{BEQ_{chem,medium}} = \sum_{i=1}^{n} \frac{\mathrm{BEQ}_{i,\,\mathrm{PS}}}{K_{\mathrm{PS/medium},i}} \xrightarrow{\text{if } K_{\mathrm{PS/medium}} \text{ is independent of } i} = \frac{1}{K_{\mathrm{PS/medium}}} \mathrm{BEQ_{chem,PS}} \tag{7}$$

If this condition holds, one can also convert the $\mathrm{BEQ_{bio,PS}}$ into a $\mathrm{BEQ_{bio,medium}}$ according to (8):

$$\mathrm{BEQ_{bio,\,medium}} = \frac{1}{K_{\mathrm{PS/medium}}} \mathrm{BEQ_{bio,\,PS}} \tag{8}$$

As mentioned above, most PS studies in water are run in the kinetic mode. Because of the dependence of $K_{\mathrm{PS/w}}$ on the hydrophobicity of the compounds, equilibrium sampling of water is difficult to apply in a quantitative way in bioassays unless dosing occurs via PD directly from the polymer [4].

For a chemical that is considered to be representative for the chemical mixture, R_s can be determined experimentally. In the water phase, chemical-specific R_s are often calibrated in situ by dissipation of performance reference compounds (PRC) which have similar physicochemical properties as the target chemicals and are dosed to the PS device before exposure in the environment. However, environmental mixtures have a very wide range of physicochemical properties so that the $\mathrm{R_s}$ of the PRCs can never be more than an approximation of the mixture sampling rate.

In the kinetic mode, the independence of R_s on the chemicals' hydrophobicity has been achieved by slowing down the water flow to a point where the diffusion across the unstirred water boundary layer is the rate-limiting resistance in the uptake process [10]. The downside of this approach is that it also lowers the overall R_s, which negatively impacts the mass of chemicals that are sampled and hence the sensitivity.

In contrast, for equilibrium sampling in sediment and biota, the condition of K_{ow}-independent $K_{\mathrm{PS/sed}}$ and $K_{\mathrm{PS/biota}}$ holds true. Therefore, a back-calculation of measured BEQs in the extract to BEQs in the original sample is possible provided the partition coefficients are known.

4 Water

Kinetic PS in water allows for sensitive in situ measurement of environmental contaminants at trace concentrations and quantification of time-weighted averaged concentrations (C_{TWA}). In numerous studies, PS has been linked with toxicological assessment of surface waters (Table 1). The result of this procedure represents a time-integrative toxicological signal of the environmental water sample, which is superior to the conventional analysis of grab water samples [52] generating snapshot information. For sampling in the water phase, PS devices such as SPMDs, LDPE, POM, and SR sheets are well developed devices for sampling of HOCs whereas they are more poorly characterized for polar compounds because the processes governing their uptake are at present not well understood [53]. To our knowledge, Sabaliunas et al. [15] presented one of the first studies combining PS in water and an in vivo bioassay with *Daphnia pulex* for toxicity assessment in a laboratory study. The authors showed that PS in SPMDs was biomimetic because the uptake of pesticides by samplers and mussels was similar. However, applying SPMD extracts in bioassays can give false-positive results because oleic acid, the hydrolysis product of the major impurity of triolein, is toxic to aquatic organisms [16].

The majority of studies (Table 1) linking PS with effect assessment of waters have used POCIS as PS devices [54]. POCIS extracts have been widely applied in receptor-mediated bioassays, in particular in the yeast estrogen screen (YES) to assess aquatic contamination with endocrine-disrupting contaminants and the effects of wastewater effluents on river water quality (e.g., [25, 26, 29, 40]). For such contaminants with highly variable aqueous concentrations, time-integrative PS is particularly useful in aquatic environments that can undergo frequent hydrological changes, e.g., small rivers. POCIS have been shown to accumulate more than 300 compounds, including pesticides, pharmaceuticals, hormones, and industrial chemicals. From the two POCIS types that are commercially available, POCIS "pharmaceuticals" and POCIS "pesticides", the latter configuration provided superior uptake for a range of polar compounds including pesticides and hormones [53].[1] Effects were typically expressed as BEQ per PS or per g PS polymer (Table 1).

Vermeirssen et al. [29] showed that accumulation of estrogens in POCIS was biomimetic because measured estradiol equivalents in PS extracts and fish were similar. By comparing PS data with repeated grab water samples, the authors further demonstrated that POCIS accumulate estrogens in a time-integrative manner. Bioanalytical assessment of POCIS extracts showed good agreement with chemical (target) analysis because the estrogenicity in the samples correlated well with chemically derived estradiol equivalents of the same POCIS extracts [29, 30].

[1] POCIS "pharmaceuticals": OASIS HLB as sorbent; POCIS "pesticides": triphasic adsorbent mixture of hydroxylated polystyrene-divinylbenzene resin and a carbonaceous sorbent dispersed on S-X3 biobeads

Table 1 Selection of studies using in situ passive sampling in water in kinetic mode combined with bioanalytical assessment in bioassays

Compartment	PS device	Chemicals targeted	Dosing type	Bioassays used	Effects reported as	Reference
Freshwater, laboratory test	SPMD	Pesticides	Solvent extract	Microtox,[a] Daphtoxkit F™,[b] Rotoxkit F™,[c] sister chromatid exchange (SCE)	LC_{50} values (relative to the reference, potassium bichromate, with known concentration in mg/L)	[15]
River water	SPMD	HOCs (qualitative analysis)	Solvent dialysate	Microtox[a]	Bioassay-directed fractionation-1/EC_{50} (without units, relative)	[16]
River, lake, paper mill effluent	SPMD	Organic chemicals (qualitative analysis)	Solvent extract	YES[d]	Bioassay-directed fractionation-"estrograms" (qualitative)	[17]
Municipal wetlands	POCIS	Surfactants, pharmaceuticals, pesticides, phthalates, fragrance-related terpenes and essential oils	Solvent extract	YES[d]	Adjusted absorbance in relation to extract volume (μL)	[18]
Sewage effluent	POCIS-Pharm	Photosystem II inhibitors and other chemicals	Solvent extract	Algal (*Pseudokirchneriella subcapitata*) growth and photosystem II inhibition (combined algae assay); Microtox[a]	BEQ (mg or μg reference compound per POCIS)	[19]
River water	POCIS-Pharm	Pharmaceuticals, alkylphenols, pesticides, PAHs, hormones, PCBs, OCPs	Solvent extract	HG5LN-hPXR,[e] MELN,[f] MDA-kb2,[g] PLHC-1[h]	Effect concentrations (POCIS/L), reference compound equivalents (μg/POCIS)	[20]
River water	POCIS-Pharm	PAHs, pharmaceuticals, alkylphenols, hormones, pesticides	Solvent extract	MELN,[f] MDA-kb2,[g] PLHC-1[h]	BEQ (ng reference compound per POCIS)	[21]
River water	POCIS-Pharm	Pharmaceuticals, steroids	Solvent extract	MELN,[f] MDA-kb2,[g] PLHC-1,[h] HG5LN-hPR,[e] HG5LN-hMR[i]	BEQ (ng reference compound per g POCIS)	[22]
River water	POCIS-Pharm	Pesticides	Solvent extract	Chlorophyll fluorescence and photosystem II inhibition in phototrophic community, phototrophic community structure	Serial dilution (dx) of the extracts	[23]
River water	POCIS-Pharm	Estrogens	Solvent extract	YES,[d] T47D-KBluc,[j] E-Screen[k]	Reference compound equivalent (ng/POCIS)	[24]
River water	POCIS-Pest	Estrogens	Solvent extract	YES[d]	EEQ[l] (ng/POCIS)	[25]

River water	POCIS-Pest	Estrogens	Solvent extract	YES[d]	EEQ[l] (ng/L), R_s used for calculations not documented	[26]
Wastewater treatment plant (WWTP) influent and effluent	POCIS-Pest	Nonylphenol, bisphenol A, estrone, 17b-estradiol, 17a-estradiol	Solvent extract	YES[d]	DEQ_{biota} (ng/sampler)	[27]
River, sewage treatment plant (STP) effluent	POCIS with STRATA-X as sorbent	Steroidal estrogens	Solvent extract	YES[d]	EEQ[l] (ng/L), R_s used for calculations not documented	[28]
River and lake water	POCIS-Pest, POCIS-Pharm	Steroidal estrogens	Solvent extract	YES[d]	EEQ[l] (ng/POCIS)	[29]
WWTP influent and effluent	POCIS-Pest	Nonylphenol, bisphenol A, estrone, 17β-estradiol	Solvent extract	YES[d]	EEQ[l] (ng/POCIS)	[30]
River water	POCIS-Pest, POCIS-Pharm	Pharmaceuticals, pesticides, perfluorinated organic compounds	Solvent extract	MVLN,[m] H4IIE-luc,[p] yeast cell-based assay for cytotoxicity, human androgen-receptor yeast cell-based assay	EEQ[l] (ng/L) calculated with R_s of E2; ng reference compound/POCIS	[31]
Surface waters	SPMD, POCIS-Pest	Pesticides, PAHs, PCBs	Solvent extract	Microtox,[a] YES[d]	Reference compound equivalent EC_{50} values, EEQ[l] in ng/sampler	[32]
Marine water	SPMD, POCIS-Pharm	PAHs, alkylphenols	Solvent dialysate / solvent extract	EROD[n] activity, vitellogenin (VTG) induction and cytotoxicity[o] in rainbow trout (*Oncorhynchus mykiss*) hepatocytes	% of control along a spatial gradient	[33]
River water	SPMD, POCIS	SPMD: PAHs, PCBs, OCP, HCB, HCHs, DDT metabolites, triclosan, methyltriclosan, PBDEs; POCIS: pesticides, pharmaceuticals, PFCs	Solvent dialysate / solvent extract	H4IIE-luc,[p] MVLN,[m] MDA-kb2,[g] yeast-cell based assay for cytotoxicity	SPMD: pg reference equivalent/L (determination of R_s by in situ calibration with PRCs); POCIS: ng reference equivalent/POCIS	[34]
River water	SPMD, POCIS-Pharm	Not analyzed	Solvent dialysate/ solvent extract	MELN,[f] MDA-kb2,[g] PLHC-1,[h] HG5LN-hPXR[e]	ng reference equivalent/g of sorbent, % of control	[35]

(continued)

Table 1 (continued)

Compartment	PS device	Chemicals targeted	Dosing type	Bioassays used	Effects reported as	Reference
River water, WWTP influent and effluent	SPMD, POCIS	Pesticides, sulfonamides, other antibiotics and pharmaceuticals, PFCs	Solvent extract	H4IIE-luc,[p] MVLN,[m] MDA-kb2,[g] human androgen-receptor yeast-cell based assay in mammalian cell lines, NR[q] uptake for cyto-toxicity, yeast-cell based assay for cytotoxicity	SPMD: ng or pg reference equiv-alent/L (determination of R_s by in situ calibration with PRCs); POCIS: ng reference equivalent/POCIS	[36]
Estuarine water	POCIS, SR sheets	PAHs, PCBs, pesticides	Solvent extract	Algal (*Dunaliella tertiolecta*) photosystem II inhibition (PAM)	% inhibition, semiquantitative	[37]
River water	POCIS-Pest, POCIS-Pharm, LDPE, SI strips	Pharmaceuticals, fungicides, germicides, flame retardants	Solvent extract	YAS[r]	mg reference equivalents/sampler	[38]
WWTP	Empore™ SDB-RPS disks	Endocrine disruptive chemicals	Solvent extract	E-Screen[k]	EEQ (ng/L), calculated as the quotient of the EC_{50} values of 17β-estradiol and the sample	[39]
River water and WWTP effluent tested in channel system	Empore™ SDB-RPS disks	Sulfamethoxazole, carbamaze-pine, clarithromycin	Solvent extract	YES[d]	EEQ[l] (ng/sampler)	[40]
Tidal river water	Empore™ SDB-RPS disks	Herbicides	Solvent extract	Algal (*Phaeodactylum tricornutum*, *Chlorella vulgaris*) photosytem II inhibition	DEQ[s] (ng/L) calculated with diu-ron R_s	[41]
River and estua-rine water	Empore™ SDB-RPS disks	Herbicides	Solvent extract	Algal (*Desmodesmus subspicatus*, *Phaeodactylum tricornutum*) photosytem II inhibition	DEQ[s] (ng/L), calculated with diu-ron R_s, EC $(REF_D)_{10}$[t]	[42]
Marine water	Empore™ SDB-RPS disks	Herbicides	Solvent extract	Coral (*Acropora millepora*) larval settlement, sea urchin (*Heliocidaris tuberculate*) larval development, Microtox,[a] algal (*Phaeodactylum tricornutum*) photosystem II inhibition	EC_{10} (% control), DEQ[s] (ng/L) calculated with an average R_s of 0.8 L/day	[43]

STP effluent	Empore™ RPS-PES disks	Organic chemicals	Solvent extract	Algal (*Phaeodactylum tricornutum*) photosytem II inhibition, Microtox,[a] umuC[w]	DEQ[s](μg/L) (calculated with average R_s for diuron) EC $(REF_D)_{50}$ or LOEC (REF_D)[t]	[44]
Water treatment plant effluent	Empore™ SDB-RPS disks	Pesticides, personal care products, endocrine disruptive compounds	Solvent extract	Microtox,[a] E-Screen,[k] algal (*Pseudoekirchneriella subcapitata*) growth and photosystem II inhibition	Baseline-TEQ[u]$_{biota}$ (mg/L), EEQ[l]$_{biota}$ (ng/L), DEQ[s]$_{biota}$ (μg/L), calculation with a hydropho-bicity-independent average R_s	[10]
STP effluent	Empore™ SDB-XC, SDB-RPS and RPS-PES disks	Pharmaceuticals, biocides	Solvent extract	Algal (*Pseudoekirchneriella subcapitata*) photosystem II inhibition	DEQ[s]$_{biota}$ (ng/sampler)	[27]
Estuarine and coastal water	SR sheets	PAHs, PCBs, pesticides, herbicides	Solvent extract	Algal (*Diacronema lutheri*) growth inhibition	EC_{10} expressed relative to the equivalent mass of PS polymer per mL of extract or assay medium (mg PS-EQ/mL)	[45]
River and estuarine water	SR sheets	PAHs, PCBs, pesticides, herbicides	Solvent extract	NR[q] uptake for cytotoxicity and EROD[n] induction in rainbow trout (*Oncorhynchus mykiss*) liver RTL-W1 cells	Activity expressed relative to the equivalent mass of PS polymer per mL extract (mg PS-EQ/mL), TEQ_{biota}	[46]
River and estuarine water	SR sheets	Determined in parallel study [46]	Solvent extract	Algal (*Diacronema lutheri*) growth inhibition, FET[v]	EC_{50} values expressed relative to the equivalent mass of SR extracted per mL of extract (mg PS-EQ/mL)	[47]
Marine water	PDMS strips	PAHs	Solvent extract	E-Screen,[k] umuC,[w] Microtox,[a] CAFLUX[x]	Reference EQ (ng/L or ng/sampler)	[48]
River water	LDPE	PAHs, screening for 1,201 chemicals	Solvent extract	FET[v]	Developmental toxicity score	[49]
Marine water	SR sheets	PCBs, PAHs, PBDEs, organotins, organonitrogen pesticides, pharmaceuticals, phenols, phthalates, PFCs	PD	Algal (*Phaeodactylum tricornutum*) growth inhibition	Hazard index	[4]

(continued)

Table 1 (continued)

Compartment	PS device	Chemicals targeted	Dosing type	Bioassays used	Effects reported as	Reference
Groundwater	Ceramic toximeter	PAHs	PD	EROD[n] activity in RTL-W1 cells	EROD induction (fold control)	[50]
Wastewater	Ceramic toximeter	Dioxin-like PCDD/Fs	Solvent extract	CALUX[y]	pg BEQ/ toximeter	[51]

[a]Bioluminescence inhibition of *Vibrio fischeri* (Microtox)
[b]Standard toxicity and genotoxicity tests with freshwater invertebrate *Daphnia pulex* (Daphtoxkit F^{TM})
[c]Standard toxicity and genotoxicity tests with freshwater invertebrate *Brachionus calyciflorus* (Rototoxkit F^{TM})
[d]Yeast estrogen screen (YES)
[e]Human pregnane X receptor reporter gene bioassay (HG5LN-hPXR)
[f]Reporter gene assay on the activation of the estrogen receptor alpha (MELN)
[g]Reporter gene assay on the activation of the androgen receptor (MDA-kb2)
[h]7-Ethoxyresorufin-*O*-deethylase (EROD) activity of a fish hepatoma cell line (PLHC-1)
[i]Human mineralocorticoid receptor reporter gene bioassay (HG5LN-hMR)
[j]Assay with estrogen-responsive luciferase reporter for the detection of estrogen receptor agonist and antagonist in human breast cancer cell line (T47D-KBluc)
[k]MCF-7 Estrogenicity Screen (E-Screen)
[l]Ethinylestradiol equivalent (EEQ)
[m]Bioluminescent MCE-7-derived cell line to study the modulation of estrogenic activity (MVLN)
[n]7-Ethoxyresorufin-*O*-deethylase (EROD)
[o]Cytotoxicity measured with alamar blue for metabolic activity and 5-carboxyfluorescein diacetate acetoxymethyl ester for membrane integrity
[p]Activation of the aryl hydrocarbon receptor (AhR) in rat hepatoma cell line (H4IIE-luc)
[q]Neutral red (NR)
[r]In vitro androgen receptor antagonist screen (YAS)
[s]Diuron equivalents (DEQ)
[t]Diuron-based relative enrichment factor (REF_D)
[u]Toxic equivalent concentration (TEQ)
[v](Zebra)fish (*Danio rerio*) embryo toxicity assay (FET)
[w]Genotoxicty assay (umuC)
[x]Chemically Activated FLUorescent gene eXpression assay (CAFLUX)
[y]Chemically Activated LUciferase gene eXpression assay (CALUX)

In further studies, POCIS extracts were tested in a bacterial bioluminescence inhibition assay and in an algal growth and photosystem II inhibition test after exposing the PS devices to treated sewage effluents [19]. The authors showed that photosynthesis inhibition correlated well with chemical equivalent concentration of the POCIS extracts. In contrast, data from tests of nonspecific toxicity did not correlate well whereby the algal inhibition test was strongly affected by specific toxicity of photosystem II inhibitors. Because the flow rate had only minor effects on chemical accumulation in POCIS [19], the authors concluded that chemical concentrations in the samplers are integrative measures of the loads of micropollutants in the effluents. Still, they did not apply any conversion and reported BEQ per PS.

More recently, PS extracts have been investigated simultaneously with a battery of in vitro or in vivo bioassays with various endpoints for a more comprehensive toxicity assessment of aquatic environments (Table 1). For example, POCIS extracts were tested in different in vitro bioassays for non-specific cytotoxicity, endocrine disruptive potential and dioxin-like toxicity after exposing PS devices in rivers (e.g., [21, 31, 36]). Interestingly, PS was even combined with community-level toxicity testing by exposing phototrophic biofilms to POCIS extracts [23].

Another PS device for polar organic chemicals is the Empore™ sulfonated styrenedivinylbenzene copolymer (SDB)-RPS disk. In several studies, "naked" Empore SDB-RPS disks (i.e., disks without any protective membrane) being suitable for time-integrative sampling of a few days were used to investigate inhibition of photosystem II in algae induced by herbicides in environmental samples. Thereby, diuron equivalent concentrations from chemical analysis and phytotoxicity of PS device extracts correlated well [42, 44]. Here, the BEQs were back-calculated from the PS device to the aqueous phase using the measured R_s of diuron.

Another configuration of the sampler, namely SDB-RPS Empore disks with protective polysulfone ethoxylate membranes, was applied in sewage treatment plant effluents for long-term integrative sampling (>1 month) of polar organic chemicals combined with toxicity assessment [41]. In contrast to POCIS, Empore™ SDB-RPS disks could only provide an integrative measure of environmental estrogens and polar pharmaceuticals under controlled conditions, i.e., in a channel with constant flow rate [40].

For toxicity assessment of HOCs in aquatic environments, PS using SR, LDPE, and SPMD has been coupled to different bioassays (e.g., [33, 38, 45, 46, 55]). Emelogu and coworkers [45, 46], for example, showed in several studies that this approach is feasible and can serve as a cost-effective early warning signal on water quality deterioration.

When applying absorption-based samplers such as LDPE and SR sheets, R_s are usually determined by in situ calibration using PRCs. A set of PRCs that have similar physicochemical properties as the target analytes and do not naturally occur in the environment, e.g., deuterated analogues of selected target analytes, are dosed to the sampler before exposure in the aquatic environment. Because dissipation is considered to have isotropic kinetics analogous to uptake, R_s of the target analytes are determined as the inverse of PRC release into the water during exposure [56, 57]. For adsorption-based samplers, R_s are generally studied in the laboratory by exposing samplers to water spiked with known contaminant concentrations under controlled conditions, i.e., temperature and flow rate. Because R_s for adsorption-based samplers are often not yet robust enough, contaminant concentrations are usually expressed in amounts per sampler and not in C_{TWA} [53]. When R_s are calibrated in situ by PRC dissipation, aqueous bioanalytical equivalent concentrations (BEQ_w) can be obtained by dividing bioanalytical equivalent concentrations in the PS device (BEQ_{PS}) by R_S for molecules with medium molecular weight and exposure time (t) of the samplers (according to (6)). This approach has been applied by Jálová et al. [36] and Booij et al. [58] for investigating SPMD extracts in different in vitro bioassays.

For polar PS devices in conjunction with an algal toxicity bioassay, the R_S of the reference herbicide diuron was used as average R_s [42]. Because algae are specifically affected by herbicides which inhibit the photosystem II and the hydrophobicity of herbicides varies over a rather small range, this approach might be justifiable but it must be kept in mind that it is only semiquantitative.

The approach has been extended for toxicity assessment of sewage treatment plant effluents and marine waters by combining PS for polar organic compounds with additional bioassays [42–44]. In particular for cytotoxicity and mortality assays, the assumption of constant R_s might not be justified because a large range of chemicals with broadly varying physicochemical properties contribute to these non-specific effects.

One way to overcome the problem of the K_{ow}-dependence of R_s is to apply PD by directly transferring the PS polymer into the bioassay (Fig. 3a). Here, a reciprocal change in mixture composition occurs that is opposite to the profile change during PS, leading to the organisms or cells "experiencing" exposure to the same mixture composition in the aqueous phase of the medium as in the environmental water phase. However, at the same time, a large fraction of the chemicals remains in the PS polymer based on its considerably elevated capacity for HOCs compared to the water, leading to reduced test sensitivity.

There are only a few studies that link PS in water and bioassays by applying PD which controls exposure concentrations by partitioning of the test substances from a dominating reservoir into the test medium; for a general discussion of PD see also the [5]. Following exposure in the marine environment, Claessens et al. [4] transferred the exposed SR sheets as partitioning donor for HOCs directly into the test medium to measure algal growth inhibition. The authors pointed out that this approach only works properly for chemicals within a certain hydrophobicity range because R_s of HOCs are slow during PS and depletion of more polar substances can occur during PD.

For chemical and toxicological monitoring of polycyclic aromatic hydrocarbons (PAHs) in groundwater, Bopp et al. [50] have developed a ceramic membrane-based, solid-sorbent sampler, the Ceramic Dosimeter [59, 60]. Thereby, ceramic tubes with a length of a few centimeters served both as a container for the solid sorbent material and as the dominant diffusion barrier. As sorbent material, polystyrene beads were used which had been found to be compatible with an in vitro fish cell line bioassay [59]. Ceramic toximeters yielded C_{TWA} of HOCs and induced cytochrome CYP1A activity in a rainbow trout liver cell line after long-term (1 year) exposure in groundwater [50].

5 Sediment and Sediment Pore Water

The traditional dose metric used for toxicity assessment of sediments is the total concentration of chemicals determined by exhaustive solvent extraction, even though it is not equivalent to the amount available for (bio)uptake and diffusion. PS devices deployed in situ in sediment do not give a measure of the total sediment concentration but of C_{free} in the sediment interstitial pore water [61] available for biouptake associated with environmental receptors. C_{free} is responsible for exposure, bioaccumulation, and effects and, thus, represents the potential risk to organisms [62]. Most PS applications have focused on assessing the concentrations and effects in pore water, with either laboratory-spiked sediments or field-contaminated sediments [63].

PS can be applied in a non-depletive or depletive mode. In non-depletive mode, samplers with a limited capacity are used to determine C_{free} without removing more than 5% of this fraction (i.e., negligible depletion). In depletive mode, high-capacity samplers which reduce the free fraction substantially lead to C_{free} being replenished, ultimately collecting the entire bioaccessible fraction that can become available. Such depletive sampling is carried out using Tenax or large-volume polymers [57]. According to the sampling site and environmental matrix, various formats of PS devices have been developed (Table 2); studies based on whole sample toxicity testing that correlated the observed data with C_{free} from PS methods are also discussed (Table 3), even if they do not fall strictly within the topic of this chapter.

Lydy et al. [63] compiled a literature survey of the available PS literature, including a wide range of papers on PS methods in contaminated sediment assessments. This survey summarized the methods in current use, materials, and applications for the assessment of sediment-associated HOCs. Of these investigations, more than 80% were based on allowing the PS polymer to come to equilibrium partitioning with the sediment interstitial pore water. In addition, kinetic sampling approaches have been used. These employed an apparent first-order model of chemical uptake into the sampler and used PRCs to estimate the equilibrium condition of the PS relative to the pore water. As PS devices, sheets of PE, POM, or SR were used or polymer-coated glass fibers (solid-phase microextraction, SPME) for which the dominant polymer phase was the silicone PDMS. The PS methods comprised, for example, low-density PE sheets in activated carbon-amended sediment [85] and thin POM sheets applied in Baltic Sea sediments [86].

Table 2 Passive sampling of sediment combined with in vivo toxicity assessment or in vitro bioanalytical tools

Compartment	PS device	PS mode	Chemicals	Dosing type	Test species or bioassays used	Endpoint	Reference
Spiked field sediments	100-μm PE sheets (RePES)[a]	Equilibrium	PAHs, PCBs	PD	*Ampelisca abdita*, *Americamysis bahia*	Survival after 7 days of exposure	[64]
River sediments	Silicone elastomer stir bars (silicone rods, $d = 5$ mm)	Equilibrium	HOCs	PD	*Scenedesmus vacuolatus*	Growth inhibition	[65]
Field-contaminated sediments, freshwater	100-μm PE sheets (RePES)	Equilibrium	PAHs	PD	*A. abdita*, *A. bahia*	Survival after 7 days of exposure	[66]
Natural and spiked sediment, river water	50-μm PE sheets	Equilibrium	2,4-Dinitrophenol, diuron, fluoranthene, nonylphenol, parathion, pentachlorophenol	PD of sediment extracts	*Danio rerio*	Embryo toxicity (48 h LC_{50} values)	[67]
Surface field sediment	SPME (10-μm PDMS fibers)	Equilibrium	OCPs, PAHs, PCBs	Dosing of SPME extracts	*Neanthes arenaceodentata*, *Musculista senhousia*, *Eohaustorius estuarius*, *Neanthes arenaceodentata*	Sublethal and lethal effects	[68]

River sediment	SPME (PDMS disk; 16 mm); Tenax porous polymer; 60 mesh to 80 mesh	Equilibrium	HOCs	Dosing extracts after exhaustive solvent extraction and after passive sampling with PDMS	Bioluminescence inhibition of *Vibrio fischeri*, combined algae test, AhR-CAFLUX, E-CALUX, Antioxidant Response Element (AREc32)[b]	Cytotoxicity, phytotoxicity that inhibits photosystem II Ah-receptor induction, estrogenic receptor induction, acting on Nrf2 pathway	[69]
Marine sediment	SPME (10-μm PDMS)	Equilibrium	Six PAHs and five PCBs	PD of recreated C_{free} mixtures	*Phaeodactylum tricornutum*	Algae growth inhibition	[70]
River sediment	SPME (10 μm PDMS fiber)	Equilibrium	HOCs	PD of sediment extract	*Danio rerio*	Embryo toxicity	[71]

[a]RePES: reverse polyethylene samplers
[b]AREc32: induction of the Nrf2-mediated oxidative stress response pathway

Table 3 Passive sampling of sediment with the measured C_{free} correlated with the results of bioanalytical assessment performed in whole sediment

Compartment	PS device	PS mode	Chemicals	Extraction method for C_{free}	Test species or bioassays used	Endpoint	Reference
Freshwater sediment; treatment with activated carbon	76-μm POM sheets	Equilibrium	PAHs	POM extractable concentration	*Gammarus pulex*, *Asellus aquaticus*	Effects of activated carbon addition on behavior (locomotion, ventilation, sediment avoidance) and mortality (LC_{50}, 8 days), and growth	[72]
Spiked freshwater sediment	SPME (30-μm PDMS fibers)	Kinetic	Bifenthrin, cyfluthrin, fenpropathrin	SPME	*Chironomus dilutus*	10-day median lethal concentrations (LC_{50})	[73]
Freshwater sediment	SPME (10-μm PDMS)	Equilibrium	Bifenthrin	SPME; Tenax extractable concentration	*Hyalella azteca*, *C. dilutus*	10-day mortality and immobilization, LC_{50}, EC_{50}	[74]
Freshwater sediment	SPME (10-μm PDMS)	Equilibrium	Pyrethroids	Ex situ SPME and 24-h extractable concentration	*H. azteca*, *C. dilutus*	10-day mortality	[75]
Field sediment, shallow estuarine wetland	SPME (10-μm PDMS)	Equilibrium	VOCs, PAHs, pesticides	SPME	*L. plumulosus*, *Mercenaria mercenaria*	Survival	[76]
Freshwater sediment	SPME (10-μm PDMS)	Equilibrium	Permethrin, DDT	SPME	*H. azteca*, *C. dilutus*	Toxic unit-based fiber LC_{50}	[77]
Field sediment, manufactured-gas plants, aluminum smelter sites	SPME (7 μm PDMS)	Equilibrium	34 parent and alkylated PAHs	SPME	*H. azteca*	28 day survival based on equilibrium partitioning and hydrocarbon narcosis model	[78]

Field sediment	SPME (85 µm polyacrylate, PA)	Equilibrium	2,4,6-Trinitrotoluene (TNT), 2-amino-4,6-dinitrotoluene (2ADNT), 4-amino-2,6-dinitrotoluene (4ADNT)	SPME	*Tubifex tubifex*	Metabolism of TNT	[79]
Sediment	SPME (85-µm PA)	Equilibrium	TNT, 2ADNT, 4ADNT	SPME, in situ organism SPME	*T. tubifex*	Lethality	[80]
River sediment	SPME (PDMS 10-µm)	Equilibrium	34 parent and alkylated PAHs	SPME	*H. azteca*	Survival and growth in 28-days bioassay	[81]
Drontermeer sediment (Dutch reference sediment)	SPME (28.5-µm PDMS fiber)	Equilibrium	34 PAHs	SPME	*Lumbriculus variegatus*	28-days reproduction, EC_{10}, EC_{50}	[82]
Sediment	SPME fiber, Tenax	Equilibrium	16 PAHs	SPME and 24 h-Tenax extraction	*H. azteca*	28-days survival	[83]
Freshwater and marine sediment	SPME (10 µm PDMS fiber)	Equilibrium	34 parent and alkylated PAHs	SPME	*H. azteca*, *L. plumulosus*	28-days survival	[84]

Further developments included SR-coated SPME fibers or SR-coated glass jars which have been used for equilibrium sampling of sediments in the laboratory (e.g., [87–91]). As main contaminant classes in sediments, PAHs and polychlorinated biphenyls (PCBs), polychlorinated dibenzo-*p*-dioxins and -furans (PCDD/Fs), and chlorinated pesticides such as dichlorodiphenyltrichloroethane (DDT) and metabolites have recently been investigated [63].

The majority of the investigations (more than 60% [63]) were conducted ex situ, whereby sediments were brought to the laboratory and tested; nevertheless, some papers have reported in situ experiments (20%, e.g., [92, 93]). Regarding the question as to whether in situ or ex situ experiments lead to more reliable results, there are a number of reasons for conducting equilibrium sampling in situ: C_{free} are often at steady state under field conditions. Processes that occur naturally in the environment, such as groundwater infiltration, are usually not fully reproducible in the laboratory, which can result in deviations between laboratory measurements and actual field levels. Furthermore, laboratory experiments are mostly unable to simulate the natural physicochemical conditions (e.g., temperature and salinity) in field sediments [94, 95]. Especially for larger rigid apolar compounds such as PAHs and PCBs, the effect of temperature on solubility is substantial [96, 97]. Because the presence of inorganic dissolved ions decreases the solubility of nonpolar or weakly polar compounds [97], R_s depend on the salinity and the ionic strength of the sample. For laboratory measurements, biocides must often be added to inhibit biodegradation. These factors can affect C_{free}, thus leading to over- or underestimation of the effective concentrations in ex situ experiments. Such problems are avoided when measuring in situ because the natural environmental conditions are preserved [93]. However, for practical reasons, ex situ sampling is more commonly applied and has generated valuable data, in particular for sediment monitoring [8, 87, 88, 90].

Fewer studies have linked PS measurements in sediments to toxicological or specific biological endpoints (Tables 2 and 3) than for water. Most of these studies targeted pore water as a proxy for the C_{free} available for uptake and diffusion in in vivo bioassays. Maruya et al. [98] compared measurements of C_{free} in pore water with water quality criteria. This approach was applied for the toxicity evaluation of sediment-associated PAHs, and the toxic response of *Hyalella azteca* was compared with the number of toxic units calculated from the SPME-derived C_{free}, using the US Environmental Protection Agency sum toxic unit model [81]. Similarly, the toxicity of pyrethroids was observed to be independent of sediment characteristics when the toxic response was based on C_{free} as determined by SPME [73, 75]. Moreover, a direct application of the concentration in the PS device was developed by determining the relationship between the sampler concentration and the response endpoint, e.g., mortality for individual species [67, 74, 80, 99].

Harwood et al. [74], for instance, showed that there was a significant relationship between the concentrations of Tenax-extractable pyrethroid insecticides and the mortality of *Hyalella azteca* and *Chironomus dilutus*. Ding et al. [100] discussed using SPME fibers to estimate the toxicity of hydrophobic pesticides such as DDT on *Hyalella azteca* and *Chironomus dilutus* by showing that SPME fibers accurately reflected the external dose. Furthermore, the authors investigated the concentrations in the SPME fiber related to the organisms' body residues with a focus on the influence of biotransformation [101] to establish the link between concentrations in the PS polymer and the dose absorbed by the organisms. In a follow-up study, the authors demonstrated the applicability of matrix-SPME in spiked and aged sediment to mimic bioaccumulation and to estimate body residues in the organisms [77].

For in vivo bioassays, PS in sediments should be combined with PD to ensure that the exposure conditions from the environment are re-established for the toxicity assessment (Fig. 3). One approach to achieve this is to use field-deployed PS devices as a PD device in bioassays [4]. For example, Witt et al. [93] used SR hollow fibers that had been deployed in an in situ equilibrium sampler as PD device in the algae growth inhibition test with the marine alga *Phaeodactylum tricornutum*. Another approach re-established the mixture composition of target HOCs in sediment pore water which had been sampled by PS methods by applying PD in toxicity tests. In a pioneering study, Lörks [71] applied PD to re-establish C_{free} of PAHs of River Elbe sediments in the fish embryo toxicity test with *Danio rerio*. The levels and composition of real mixtures of PAHs were realistically reproduced in the toxicity test as confirmed by chemical analysis. Constant exposure concentrations as maintained by PD led to higher reproducibility as well as a relative increase in toxicity compared to standard dosing procedures, which can be explained by C_{free} being replenished from the PD device.

Even fewer studies have applied bioassays to sense the actual effect equivalents in the sediment itself. Here, the difference to total extraction is again only that C_{free} are accessible to the PS device because the transfer of compounds from the organic carbon (oc) of the sediment to the sampler is mediated via the pore water and not via direct transfer. However, in this case, it is possible to back-calculate from BEQ_{PS} to BEQ_{oc} because the partition coefficient $K_{\text{PS/oc}}$ is independent of hydrophobicity. If equilibrium is attained between all phases (sediment, water, PS), then by the laws of thermodynamics the partition coefficient between PS and sediment oc equals the ratio of the partition coefficients between both phases and water (9) as demonstrated in Fig. 4.

$$K_{\text{PS/oc}} = \frac{K_{\text{PS/w}}}{K_{\text{oc/w}}} \tag{9}$$

Li et al. [69] combined passive equilibrium sampling with solvent spiking of extracts in five in vitro bioassays (Table 2). They applied (9) to relate the measured BEQ_{PS} to BEQ_{oc} and demonstrated that BEQs from exhaustive extraction agreed quite well with values estimated from PDMS-based PS extracts via the constant oc/PDMS partition coefficient. In their case study, the authors could demonstrate the feasibility of linking PS of freely dissolved chemicals in sediments with different toxic endpoints.

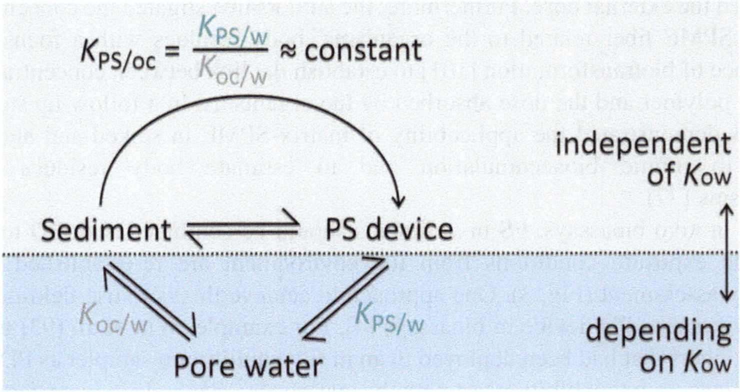

Fig. 4 Although both $K_{oc/w}$ and $K_{PS/w}$ are K_{ow}-dependent, $K_{PS/oc}$ is independent of hydrophobicity

The measured BEQ_{PS} can be used to derive BEQ_{oc} using (10). Provided that the partitioning to oc is the main process determining bioavailability, BEQ_{oc} can also be related to BEQ_{sed} by means of (11).

$$BEQ_{oc} = \frac{BEQ_{PS}}{K_{PS/oc}} \quad (10)$$

$$BEQ_{sed} = f_{oc} BEQ_{oc} \quad (11)$$

6 Biota

Biota extraction can be combined with (eco)toxicological profiling by exhaustive solvent extraction of tissue homogenates followed by pipetting of the extract into bioassays [2]. However, amongst other issues, exhaustive extraction techniques involve considerable co-extraction of matrix constituents which can hamper chemical analysis or have an impact on the implementation of bioanalytical methods, e.g., by influencing the kinetics [102]. Hence, laborious and cost-intensive cleanup procedures become necessary which can additionally change the mixture composition in the extracts because some of the chemicals may, for example, sorb to the media used for removing matrix constituents or may not be stable in, e.g., acid treatment. Two important advantages of PS approaches in this context are that they (1) achieve largely unbiased sampling of a broad range of chemicals into the PS polymer and (2) leave most of the matrix behind. Hence, compared to exhaustive extraction, PS may represent a superior approach for extraction in biota in combination with chemical analysis and mixture toxicity assessment. Furthermore, PS in biological fluids and tissues provides a way of accessing the internal exposure of organisms and is closely linked with toxicity.

Studies on bioanalytical assessment of mixtures of pollutants in higher organisms [103] are much scarcer than applications in water (Sect. 4) and sediment (Sect. 5). Therefore, in this section, we discuss applications of PS in biota that are limited to chemical analysis but have the potential of being linked with bioanalytical assessment before we review the small number of studies combining PS with bioanalysis (Table 4).

SPME, with mostly in vitro but also a few in vivo applications, was one of the first techniques to allow the direct extraction and measurement of analytes from complex matrices such as blood, urine, and lipid-rich tissues, e.g., Ossiander et al. [105]. However, because of the small volumes of the polymer coatings and related issues with sensitivity, a combination with bioanalytical profiling is generally not possible. To overcome the inherent low sensitivity of SPME approaches, polymer-based PS methods with larger polymer volumes which are not necessarily non-depletive were explored. For example, Jahnke et al. [106] used PDMS thin-films as the sampling material for lipid-rich tissues, whereas sampling kinetics were too slow in lean fish.

Mäenpää et al. [107] developed a method for distinguishing the equilibrium partitioning of halogenated organic compounds in fish storage lipids, membrane lipids, and proteins. In addition, the authors showed that it is of relevance for the partitioning in tissues which of these constituents represents the most relevant sorptive phase on a fresh weight basis.

As for in vivo sampling, Zhou et al. [108] inserted SPME fibers in the dorsal muscle of rainbow trout after exposure to water contaminated with different pharmaceuticals. Furthermore, Adolfsson-Erici et al. [109] used acupuncture needles covered with PDMS tubing to study the elimination kinetics of hydrophobic chemicals in rainbow trout, but even here the mass of chemicals enriched in the polymer was too low to allow a toxicological characterization using bioassays. Later, an in vivo PS approach was developed that included implanting silicone tubing in brown trout [110], thereby providing a substantially enhanced mass of polymer and correspondingly the sampled mixture of chemicals. Mayer et al. [111] used PDMS microtubes for sampling of PAHs in lipid-rich mussel samples.

Pioneering work in the coupling of PS with mixture toxicity assessment in bioassays has been carried out by Jin et al. [102]. The authors used PDMS samplers to enrich PCDDs from lipid-rich dugong blubber after spiking the tissues with PCDDs (Table 4). Afterwards, the obtained PS extracts were dosed into the CAFLUX in vitro bioassay, which specifically responds to dioxin-like chemicals. Because the partition coefficients for dioxins between PDMS and blubber were independent of hydrophobicity, consistent with the considerations in Fig. 2, it was possible to back-calculate from the measured BEQ_{PS} in the PS extracts to the $BEQ_{blubber}$ using (8). The authors found that the small quantities of co-extracted lipids (up to 0.3% of the PDMS weight) did not affect the response. However, the cells dosed with the reference mixtures (i.e., without lipids) needed three times longer than those dosed with the sample extracts to reach the same induction level, with comparable effect levels after 72 instead of 24 h of exposure, potentially because of facilitated chemical uptake by cells mediated through the lipid matrix

Table 4 Passive sampling of biota combined with bioanalytical assessment

Compartment	PS device	PS mode (eq./kin.)	Chemicals	Dosing type	Bioassays used	Reference
Dugong blubber	PDMS thin-films (16 mm diameter, 1 mm thickness)	Equilibrium	PCDDs	Solvent spike	CAFLUX[25]	[102]
Dugong blubber	PDMS thin-films (16 mm diameter, 1 mm thickness)	Equilibrium	PCDDs, PCBs, PBDEs, PAHs, OCPs, pyrethroid pesticides	Solvent spike	CAFLUX, AREc32, transcription factor p53-mediated apoptosis (P53-bla)[a], nuclear factor-kappaB (NF-κB-bla)[b]	[103]
Turtle blood	PDMS thin-films (16 mm diameter, 1 mm thickness)	Equilibrium	PCDDs, PCBs, PBDEs	Solvent spike	CAFLUX, AREc32	[104]

[a]p53 activation indicative of tumor suppressor gene
[b]NF-κB activation indicative of response to inflammation

[102]. Hence, it is important to address matrix effects in every study. The method was subsequently applied to unspiked dugong blubber samples, and the comparison of bioassay response with quantified PCDDs demonstrated that all mixture effects in the dugong blubber samples could be explained by PCDDs [102].

Furthermore, Jin et al. [103] used the PDMS-based PS method for dugong blubber to obtain extracts that were subjected to a range of in vitro bioassays indicative for aryl hydrocarbon receptor (AhR)-mediated activity and key stress response pathways such as oxidative stress, DNA damage, and inflammation. Effects were detected for all bioassays but the activation of the AhR and the Nrf2-mediated oxidative stress response accounted for the most explicit effects in the tests, whereas assays for DNA damage response and inflammation gave no significant effect (less than 5% of the total stress response for PCDDs).

The same research group developed a PDMS/blood partitioning system and determined partition coefficients for model chemicals including PCDDs, PCBs, and polybrominated diphenylethers (PBDEs [104]). PDMS extracts of marine turtle blood samples with known concentrations of persistent organic pollutants were tested in CAFLUX and Antioxidant Response Element (AREc32) bioassays. The BEQ_{biota} were calculated from the bioassay responses (5) and compared with BEQ_{chem} calculated from results of chemical analyses and with single chemicals' relative effect potencies (3). The concentrations of the dioxin-like chemicals in the turtle blood could explain all effects observed in the CAFLUX assay for the activation of the arylhydrocarbon receptor. In contrast, the activity in the oxidative stress response assay AREc32 proved to be higher than could be explained by the detected chemicals indicating that unknown bioactive chemicals were present in the blood samples. The work shows the high potential of linking *in tissue* passive sampling with toxicological characterization, because Jin et al. [104] succeeded in a near-exhaustive extraction (62–84%), a yield comparable with conventional solvent extractions. By enhancing the sampling efficiency, the probability for receiving complete dose–response curves is elevated. Low mass transfer of chemicals into PDMS is always a limiting factor for combining chemical analysis and toxicological characterization, and may be overcome by a higher surface area/volume ratio of the sampler to enhance uptake rates, by more PS material or new polymer devices with higher affinities for the contaminants. However, sampling phases with higher capacities usually go along with prolonged sampling times which can be problematic with regards to sample stability.

7 Opportunities and Challenges for Combining Passive Sampling and Bioassays

Based on the current research summarized above, the following conclusions can be drawn: PS provides information on the concentrations of environmental pollutants in diverse media that are available for partitioning and (bio)uptake. Solvent extracts of PS devices can easily be dosed into in vitro and in vivo bioassays but the interpretation of results is not straightforward and depends on the scientific questions to be tackled.

- For the water phase, numerous methods have been described (see Sect. 4 and Table 1), but equilibration of the PS devices is usually not achieved because of the low levels of chemicals and the unstirred boundary layer hampering uptake of hydrophobic chemicals into the sampler. An approach to overcome this issue may be "active passive sampling" as suggested by Vrana et al. [112], whereby water is pumped through a box containing the PS devices to enhance the sampling rates. Using this approach may be a way to overcome the limitations posed by too low chemical amounts being sampled to meet the detection limits of bioassays. This approach does not, however, solve the hydrophobicity-dependent R_s but even aggravates the problem because R_s are increased. Only for very low R_s does the sampling of typical water-borne pollutants become hydrophobicity-independent [10], with the disadvantage that low R_s result in low quantities of sampled chemicals and, hence, low sensitivity.
- For sediment (Sect. 5 and Table 2), a number of suitable PS methods exist for sensitive in situ or ex situ sampling. In undisturbed sediment in the field, equilibration may take between weeks and months, whereas agitation in the laboratory allows shorter sampling times (a few weeks). PS in sediments can target two aspects: (1) the combination of PS with PD into in vitro bioassays allows the assessment of the freely dissolved and toxicologically active fraction of the chemicals and (2) the combination of PS with spiking the extracts into in vitro bioassays allows a quantitative calculation of BEQ in the sediment phase because of the hydrophobicity-independent PS/sediment (oc) partition coefficient. A challenge could be to sample sufficient amounts of chemicals for the assessment of a number of toxicological endpoints or bioassays, and hence a larger polymer mass may be required as described by Li et al. [69].
- Biota is the least widely studied medium in PS research so far (Sect. 6 and Table 4). This fact may be explained in part by methodological challenges, i.e., slow sampling kinetics in lean tissues combined with limited sample stability. Thin polymers as, for example, in SPME fibers allow fast equilibration but lead to issues regarding sensitivity for chemical analysis, above all in combination with toxicological assessment. New approaches such as using thinner polymers with larger surface to volume ratios as discussed above have the potential to enhance substantially the applicability of this new approach.

Huge potential lies in the application of these improved PS methods in combination with (eco)toxicological profiling to achieve a substantial refinement of environmental hazard and risk assessment and management of water bodies, sediments, and biota. General challenges in the combination of PS with bioanalytical assessment lie in the sensitivity, i.e., the extraction of sufficient amounts of chemicals, in transferring or re-establishing the complex mixtures of chemicals from the environmental medium of interest to the bioassay, e.g., by combining PS and PD for aqueous media, and in the co-extraction of matrix constituents which can hamper both chemical analysis and the characterization of effects. Cleanup procedures are not only laborious and cost-intensive; they also easily involve the risk of changing the chemical mixture composition and hence should largely be avoided.

Finally, interesting new approaches include (1) simultaneous sampling of polar and non-polar chemicals that usually require different PS devices (e.g., [38]), (2) if sufficient amounts of chemicals can be sampled, effect-directed analysis of PS extracts could be performed to identify biologically active compounds in complex environmental mixtures (e.g., [22, 38]), and (3) toxicity testing of PS extracts in the laboratory could be combined with in situ testing of biomarkers or similar indicators of animal health for integrated assessment of aquatic ecosystems [34].

References

1. Escher B, Leusch F (2012) Bioanalytical tools in water quality assessment. IWA, London
2. Jahnke A, Mayer P, Schäfer S, Witt G, Haase N, Escher BI (2016) Strategies for transferring mixtures of organic contaminants from aquatic environments into bioassays. Environ Sci Technol. doi:10.1021/acs.est.5b04687
3. Mayer P, Tolls J, Hermens JLM, Mackay D (2003) Equilibrium sampling devices: an emerging strategy for monitoring exposure to hydrophobic organic chemicals. Environ Sci Technol 37:184A–191A
4. Claessens M, Monteyne E, Wille K, Vanhaecke L, Roose P, Janssen CR (2015) Passive sampling reversed: coupling passive field sampling with passive lab dosing to assess the ecotoxicity of mixtures present in the marine environment. Mar Pollut Bull 93:9–19
5. Jahnke A, Witt G, Schäfer S, Haase N, Escher BI (2015) Combining passive sampling with toxicological characterization of complex mixtures of pollutants from the aquatic environment. Adv Biochem Eng Biotechnol. doi:10.1007/10_2015_5014
6. Mills GA, Fones GR, Booij K, Greenwood R (2011) Passive sampling technologies. In: Quevauviller P, Roose P, Varreet G (eds) Chemical marine monitoring: policy framework and analytical trends. Wiley, Hoboken
7. ter Laak TL, ter Bekke MA, Hermens JLM (2009) Dissolved organic matter enhances transport of PAHs to aquatic organisms. Environ Sci Technol 43:7212–7217
8. Jahnke A, Mayer P, McLachlan MS, Wickstrom H, Gilbert D, MacLeod M (2014) Silicone passive equilibrium samplers as "chemometers" in eels and sediments of a Swedish lake. Environ Sci Process Impacts 16:464–472
9. Leslie HA, Oosthoek AJP, Busser FJM, Kraak MHS, Hermens JLM (2002) Biomimetic solid-phase microextraction to predict body residues and toxicity of chemicals that act by narcosis. Environ Toxicol Chem 21:229–234
10. Escher BI, Lawrence M, Macova M, Mueller JF, Poussade Y, Robillot C, Roux A, Gernjak W (2011) Evaluation of contaminant removal of reverse osmosis and advanced oxidation in full-scale operation by combining passive sampling with chemical analysis and bioanalytical tools. Environ Sci Technol 45:5387–5394
11. Armitage JM, Wania F, Arnot JA (2014) Application of mass balance models and the chemical activity concept to facilitate the use of in vitro toxicity data for risk assessment. Environ Sci Technol 48:9770–9779
12. Heringa MB, Schreurs R, Busser F, Van Der Saag PT, Van Der Burg B, Hermens JLM (2004) Toward more useful in vitro toxicity data with measured free concentrations. Environ Sci Technol 38:6263–6270

13. Wagner M, Vermeirssen ELM, Buchinger S, Behr M, Magdeburg A, Oehlmann J (2013) Deriving bio-equivalents from in vitro bioassays: assessment of existing uncertainties and strategies to improve accuracy and reporting. Environ Toxicol Chem 32:1906–1917
14. Villeneuve DL, Blankenship AL, Giesy JP (2000) Derivation and application of relative potency estimates based on in vitro bioassay results. Environ Toxicol Chem 19:2835–2843
15. Sabaliunas D, Lazutka J, Sabaliuniene I, Sodergren A (1998) Use of semipermeable membrane devices for studying effects of organic pollutants: comparison of pesticide uptake by semipermeable membrane devices and mussels. Environ Toxicol Chem 17:1815–1824
16. Sabaliunas D, Ellington J, Sabaliuniene I (1999) Screening bioavailable hydrophobic toxicants in surface waters with semipermeable membrane devices: role of inherent oleic acid in toxicity evaluations. Ecotoxicol Environ Saf 44:160–167
17. Rastall AC, Getting D, Goddard J, Roberts DR, Erdinger L (2006) A biomimetic approach to the detection and identification of estrogen receptor agonists in surface waters using semipermeable membrane devices (SPMDs) and bioassay-directed chemical analysis. Environ Sci Pollut Res Int 13:256–267
18. Petty JD, Huckins JN, Alvarez DA, Brumbaugh WG, Cranor WL, Gale RW, Rastall AC, Jones-Lepp TL, Leiker TJ, Rostad CE, Furlong ET (2004) A holistic passive integrative sampling approach for assessing the presence and potential impacts of waterborne environmental contaminants. Chemosphere 54:695–705
19. Vermeirssen ELM, Hollender J, Bramaz N, van der Voet J, Escher BI (2010) Linking toxicity in algal and bacterial assays with chemical analysis in passive samplers deployed in 21 treated sewage effluents. Environ Toxicol Chem 29:2575–2582
20. Creusot N, Kinani S, Balaguer P, Tapie N, LeMenach K, Maillot-Marechal E, Porcher JM, Budzinski H, Ait-Aissa S (2010) Evaluation of an hPXR reporter gene assay for the detection of aquatic emerging pollutants: screening of chemicals and application to water samples. Anal Bioanal Chem 396:569–583
21. Tapie N, Devier MH, Soulier C, Creusot N, Le Menach K, Ait-Aissa S, Vrana B, Budzinski H (2011) Passive samplers for chemical substance monitoring and associated toxicity assessment in water. Water Sci Technol 63:2418–2426
22. Creusot N, Ait-Aissa S, Tapie N, Pardon P, Brion F, Sanchez W, Thybaud E, Porcher J-M, Budzinski H (2014) Identification of synthetic steroids in river water downstream from pharmaceutical manufacture discharges based on a bioanalytical approach and passive sampling. Environ Sci Technol 48:3649–3657
23. Pesce S, Morin S, Lissalde S, Montuelle B, Mazzella N (2011) Combining polar organic chemical integrative samplers (POCIS) with toxicity testing to evaluate pesticide mixture effects on natural phototrophic biofilms. Environ Pollut 159:735–741
24. Alvarez DA, Shappell NW, Billey LO, Bermudez DS, Wilson VS, Kolpin DW, Perkins SD, Evans N, Foreman WT, Gray JL, Shipitalo MJ, Meyer MT (2013) Bioassay of estrogenicity and chemical analyses of estrogens in streams across the United States associated with livestock operations. Water Res 47:3347–3363
25. Vermeirssen ELM, Suter MJF, Burkhardt-Holm P (2006) Estrogenicity patterns in the Swiss midland river Lutzelmurg in relation to treated domestic sewage effluent discharges and hydrology. Environ Toxicol Chem 25:2413–2422
26. Burki R, Vermeirssen ELM, Korner O, Joris C, Burkhardt-Holm P, Segner H (2006) Assessment of estrogenic exposure in brown trout (Salmo trutta) in a Swiss midland river: integrated analysis of passive samplers, wild and caged fish, and vitellogenin mRNA and protein. Environ Toxicol Chem 25:2077–2086
27. Vermeirssen ELM, Bramaz N, Hollender J, Singer H, Escher BI (2009) Passive sampling combined with ecotoxicological and chemical analysis of pharmaceuticals and biocides - evaluation of three Chemcatcher (TM) configurations. Water Res 43:903–914
28. Balaam JL, Grover D, Johnson AC, Jurgens M, Readman J, Smith AJ, White S, Williams R, Zhou JL (2010) The use of modelling to predict levels of estrogens in a river catchment: How

does modelled data compare with chemical analysis and in vitro yeast assay results? Sci Total Environ 408:4826–4832
29. Vermeirssen ELM, Körner O, Schönenberger R, Suter MJF, Burkhardt-Holm P (2005) Characterization of environmental estrogens in river water using a three pronged approach: active and passive water sampling and the analysis of accumulated estrogens in the bile of caged fish. Environ Sci Technol 39:8191–8198
30. Liscio C, Magi E, Di Carro M, Suter MJF, Vermeirssen ELM (2009) Combining passive samplers and biomonitors to evaluate endocrine disrupting compounds in a wastewater treatment plant by LC/MS/MS and bioassay analyses. Environ Pollut 157:2716–2721
31. Jarosova B, Blaha L, Vrana B, Randak T, Grabic R, Giesy JP, Hilscherova K (2012) Changes in concentrations of hydrophilic organic contaminants and of endocrine-disrupting potential downstream of small communities located adjacent to headwaters. Environ Int 45:22–31
32. Alvarez DA, Cranor WL, Perkins SD, Clark RC, Smith SB (2008) Chemical and toxicologic assessment of organic contaminants in surface water using passive samplers. J Environ Qual 37:1024–1033
33. Harman C, Farmen E, Tollefsen KE (2010) Monitoring North Sea oil production discharges using passive sampling devices coupled with in vitro bioassay techniques. J Environ Monit 12:1699–1708
34. Zounkova R, Jalova V, Janisova M, Ocelka T, Jurcikova J, Halirova J, Giesy JP, Hilscherova K (2014) In situ effects of urban river pollution on the mudsnail Potamopyrgus antipodarum as part of an integrated assessment. Aquat Toxicol 150:83–92
35. Creusot N, Tapie N, Piccini B, Balaguer P, Porcher JM, Budzinski H, Ait-Aissa S (2013) Distribution of steroid- and dioxin-like activities between sediments, POCIS and SPMD in a French river subject to mixed pressures. Environ Sci Pollut Res Int 20:2784–2794
36. Jálová V, Jarosova B, Blaha L, Giesy JP, Ocelka T, Grabic R, Jurcikova J, Vrana B, Hilscherova K (2013) Estrogen-, androgen- and aryl hydrocarbon receptor mediated activities in passive and composite samples from municipal waste and surface waters. Environ Int 59:372–383
37. Booij P, Sjollema SB, Leonards PEG, de Voogt P, Stroomberg GJ, Vethaak AD, Lamoree MH (2013) Extraction tools for identification of chemical contaminants in estuarine and coastal waters to determine toxic pressure on primary producers. Chemosphere 93:107–114
38. Liscio C, Abdul-Sada A, Al-Salhi R, Ramsey MH, Hill EM (2014) Methodology for profiling anti-androgen mixtures in river water using multiple passive samplers and bioassay-directed analyses. Water Res 57:258–269
39. Tan BLL, Hawker DW, Muller JF, Leusch FDL, Tremblay LA, Chapman HF (2007) Comprehensive study of endocrine disrupting compounds using grab and passive sampling at selected wastewater treatment plants in South East Queensland, Australia. Environ Int 33:654–669
40. Vermeirssen ELM, Asmin J, Escher BI, Kwon JH, Steimen I, Hollender J (2008) The role of hydrodynamics, matrix and sampling duration in passive sampling of polar compounds with Empore (TM) SDB-RPS disks. J Environ Monit 10:119–128
41. Muller R, Schreiber U, Escher BI, Quayle P, Nash SMB, Mueller JF (2008) Rapid exposure assessment of PSII herbicides in surface water using a novel chlorophyll a fluorescence imaging assay. Sci Total Environ 401:51–59
42. Escher BI, Quayle P, Muller R, Schreiber U, Mueller JF (2006) Passive sampling of herbicides combined with effect analysis in algae using a novel high-throughput phytotoxicity assay (maxi-imaging-PAM). J Environ Monit 8:456–464
43. Shaw M, Negri A, Fabricius K, Mueller JF (2009) Predicting water toxicity: pairing passive sampling with bioassays on the Great Barrier Reef. Aquat Toxicol 95:108–116
44. Muller R, Tang JYM, Thierb R, Mueller JF (2007) Combining passive sampling and toxicity testing for evaluation of mixtures of polar organic chemicals in sewage treatment plant effluent. J Environ Monit 9:104–109

45. Emelogu ES, Pollard P, Dymond P, Robinson CD, Webster L, McKenzie C, Dobson J, Bresnan E, Moffat CF (2013) Occurrence and potential combined toxicity of dissolved organic contaminants in the Forth estuary and Firth of Forth Scotland assessed using passive samplers and an algal toxicity test. Sci Total Environ 461:230–239
46. Emelogu ES, Pollard P, Robinson CD, Smedes F, Webster L, Oliver IW, McKenzie C, Seiler TB, Hollert H, Moffat CF (2013) Investigating the significance of dissolved organic contaminants in aquatic environments: coupling passive sampling with in vitro bioassays. Chemosphere 90:210–219
47. Emelogu ES, Seiler T-B, Pollard P, Robinson CD, Webster L, McKenzie C, Heger S, Hollert H, Bresnan E, Best J, Moffat CF (2014) Evaluations of combined zebrafish (Danio rerio) embryo and marine phytoplankton (Diacronema lutheri) toxicity of dissolved organic contaminants in the Ythan catchment, Scotland, UK. Environ Sci Pollut Res Int 21:5537–5546
48. Bi HP, Rissik D, Macova M, Hearn L, Mueller JF, Escher B (2011) Recovery of a freshwater wetland from chemical contamination after an oil spill. J Environ Monit 13:713–720
49. Allan SE, Smith BW, Tanguay RL, Anderson KA (2012) Bridging environmental mixtures and toxic effects. Environ Toxicol Chem 31:2877–2887
50. Bopp SK, McLachlan MS, Schirmer K (2007) Passive sampler for combined chemical and toxicological long-term monitoring of groundwater: the ceramic toximeter. Environ Sci Technol 41:6868–6876
51. Addeck A, Croes K, Van Langenhove K, Denison MS, Elhamalawy A, Elskens M, Baeyens W (2014) Time-integrated monitoring of polychlorinated dibenzo-p-dioxins and polychlorinated dibenzofurans (PCDD/Fs) in urban and industrial wastewaters using a ceramic toximeter and the CALUX bioassay. Chemosphere 94:27–35
52. Vrana B, Mills GA, Allan IJ, Dominiak E, Svensson K, Knutsson J, Morrison G, Greenwood R (2005) Passive sampling techniques for monitoring pollutants in water. Trends Analyt Chem 24:845–868
53. Harman C, Allan IJ, Vermeirssen ELM (2012) Calibration and use of the polar organic chemical integrative sampler – a critical review. Environ Toxicol Chem 31:2724–2738
54. Alvarez DA, Petty JD, Huckins JN, Jones-Lepp TL, Getting DT, Goddard JP, Manahan SE (2004) Development of a passive, in situ, integrative sampler for hydrophilic organic contaminants in aquatic environments. Environ Toxicol Chem 23:1640–1648
55. Emelogu ES, Pollard P, Robinson CD, Webster L, McKenzie C, Napier F, Steven L, Moffat CF (2013) Identification of selected organic contaminants in streams associated with agricultural activities and comparison between autosampling and silicone rubber passive sampling. Sci Total Environ 445:261–272
56. Booij K, Smedes F (2010) An improved method for estimating in situ sampling rates of nonpolar passive samplers. Environ Sci Technol 44:6789–6794
57. Smedes F, van Vliet LA, Booij K (2013) Multi-ratio equilibrium passive sampling method to estimate accessible and pore water concentrations of polycyclic aromatic hydrocarbons and polychlorinated biphenyls in sediment. Environ Sci Technol 47:510–517
58. Booij K, Smedes F, van Weerlee EM, Honkoop PJC (2006) Environmental monitoring of hydrophobic organic contaminants: the case of mussels versus semipermeable membrane devices. Environ Sci Technol 40:3893–3900
59. Bopp SK, Bols NC, Schirmer K (2006) Development of a solvent-free, solid-phase in vitro bioassay using vertebrate cells. Environ Toxicol Chem 25:1390–1398
60. Schirmer K, Bopp S, Gehrhardt J (2007) Use of passive sampling devices in toxicity assessment of groundwater. In: Greenwood R, Mills G, Vrana B (eds) Passive sampling techniques in environmental monitoring. Wilson & Wilson's, Amsterdam, pp 393–405
61. Mayer P, Parkerton TF, Adams RG, Cargill JG, Gan J, Gouin T, Gschwend PM, Hawthorne SB, Helm P, Witt G, You J, Escher BI (2014) Passive sampling methods for contaminated sediments: scientific rationale supporting use of freely dissolved concentrations. Integr Environ Assess Manag 10:197–209

62. Reichenberg F, Mayer P (2006) Two complementary sides of bioavailability: accessibility and chemical activity of organic contaminants in sediments and soils. Environ Toxicol Chem 25:1239–1245
63. Lydy MJ, Landrum PF, Oen AMP, Allinson M, Smedes F, Harwood AD, Li H, Maruya KA, Liu J (2014) Passive sampling methods for contaminated sediments: state of the science for organic contaminants. Integr Environ Assess Manag 10:167–178
64. Perron MM, Burgess RM, Ho KT, Pelletier MC, Friedman CL, Cantwell MG, Shine JP (2009) Development and evaluation of reverse polyethylene samplers for marine phase II whole-sediment toxicity identification evaluations. Environ Toxicol Chem 28:749–758
65. Bandow N, Altenburger R, Streck G, Brack W (2009) Effect-directed analysis of contaminated sediments with partition-based dosing using green algae cell multiplication inhibition. Environ Sci Technol 43:7343–7349
66. Perron MM, Burgess RM, Ho KT, Pelletier MC, Friedman CL, Cantwell MG, Shine JP (2011) Limitations of reverse polyethylene samplers (RePES) for evaluating toxicity of field contaminated sediments. Chemosphere 83:247–254
67. Zielke H, Seiler TB, Niebergall S, Leist E, Brinkmann M, Spira D, Streck G, Brack W, Feiler U, Braunbeck T, Hollert H (2011) The impact of extraction methodologies on the toxicity of sediments in the zebrafish (Danio rerio) embryo test. J Soil Sediment 11:352–363
68. Burton GA Jr, Rosen G, Chadwick DB, Greenberg MS, Taulbee WK, Lotufo GR, Reible DD (2012) A sediment ecotoxicity assessment platform for in situ measures of chemistry, bioaccumulation and toxicity. Part 1: system description and proof of concept. Environ Pollut 162:449–456
69. Li J-Y, Tang JYM, Jin L, Escher BI (2013) Understanding bioavailability and toxicity of sediment-associated contaminants by combining passive sampling with in vitro bioassays in an urban river catchment. Environ Toxicol Chem 32:2888–2896
70. Witt G, Niehus NC, Konopka KF, Mayer P, Floeter C (2015) Comparison of passive and standard dosing of polycyclic aromatic hydrocarbons to the marine algae Phaeodactylum tricornutum. SETAC Europe 25th annual meeting, Barcelona, Spain
71. Lorks J (2010) Passive dosing of extracts using PDMS layers in the fish embryo test as an alternative to direct contact exposure with marine Baltic Sea sediments. Diploma thesis, RWTH Aachen
72. Kupryianchyk D, Reichman EP, Rakowska MI, Peeters ETHM, Grotenhuis JTC, Koelmans AA (2011) Ecotoxicological effects of activated carbon amendments on macroinvertebrates in nonpolluted and polluted sediments. Environ Sci Technol 45:8567–8574
73. Xu Y, Spurlock F, Wang Z, Gan J (2007) Comparison of five methods for measuring sediment toxicity of hydrophobic cantaminants. Environ Sci Technol 41:8394–8399
74. Harwood AD, Landrum PF, Lydy MJ (2013) Bioavailability-based toxicity endpoints of bifenthrin for *Hyalella azteca* and *Chironomus dilutus*. Chemosphere 90:1117–1122
75. Harwood AD, Landrum PF, Weston DP, Lydy MJ (2013) Using SPME fibers and Tenax to predict the bioavailability of pyrethroids and chlorpyrifos in field sediments. Environ Pollut 173:47–51
76. Neale PA, Antony A, Bartkow ME, Farre MJ, Heitz A, Kristiana I, Tang JYM, Escher BI (2012) Bioanalytical assessment of the formation of disinfection byproducts in a drinking water treatment plant. Environ Sci Technol 46:10317–10325
77. Ding Y, Landrum PF, You J, Lydy MJ (2013) Assessing bioavailability and toxicity of permethrin and DDT in sediment using matrix solid phase microextraction. Ecotoxicology 22:109–117
78. Hawthorne SB, Azzolina NA, Neuhauser EF, Kreitinger JP (2007) Predicting bioavailability of sediment polycyclic aromatic hydrocarbons to *Hyalella azteca* using equilibrium partitioning, supercritical fluid extraction, and pore water concentrations. Environ Sci Technol 41:6297–6304
79. Bowen AT, Conder JM, La Point TW (2006) Solid phase microextraction of aminodinitrotoluenes in tissue. Chemosphere 63:58–63

80. Conder JM, La Point TW, Steevens JA, Lotufo GR (2004) Recommendations for the assessment of TNT toxicity in sediment. Environ Toxicol Chem 23:141–149
81. Kreitinger JP, Neuhauser EF, Doherty FG, Hawthorne SB (2007) Greatly reduced bioavailability and toxicity of polycyclic aromatic hydrocarbons to *Hyalella azteca* in sediments from manufactured-gas plant sites. Environ Toxicol Chem 26:1146–1157
82. Paumen ML, Stol P, Ter Laak TL, Kraak MHS, van Gestel CAM, Admiraal W (2008) Chronic exposure of the oligochaete *Lumbriculus variegatus* to polycyclic aromatic compounds (PACs): bioavailability and effects on reproduction. Environ Sci Technol 42:3434–3440
83. Pang J, Sun B, Li H, Mehler WT, You J (2012) Influence of bioturbation on bioavailability and toxicity of PAHs in sediment from an electronic waste recycling site in South China. Ecotoxicol Environ Saf 84:227–233
84. Arp HPH, Azzolina NA, Cornelissen G, Hawthorne SB (2011) Predicting pore water APA-34 PAH concentrations and toxicity in pyrogenic-impacted sediments using pyrene content. Environ Sci Technol 45:5139–5146
85. Oen AMP, Janssen EML, Cornelissen G, Breedveld GD, Eek E, Luthy RG (2011) In situ measurement of PCB pore water concentration profiles in activated carbon-amended sediment using passive samplers. Environ Sci Technol 45:4053–4059
86. Cornelissen G, Wiberg K, Broman D, Arp HPH, Persson Y, Sundqvist K, Jonsson P (2008) Freely dissolved concentrations and sediment-water activity ratios of PCDD/Fs and PCBs in the open Baltic Sea. Environ Sci Technol 42:8733–8739
87. Jahnke A, Mayer P, McLachlan MS (2012) Sensitive equilibrium sampling to study polychlorinated biphenyl disposition in Baltic Sea sediment. Environ Sci Technol 46:10114–10122
88. Mäenpää K, Leppänen MT, Reichenberg F, Figueiredo K, Mayer P (2011) Equilibrium sampling of persistent and bioaccumulative compounds in soil and sediment: comparison of two approaches to determine equilibrium partitioning concentrations in lipids. Environ Sci Technol 45:1041–1047
89. Mayer P, Vaes WHJ, Wijnker F, Legierse KCHM, Kraaij RH, Tolls J, Hermens JLM (2000) Sensing dissolved sediment porewater concentrations of persistent and bioaccumulative pollutants using disposable solid-phase microextraction fibers. Environ Sci Technol 34:5177–5183
90. Schäfer S, Antoni C, Möhlenkamp C, Claus E, Reifferscheid G, Heininger P, Mayer P (2015) Equilibrium sampling of polychlorinated biphenyls in River Elbe sediments – linking bioaccumulation in fish to sediment contamination. Chemosphere 138:856–862
91. Witt G, Liehr GA, Borck D, Mayer P (2009) Matrix solid-phase microextraction for measuring freely dissolved concentrations and chemical activities of PAHs in sediment cores from the western Baltic Sea. Chemosphere 74:522–529
92. Maruya KA, Zeng EY, Tsukada D, Bay SM (2009) A passive sampler based on solid-phase microextraction for quantifying hydrophobic organic contaminants in sediment pore water. Environ Toxicol Chem 28:733–740
93. Witt G, Lang SC, Ullmann D, Schaffrath G, Schulz-Bull D, Mayer P (2013) Passive equilibrium sampler for in situ measurements of freely dissolved concentrations of hydrophobic organic chemicals in sediments. Environ Sci Technol 47:7830–7839
94. Bao LJ, Zeng EY (2011) Passive sampling techniques for sensing freely dissolved hydrophobic organic chemicals in sediment porewater. Trends Analyt Chem 30:1422–1428
95. Difilippo EL, Eganhouse RP (2010) Assessment of PDMS-water partition coefficients: implications for passive environmental sampling of hydrophobic organic compounds. Environ Sci Technol 44:6917–6925
96. Schenker U, MacLeod M, Scheringer M, Hungerbuhler K (2005) Improving data quality for environmental fate models: a least-squares adjustment procedure for harmonizing physicochemical properties of organic compounds. Environ Sci Technol 39:8434–8441

97. Schwarzenbach RPG, Philip M, Imboden DM (2003) Environmental organic chemistry, 2nd edn. Wiley, Hoboken
98. Maruya KA, Landrum PF, Burgess RM, Shine JP (2012) Incorporating contaminant bioavailability into sediment quality assessment frameworks. Integr Environ Assess Manag 8:659–673
99. Rosen G, Chadwick DB, Burton GA, Taulbee WK, Greenberg MS, Lotufo GR, Reible DD (2012) A sediment ecotoxicity assessment platform for in situ measures of chemistry, bioaccumulation and toxicity. Part 2: integrated application to a shallow estuary. Environ Pollut 162:457–465
100. Ding Y, Landrum PF, You J, Harwood AD, Lydy MJ (2012) Use of solid phase microextraction to estimate toxicity: relating fiber concentrations to toxicity-part I. Environ Toxicol Chem 31:2159–2167
101. Ding Y, Landrum PF, You J, Harwood AD, Lydy MJ (2012) Use of solid phase microextraction to estimate toxicity: relating fiber concentrations to body residues-part II. Environ Toxicol Chem 31:2168–2174
102. Jin L, Gaus C, van Mourik L, Escher BI (2013) Applicability of passive sampling to bioanalytical screening of bioaccumulative chemicals in marine wildlife. Environ Sci Technol 47:7982–7988
103. Jin L, Gaus C, Escher BI (2015) Adaptive stress response pathways induced by environmental mixtures of bioaccumulative chemicals in dugongs. Environ Sci Technol 49:6963–6973
104. Jin L, Escher BI, Limpus CJ, Gaus C (2015) Coupling passive sampling with in vitro bioassays and chemical analysis to understand combined effects of bioaccumulative chemicals in blood of marine turtles. Chemosphere 138:292–299
105. Ossiander L, Reichenberg F, McLachlan MS, Mayer P (2008) Immersed solid phase microextraction to measure chemical activity of lipophilic organic contaminants in fatty tissue samples. Chemosphere 71:1502–1510
106. Jahnke A, Mayer P, Broman D, McLachlan MS (2009) Possibilities and limitations of equilibrium sampling using polydimethylsiloxane in fish tissue. Chemosphere 77:764–770
107. Mäenpää K, Leppänen MT, Figueiredo K, Tigistu-Sahle F, Käkelä R (2015) Sorptive capacity of membrane lipids, storage lipids, and proteins: a preliminary study of partitioning of organochlorines in lean fish from a PCB-contaminated freshwater lake. Arch Environ Contam Toxicol 68:193–203
108. Zhou SN, Oakes KD, Servos MR, Pawliszyn J (2008) Application of solid-phase microextraction for in vivo laboratory and field sampling of pharmaceuticals in fish. Environ Sci Technol 42:6073–6079
109. Adolfsson-Erici M, Akerman G, McLachlan MS (2012) In-vivo passive sampling to measure elimination kinetics in bioaccumulation tests. Chemosphere 88:62–68
110. Allan IJ, Baek K, Haugen TO, Hawley KL, Hogfeldt AS, Lillicrap AD (2013) In vivo passive sampling of nonpolar contaminants in brown trout (*Salmo trutta*). Environ Sci Technol 47:11660–11667
111. Mayer P, Toraeng L, Glaesner N, Joensson JA (2009) Silicone membrane equilibrator: measuring chemical activity of nonpolar chemicals with poly(dimethylsiloxane) microtubes immersed directly in tissue and lipids. Anal Chem 81:1536–1542
112. Vrana B, Smedes F, Rusina TP, Okonski K, Allan IJ, Grung M, Hilscherova K, Novák J, Tarábek P, Slobodník J (2015) Passive sampling: chemical analysis and toxicological profiling. ICPDR report: Joint Danube Survey 3. A comprehensive analysis of Danube Water Quality

Adv Biochem Eng Biotechnol (2017) 157: 263–292
DOI: 10.1007/10_2015_5017

Published online: 13 July 2016

Defining and Controlling Exposure During In Vitro Toxicity Testing and the Potential of Passive Dosing

Kilian E.C. Smith and Sabine Schäfer

Abstract Toxicity testing using in vitro bioassays is assuming an increasingly important role. Nevertheless, several issues remain with regard to their proper application, which mainly relate to the proper definition and control of the test chemical(s) concentrations to which the cells or tissues are exposed. This has fundamental implications for understanding the underlying relationship between the in vitro exposure regime and response, and leads to uncertainty in the resulting bioassay data. This chapter covers the definition and control of exposure of hydrophobic organic chemicals (HOCs) in in vitro bioassays aimed at measuring their toxicity. A review of the fate of HOCs in typical in vitro set-ups is followed by a discussion of how to define the test exposure. Currently applied approaches for introducing HOCs into in vitro bioassays are then related to these different definitions of test exposure. Finally, passive dosing as one possible approach for giving defined and constant dissolved concentrations of HOCs in in vitro toxicity tests is introduced, using examples taken from the literature, and how this might be better integrated into high throughput in vitro toxicity testing is discussed.

Keywords Dissolved concentration, Exposure, Hydrophobic organic chemical, In vitro toxicity bioassay, Passive dosing

---

K.E.C. Smith (✉)
RWTH Aachen University, Institute for Environmental Research, Worringerweg 1, 52074 Aachen, Germany
e-mail: kilian.smith@bio5.rwth-aachen.de

S. Schäfer
Department of Qualitative Hydrology, German Federal Institute of Hydrology, Am Mainzer Tor 1, 56068 Koblenz, Germany
e-mail: sabine.schafer@bafg.de

Contents

1 Introduction

Testing the toxicity of organic chemicals plays an important role in protecting human health as well as the environment. Whereas chemicals used in medicinal applications already have a detailed description of their toxicity because this is mandatory before their release onto the market, for many other chemicals this is not the case, despite requirements set out in regulations such as the European Union's REACH [1]. One reason is that existing approaches for toxicity testing can be rather cost and labor intensive, leading to bottlenecks in the generation of the required toxicity data [2, 3]. This is a rather critical issue given the huge number of existing (and future) organic chemicals, and is exacerbated by the requirement for data covering different toxicity endpoints because these require targeted toxicity tests.

In this context, the application of in vitro tests can play an important role [4, 5]. Their miniaturized format is amenable to automation for increasing throughput, and additionally reduces the consumables footprint. Also relevant is that the ethical concerns accompanying the use of animals for toxicity testing do not play a role in in vitro testing. Such considerations are reflected in the US National Research Council report "Toxicity testing in the 21st Century, A Vision and Strategy" [6, 7], which calls for a change from toxicity testing based on high-dose testing in animals to a strategy where in vitro bioassays using cells or tissues relevant for humans are used to evaluate adverse changes in the cellular signaling pathways. Furthermore, the EU REACH and cosmetics legislation specifically advocate the development of safety and risk assessment methods which do not rely on animal testing, and here in vitro testing also plays an increasingly important role [1, 8].

In vitro tests are highly diverse because they can be based on sub-cellular fractions, cells, or even tissues derived from different organisms and organs. They can therefore deliver species- and organ-specific toxicity information, and can also target different toxicity endpoints ranging from apical responses at the cellular level (e.g., cell viability or death) down to molecular-level events (e.g., gene or protein expression) [4, 9]. This makes them useful both for hazard identification and for unraveling the mechanisms underlying the toxicity. Molecular biology has led to the development of engineered reporter bioassays, in which the occurrence of a specific molecular event quantifiably translates into a measureable response. Such bioassays can be used to quantify specific types of biological activity, e.g., endocrine disruption or aryl hydrocarbon receptor activation [10]. In vitro tests can also be used to derive metabolic rates, which are integral for translating in vitro toxicity data into an in vivo setting via the application of approaches such as adsorption-distribution-metabolism-excretion modeling [11].

Another area where in vitro tests are increasingly important is the testing of environmental samples containing organic contaminants. Environmental toxicity arises from exposure to complex chemical mixtures which are often undefined and present at low levels. This makes it challenging to apply chemical analysis to measure the concentrations of all constituents of a sample for predicting their combined toxicity via concentration or response addition [12]. Directly measuring toxicity using in vitro bioassays takes into account all mixture constituents, as well as their interactions, and can be done either by direct sample application [13, 14] or by initial extraction and applying a portion of the extract into the test [9]. One such example where in vitro tests are used in this way is the acute toxicity testing of waste-water effluents using the zebrafish embryo toxicity test [15].

Despite the undoubted advantages of using in vitro tests, a number of issues remain with regard to their application for the toxicity testing of organic chemicals and environmental samples. These largely relate to the proper definition and control of the test chemical(s) concentrations to which the cells or tissue are exposed [16, 17], with fundamental implications for understanding the underlying relationship between exposure and response. This not only leads to uncertainty in the data resulting from such in vitro tests, but also hampers extrapolation of this data to the in vivo situation for risk assessment purposes [16]. The set-ups used for in vitro testing have been optimized to support the cell growth, but unfortunately are not always suited for organic chemicals. For example, many in vitro tests use open plastic microtiter plates for optimal cell adhesion and gas exchange, "rich" culture media to stimulate cell growth and, in the case of mammalian cell lines, elevated temperatures. However, these conditions can also lead to large losses or poor availability of the test compound [18–21]. A related issue is that the instrumentation for automating the in vitro testing process has evolved to be compatible with microtiter plates, making radical format changes difficult.

This chapter covers the definition and control of exposure of organic chemicals when testing their toxicity using in vitro bioassays. The focus is primarily on hydrophobic organic compounds (HOCs) because their low aqueous solubility and propensity to sorb to surfaces, particles, or proteins make control of their exposure particularly challenging. Nevertheless, many of the same considerations also apply to compounds with lower hydrophobicities. A discussion of the fate of HOCs in typical in vitro set-ups is followed by considerations of how best to define the test exposure. Current approaches for introducing HOCs into in vitro bioassays are then related to these various definitions of test exposure. Finally, passive dosing as one approach for giving defined and constant dissolved concentrations of HOCs in in vitro toxicity tests is introduced using examples taken from the literature, together with an outlook on how this can be better integrated into high throughput in vitro testing.

2 Fate of Test Chemicals in In Vitro Bioassays

Test chemical(s) added to an in vitro test are subject to a range of loss processes including sorption, volatilization, and (bio)transformation [19, 22, 23]. Moreover, it is not the total but rather dissolved concentrations that are considered as being bioavailable and thus the effective concentrations driving uptake and toxicity [16, 17, 24]. These dissolved levels are determined by the balance between the different loss processes (Fig. 1).

In many bioassays, protein and lipid form an integral part of the culture medium but can be a significant sorbing phase leading to a reduction in test compound bioavailability. The extent of this sorption depends on the amount of protein and lipid in the medium as well as the compound properties and concentrations because of nonlinearities in the sorption isotherms [19]. Medium sorption is particularly relevant for more hydrophobic compounds. For example, the free fraction of different PAHs measured in a typical RPMI 1640 medium supplemented with 10 vol% fetal calf serum ranged from 30% (naphthalene, $\log K_{OW}$ 3.33) down to

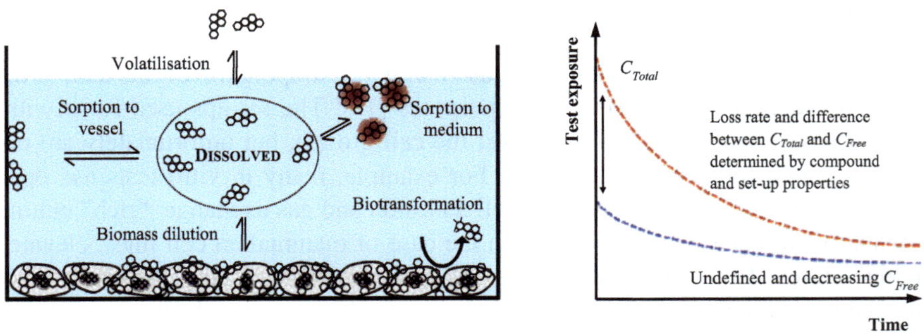

Fig. 1 Fate processes determining the total (C_{total}) and freely dissolved (C_{free}) concentrations of an organic compound under typical in vitro bioassay conditions. C_{free} is considered to represent the bioavailable fraction of a compound, and thus the effective concentration driving uptake and toxicity. The difference between C_{total} and C_{free}, as well as the magnitude of the losses driving the decrease in the medium concentrations, are determined by the compound properties and in vitro test characteristics

0.1% (benzo(*a*)pyrene, log K_{OW} 6.13) [21]. Polar and ionizable compounds might also significantly bind to the protein components in the medium, although the relevance of this effect is less well understood [25].

Sorption also takes place at the walls of the microplate wells, particularly when these are made out of plastic. Here, sorption increases in line with compound hydrophobicity. The relative contribution of this process to the overall sorptive losses depends on the ratio of the wall surface area to the volume of the exposure medium [19, 20]. However, under typical in vitro conditions using "rich" cell culture media, sorption to the plastic walls is often relatively minor because the former frequently dominates the overall sorption [19]. Nevertheless, sorption to surfaces might become significant if culture media with limited lipid and protein contents are used. Such media are being developed to overcome the problems regarding the medium sorption outlined above.

Another process affecting the dissolved test concentrations is sorption to the target cells [22]. Of course, it is this very sorption that leads to accumulation at the target site and thus results in the toxicity. Here, the critical point is that when cells accumulate a significant fraction of the added test compound, it is no longer appropriate to use nominal concentrations to describe the test exposure. Gülden et al. [22] investigated the impact of increasing cell concentrations on the EC_{50} of a range of compounds sorbing via different mechanisms to the adherent embryonic mouse cell line Balb/c 3 T3. Cell binding resulted in higher EC_{50} values based on nominal concentrations, and this effect was increased at higher cell densities and also for those compounds which were accumulated to a greater extent.

Although some in vitro tests use cell lines with an inherently low biotransformation capacity, this loss route can be another confounding issue. Biotransformation losses are compound specific, and chemicals are thus affected differently. Furthermore, the parent compound might be transformed into metabolites with higher or lower activities and/or different types of toxicity. In fact, a number of in vitro toxicity tests rely on biotransformation to "activate" the test chemical so that it can exert its toxic effect [26]. Therefore, the implications of biotransformation are complex. On the one hand, this can lead to a reduction in the availability of the chemical and thus a lower test response. On the other, it can result in the production of metabolites with increased or different toxicities which can alter the test response. The importance of biotransformation therefore depends on the chemical, its bioavailable concentration, and the biotransformation characteristics of the bioassay (e.g., cell type or the addition of exogenous enzymes).

A compound with a high air to water partition ratio can partition appreciably into the headspace above the well so that that volatile losses become significant [20, 23]. Volatile losses are further increased at higher temperatures, such as the 37°C typical of mammalian cell bioassays. Volatilization not only significantly reduces the test concentrations – the volatilized compounds can also partition into adjacent wells which leads to artifacts caused by cross-contamination. This is particularly relevant for high throughput screening, where large numbers of chemicals and mixtures are tested simultaneously using microtiter plates.

In summary, in an in vitro test the bioavailable concentration of a chemical that drives uptake and toxicity is determined by different fate processes which in turn depend on the properties of the compound (hydrophobicity, volatility, etc.) and the test system (culture medium, cell types, test temperature, etc.). When testing the toxicity of single compounds using in vitro tests, this is reflected in the dual issues of dissolved concentrations which are undefined but also decrease during the test. For mixtures there is the additional complication that these effects impact the individual constituents differently, and therefore there is also a change in the bioavailable mixture profile. This might lead to a compound being incorrectly prioritized as being the key toxicity driver in the mixture. Although it is possible to account for some of these factors for single compounds and simple mixtures by using (equilibrium) modeling approaches (e.g., [19, 22, 27]), this is much more challenging for complex mixtures and even impossible in those instances where the toxicity of an undefined environmental sample is being considered.

3 Defining Exposure in In Vitro Bioassays

The fundamental purpose of an in vitro toxicity test is to provide information on whether a chemical is hazardous or not. However, often an additional aim is to obtain information on how the toxic response varies with increasing concentrations. This might then be used, for example, to identify whether there is a threshold level below which no harm is expected for use within risk assessment. To achieve this, the measured bioassay response has to be related to the test chemical(s) exposure, and here the choice of exposure metric plays a fundamental role.

Ideally, the exposure metric should describe the exposure at the site of toxic action as this is most closely related to the initial molecular changes within the cell as induced by the chemical [16, 17]. However, in many cases the exact nature of this site is unknown or not analytically accessible, and therefore nominal or total concentrations are commonly used. The nominal concentration is given by the amount of added test chemical divided by the medium volume, and, although practical, its use is only appropriate when it accurately describes the bioavailable concentrations driving uptake and toxicity. Even when the medium concentration is determined immediately after compound addition, there are often marked differences to the nominal concentrations caused by handling losses or incomplete dissolution [28]. Furthermore, depending on the properties of the compound and the test set-up, the nominal concentration can significantly overestimate the bioavailable concentrations in the test because of the various loss processes described above. This can lead to a shift in the concentration-response curve to higher concentrations [18, 29, 30], which partly explains the low apparent sensitivity sometimes observed in in vitro tests compared to the in vivo situation [31]. At the extreme, this could mean that no test response is observed and the compound is classified as being non-toxic because of inadequate testing conditions leading to a limited bioavailability rather than its intrinsic toxic properties.

Therefore, rather than using nominal concentrations, a number of in vitro toxicity test guidelines stipulate that total concentrations at the start and end of the toxicity test be analytically confirmed. This accounts for losses such as sorption to the vessel walls and volatilization, but does not consider the all-important difference between the total and bioavailable concentrations. Furthermore, when losses are severe it is challenging how to best define this decreasing exposure. Examples of time-dependent exposure metrics include the integrated dose over time (AUC, area under the curve) or the dose divided by the time period (TWA, time weighted average) [16]. In any case, the considerations concerning bioavailability described above still apply when such time-dependent exposure metrics are based on nominal or total concentrations. For more complex exposure scenarios, such as pulsed exposures, it is possible to incorporate time effects by applying biokinetic and toxicodynamic models to give the concentration-effect relationship over time [32].

Therefore, one of the best alternatives for an exposure metric in in vitro tests is possibly the freely dissolved concentration (C_{free}). This is considered to represent that portion of the test chemical in the medium which is bioavailable for cell uptake and toxicity [16, 17, 24]. Although C_{free} describes the exposure external to the cells, it can be related to whole- or even sub-cellular concentrations via uptake and partitioning models (e.g., [19, 22, 33], and can be measured using dialysis, ultrafiltration, centrifugation, or solid phase microextraction (SPME). The latter has the advantage that it circumvents some of the artifacts associated with other methods such as filter sorption or inclusion of colloid-associated compound in the freely dissolved fraction [17].

Using C_{free} as the exposure metric has the further advantage that the test data can be used as input for the chemical activity concept. An interesting development is the application of chemical activity to describe baseline narcotic toxicity. For a wide range of organisms, chemicals, and mixtures thereof, the onset of narcotic toxicity consistently appears within a chemical activity range from 0.01 to 0.1 in the external exposure medium [34–37]. In addition, it appears that chemical activity for describing mixture toxicity is additive (e.g., [38, 39]). These observations agree with the critical membrane concept for narcotic toxicity [40], likely because of a (near) chemical equilibrium between the organism and exposure medium and because of the properties of the compounds studied so far, which all have relatively similar lipid activity coefficients and are minimally metabolized.

4 Conventional Approaches for Controlling Exposure

Compounds which are sorbed, biotransformed, or volatilized are not available for uptake by the biological target and thus cannot contribute directly to the toxic response. Increased standardization of in vitro testing protocols (e.g., the use of standard culture media and fixed cell numbers) reduces variability between similar types of tests but does not adequately consider such losses. Therefore, to improve the reliability of in vitro bioassay data it is necessary to account for, or even better to compensate for, such losses.

If chemicals are sufficiently water soluble it is possible to dissolve them directly in the culture medium, and then directly expose the cells in this medium. However, many HOCs are challenging to dissolve initially in aqueous solutions. Therefore, one widely used approach for their introduction into in vitro tests is by spiking the compounds using a co-solvent as a solubilizer. For this, a concentrated solution of the test compound(s) in a biocompatible and water-miscible solvent is prepared, and a small volume of this solution is added to the test medium. Careful consideration should be given to ensure that the volume of solvent added does not result in a measureable toxic effect. This can be determined experimentally, but does not preclude mixture toxicity caused by the combination of the spiking solvent and test chemical(s). Furthermore, solvents such as dimethylsulfoxide (DMSO) can alter the permeability of the cell membranes and thus lead to different uptake kinetics [41, 42]. Another issue with solvent spiking is that the test compound is introduced as a concentrated solution. When directly spiking the test medium in the microplate well, this can lead to a temporary inhomogeneity in the exposure as the solvent spike hits the aqueous phase. This might occur either because the solvent plus compound sinks to the bottom of the well where adherent cells are located, leading to a localized high concentration region, and/or because the compound precipitates and needs to redissolve again to become bioavailable [28]. This might be solved by indirect dosing, where the culture medium is first spiked and thoroughly mixed before using this to dose the cells [28].

However, the main criticism with solvent spiking is that the bioavailable dissolved concentrations are neither defined nor buffered. Although the various loss processes can be accounted for by using analytical [17] or modeling [19, 22] approaches to obtain a better definition of the bioavailable exposure concentrations, two important limitations nevertheless remain. First, there is the issue of a low compound bioavailability because of sorptive and other losses leading to a reduction in the apparent test sensitivity [18, 29, 30]. In part this might be solved by adding increasing amounts of test compound to compensate for such losses, but this can lead to precipitation, particularly when studying toxicity close to a compound's aqueous solubility. Second, if concentrations are not buffered and decrease significantly, it becomes challenging to find the correct time-resolved measure of exposure [16, 20].

One way to ameliorate the effects of rapidly declining concentrations is by medium renewal as is often done in toxicity tests with larger organisms. Strictly speaking, medium renewal does not result in constant concentrations but rather a sequence of pulsed exposures, the amplitude of which depends on the loss rate relative to the frequency of medium renewal. In addition to buffering test losses, this approach also removes excreted metabolic products. However, there is the potential for negative effects on the cells caused by handling stress. In any case, simply renewing the medium does not address the discrepancy between the nominal/total concentrations and C_{free}, and additional modeling or analytical efforts would be required. Although medium renewal is much easier with adhered rather than suspended cells, it is rarely applied in in vitro systems.

A more advanced approach is to deliver a continuous stream of medium with constant compound concentrations, either from a reservoir of spiked medium or by using a generator column. Despite not being so easy to interface with in vitro test set-ups, examples exist such as the flow-through set-up developed for the fish embryo toxicity test using zebrafish (*Danio rerio*) [43]. In this study, culture medium containing 4-chlorophenol was continuously supplied to the wells of a modified 24-well microplate from a large reservoir in a flow-through set up. The wells were lined with a gauze net preventing loss of the eggs. In addition to providing constant concentrations of test compound for exposures up to 120 h, other benefits included a good supply of oxygen, removal of metabolic waste products, and the possibility of collecting large medium volumes for chemical analysis. The main limitations were a lack of cheap and commercially available plates for this purpose, incompatibility with cell suspensions which are too small to be retained in the well, and the need to differentiate between the nominal/total concentrations and C_{free}.

Considerable effort has been made to solve the problems of sorption to the vessel walls or medium components. Many in vitro bioassays use cells forming confluent layers which require the plastic material for optimal cell adhesion, and here the choice of plastic can reduce the sorption of the test compound, although this does not eliminate the problem entirely [20]. For tests using cell suspensions, vessels made out of glass which minimally sorb many organic compounds might be used. A parallel development is the use of defined culture media containing minimal protein and lipid to reduce this loss pathway, although as yet these media cannot be applied in all in vitro bioassays [44].

Compounds with high air to water partition ratios are prone to volatile losses [20], a process particularly difficult to model because of its dynamic nature [19]. Stalter et al. [23] developed a sealed headspace-free set-up in plastic micro-titer plates to study the toxicity of volatile compounds in different in vitro tests: bioluminescence inhibition in the bacteria *Vibrio fischeri*, genotoxicity in the bacteria *Salmonella typhimurium* umuC, and Ames fluctuation bioassays and oxidative stress response in the mammalian AREc32 bioassay. For the different bacterial bioassays used in the study, simple culture media lacking any sorbing components were used, and here a large increase in the apparent test sensitivity was found for the sealed headspace-free set-up compared to the conventional design with an unsealed headspace. This was traced back to an analytically confirmed reduction in volatile losses in the sealed headspace-free format. Interestingly, for the AREc32 bioassay which used culture medium containing serum components, the effect was not so marked. This was likely because the constituents of the culture medium had a retaining effect on the test compounds. Although the sealed headspace-free set-up worked well for the relatively short exposures used in this study, for longer exposures oxygen supply to the cells might become limiting.

Therefore, despite the different experimental approaches discussed above, there still remains an urgent need for new approaches to control and maintain the C_{free} of organic compounds and their mixtures in in vitro toxicity tests. These should be practical to apply and also be compatible with existing test protocols and with the drive towards high throughput testing. The ability to test at a priori defined and constant C_{free} would mean the same exposure metric is used in different in vitro bioassays, irrespective of variations in the testing protocols such as differences in the medium composition or cell densities. This would be particularly important for mixture toxicity testing, as only in this way can the mixture profile be maintained between different bioassays. This would simplify interpretation and comparison of in vitro toxicity data and facilitate the application of test batteries covering different toxicity endpoints. Passive dosing is one approach that can solve these challenges.

5 Principles of Passive Dosing

Passive dosing is referred to in the literature by various names including partitioning driven administration, partition controlled delivery, and partition (ing)-based dosing [45–51].

In passive dosing, a dominating reservoir of sorbed organic compound acts as a constant partitioning source to the aqueous phase (Fig. 2). When equilibrium is established between the dissolved (C_{free}, e.g., mg L^{-1}) and polymer-sorbed (C_{sorbed}, e.g., mg L^{-1}) concentrations of organic compound, C_{free} is given by

$$C_{\mathrm{free}} = \frac{C_{\mathrm{sorbed}}}{K_{\mathrm{sorbed/free}}} \tag{1}$$

where $K_{\mathrm{sorbed/free}}$ (L L^{-1}) is the equilibrium partition ratio that depends on the properties of the compound, passive dosing phase, and aqueous medium. Losses that lead to a perturbation of C_{free} are compensated for by additional partitioning

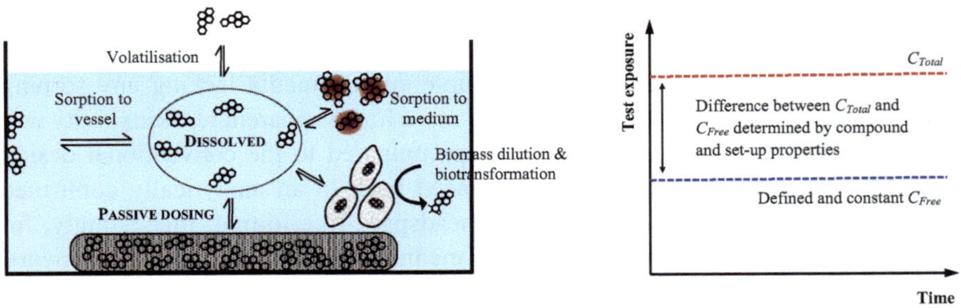

Fig. 2 Passive dosing for producing defined and constant freely dissolved concentrations (C_{free}) in in vitro toxicity bioassays. Partitioning from the dominating reservoir of compound sorbed to a polymer defines and maintains C_{free} even in the face of continuing test losses. The difference between the total medium concentrations (C_{total}) and C_{free} is determined by the compound and set-up properties

from the sorbed reservoir which is dimensioned to be large enough that any mass depletion is negligible. Therefore, C_{sorbed} and thus also C_{free} remain constant.

For the control of exposure in in vitro tests, passive dosing has a number of important features:

1. *Control of* C_{free}. Simply varying C_{sorbed} allows for the control of C_{free} to any value between zero and aqueous solubility. Furthermore, the approach is based on partitioning and when working close to aqueous solubility there are no problems with precipitation of test substance which can be an issue with solvent spiking.
2. *Defined* C_{free}. By properly dimensioning the passive dosing format, C_{free} is determined by equilibrium partitioning and thus known by applying experimentally determined values of $K_{sorbed/free}$.
3. *Constant* C_{free}. When the sorbed reservoir of organic compound is large enough to dominate any losses, significant depletion of the passive dosing phase is avoided and C_{free} remains constant.
4. *Conservation of mixture composition.* When the above features apply to all dosed compounds, the C_{free} mixture profile is defined and remains constant.
5. *Avoiding co-solvents.* The organic compounds are introduced associated with a biocompatible and inert passive dosing phase, and artifacts associated with spiking co-solvents are avoided.
6. *Analytical confirmation of exposure.* At experiment completion, C_{free} can be analytically confirmed by directly measuring the equilibrium partitioning concentrations in pure water or by measuring C_{sorbed} and applying $K_{sorbed/free}$.

6 Implementation of Passive Dosing

The development and application of equilibrium passive dosing follows a sequence of steps: format selection and dimensioning, cleaning and loading of the dosing phase, application in the experimental set-up, and exposure confirmation. The following sections give detailed information about these aspects.

6.1 Passive Dosing Format Selection and Dimensioning

The first step is selection of a format compatible with the requirements of the in vitro test. For example, considerations might include the physical size of the format or whether contact between the passive dosing phase and cells needs to be avoided. The second step involves ensuring that the passive dosing phase is not significantly depleted because of losses during the test but also that equilibrium partitioning applies.

Ideally, the passive dosing phase should be chemically inert, biocompatible, and have a high capacity for the test compounds. Furthermore, partitioning should be linear over the full concentration range with no sorption competition between the test chemicals and medium constituents. Finally, the passive dosing material should allow for a fast release of test compounds to keep up with various loss processes for ensuring that equilibrium partitioning applies. This explains why, to date, silicone polymers have most often been applied as the passive dosing phase, particularly for HOCs [18, 29, 30, 47, 49, 52, 53]. Therefore, although the following discussion focuses on silicone, the same considerations apply when considering other polymers and materials for use as the passive dosing phase. Indeed, some of these alternatives are mentioned in Sect. 7.1.

Silicone can be obtained at reasonable cost in a wide range of formats, conferring flexibility in selecting a format suited for any particular experimental set-up. Silicone is also chemically inert, meaning that it can be rigorously cleaned and then loaded with HOCs using a wide range of solvents without problems related to swelling or cracking [54]. Its biocompatibility is well documented, and it is available in highly pure forms such as food- or medical-grade silicone to allow its use as the passive dosing phase in sensitive bioassays [21]. Related to this is the fact that silicone can be easily sterilized which is a prerequisite for its application in some bioassays. Both single HOCs and their mixtures dissolve in the silicone matrix, with the sorption isotherms being linear over the full range of concentrations up to the upper limit as set by solubility [55]. Silicone also has a sufficiently high affinity for HOCs, so that the amount of silicone that needs to be introduced into a set-up to ensure it dominates as a partitioning source is practical to work with. Also important is the low mass transfer resistance of the silicone matrix towards the HOCs. Therefore, even under well-mixed conditions, HOC mass transfer into the aqueous phase is rate limited by diffusion through the unstirred water layer immediately adjacent to the silicone surface [56]. Concentration gradients thus do not develop within the silicone matrix and it can be considered as homogenous, which is critical for the correct application of equilibrium partition ratios for calculating C_{free}. Finally, extensive literature exists on the equilibrium partition ratios of a large number of HOCs between silicone and water [57, 58].

At equilibrium, a compound's distribution between the silicone and the medium is given by

$$f_{\text{silicone}} = \frac{1}{1 + \frac{1}{K_{\text{silicone/medium}} \cdot \frac{V_{\text{silicone}}}{V_{\text{medium}}}}} \tag{2}$$

where f_{silicone} is the fraction of the total compound mass in the silicone at equilibrium, $K_{\text{silicone/medium}}$ (L L^{-1}) is the equilibrium silicone to medium partition ratio, and V_{silicone} (L) and V_{medium} (L) are the volumes of silicone and medium, respectively. Equation (2) indicates that the capability of the silicone passive dosing phase to function as a non-depleted buffer depends on its volume relative to that of the culture medium (i.e., $V_{\text{silicone}}/V_{\text{medium}}$) and also to the compound affinity of the silicone relative to the medium (i.e., $K_{\text{silicone/medium}}$). Therefore, high $K_{\text{silicone/free}}$

compounds are more effectively buffered because of the larger mass fraction in the silicone.

When the culture medium does not contain appreciable sorbing components, then $K_{\text{silicone/medium}}$ approaches the silicone to water partition ration ($K_{\text{silicone/free}}$, L L^{-1}) and indeed this can be used as a first approximation to estimate whether the chosen format is depleted upon equilibration with the medium. This has the advantage that values of $K_{\text{silicone/free}}$ are available for many compounds [57, 58], but two things need to be kept in mind. First, when the culture medium contains serum this has a high sorption propensity for HOCs. Second, volatile and other losses are not included in this calculation and lead to additional depletion.

The silicone surface area relative to the medium volume plays a determining role in how quickly a partitioning equilibrium is reached. However, the actual importance of this depends on the context of the study. In longer term bioassays (e.g., 24 or 48 h) the few hours needed to reach equilibrium partitioning are not so critical. However, in short term bioassays, steps might be required to ensure that equilibration is more rapid. This might involve selecting a format with a higher surface area, increasing mixing in the medium, or incorporating a pre-equilibration step between the silicone and medium prior to addition of the test organism or cells [21, 30, 59].

6.2 *Initial Cleaning of the Silicone*

Unpolymerized oligomers and other impurities in the silicone can lead to problems, but can be relatively simply avoided by selecting high quality silicones and incorporating an initial cleaning step. This is particularly important for in vitro toxicity testing because the value of the toxicity data depends heavily on the control treatments showing limited toxicity. A rigorous cleaning process of extracting the silicone with analytical grade and water-miscible solvents such as ethanol or methanol, followed by sequential rinsing with distilled water to remove remaining solvent traces, has been shown to be effective [18, 21].

6.3 *Loading of the Silicone*

The silicone concentrations directly determine C_{free} via equilibrium partitioning (see (1)), and the correct loading of the passive dosing phase is thus a critical step. Loading approaches can be grouped into those based on partitioning from a loading solution and "spiking" from a concentrated solution. In both cases, careful consideration should be given to the solvent used to make up the loading solution. One important aspect is miscibility with water, because this can greatly simplify removal of any remaining solvent after loading of the silicone with compound.

Good experience has been obtained when using, for example, pure methanol or methanol:water mixtures as the solvent [18, 21, 49, 52].

For the partitioning approaches, a loading solution of the test compound(s) is prepared in a water-miscible solvent and brought into contact with the silicone such that the compound partitions between the two. The simplest approach is to allow an equilibrium to be reached, and then to determine the fraction of test compound in the silicone, e.g., by measuring the solvent concentration post-loading and assuming a 100% mass balance. Another approach is to avoid a significant depletion of the loading solution because of partitioning into the silicone so that the concentration can be assumed constant and equal to the nominal concentration [52, 60]. An equation analogous to (2) can be used to calculate the loading volume required to avoid depletion. Unfortunately, the required equilibrium partition ratios between solvent and silicone are still relatively scarce. However, examples include the PAH partitioning ratios between silicone and methanol from Reichenberg et al. [61] or Smith et al. [21], and the PAH and PCB partition ratio between silicone and methanol/water (80:20 v/v) from Booij et al. [62]. A variation of the above negligible-depletion approach is to use sequential aliquots of loading solution, each with a smaller volume. This reduces the amount of solvent and compound required. A special case of negligible-depletion loading occurs when loading the silicone to saturation, which translates into equilibrium C_{free} levels in the medium that are at aqueous solubility. Here, a loading solution containing excess compound is used to load the silicone dosing phase. For solid compounds, the crystals present in the loading suspension dissolve to replenish any depletion caused by partitioning into the silicone. At the end of the loading step, the continued presence of crystals indicates that the loading solution, and thus also the silicone, is saturated [49]. If required, the loading suspension can then be decanted and re-used.

The other loading approach is by "spiking" from a concentrated solution. Here, a defined mass of test compound is quantitatively transferred into the silicone from a small volume of concentrated spike solution. This is useful when working with HOCs which are available in small quantities or are costly. Two variations are possible: (1) partitioning from a water-miscible solvent and (2) partitioning from a volatile solvent followed by its evaporation. In the first approach, a concentrated spiking solution of the test compound is made up in a water-miscible solvent, and the appropriate volume (and thus the required mass of compound) is added to the silicone. The HOC is allowed to partition into the silicone, and small volumes of water are then added incrementally, allowing sufficient time between each addition for a new equilibrium partitioning to be reached. By gradually increasing the water to solvent ratio, partitioning into the silicone is favored and the HOC is thereby quantitatively "forced" into the silicone [59]. The water increments should be small enough to avoid the formation of crystals in the solvent:water mixture. In the second approach, a volatile solvent is used to make up the loading spike, and the compound is then forced into the silicone by slow evaporation of the solvent [29]. When using these spiking approaches, C_{silicone} is known and C_{free} can be calculated using the appropriate partition ratio.

For all approaches, sufficient time should be allowed during the loading step to ensure equilibrium partitioning between the loading solution and silicone is reached. Fortunately, when using methanol, or even methanol:water mixtures, this is relatively quick. For example, under quiescent conditions, PAH equilibrium between either pure methanol or methanol:water and silicone is completed within a few hours [18, 21, 52].

6.4 Removal of Loading Solution Traces

After loading is completed, remaining solvent traces have to be removed. When loading using water-miscible solvents, after pouring off the loading solution the silicone surfaces should be immediately wiped using lint-free tissue to soak up adhering solvent and thus prevent formation of a film of microcrystals. This is followed by sequential rinses with small volumes of water to remove any remaining solvent. It is important here to use small volumes of water to minimize removal of the test compound and to leave sufficient contact time for each rinse to ensure complete removal of the loading solvent. Of course this rinsing step necessitates the use of a water-miscible solvent for making up the loading solution. In the case of spiking with a volatile solvent, the evaporation step should be optimized to remove the solvent, minimize losses of the target compound, and avoid the formation of crystals at the polymer surface [45].

6.5 Exposure Confirmation

At experiment completion, the confirmation of exposure demonstrates that the silicone passive dosing phase was loaded to the correct level and negligibly depleted. This is particularly important for toxicity testing, where a criterion for further considering the data is often that exposure has been analytically confirmed. Exposure can be confirmed in different ways. The post-test silicone concentrations can be analyzed following a simple solvent extraction and compared to the initial values [21]. This approach has the advantage that concentrations of HOCs in silicone are relatively high, which simplifies the analysis. Alternatively, the silicone can be equilibrated with a small volume of pure water at the test temperature [49]. Measuring the equilibrium C_{free} values provides a direct measure of the dissolved exposure concentrations in the experimental set-up, but is challenging when working with very low dissolved concentrations.

7 Application of Passive Dosing for Toxicity Testing

In the following sections, a brief overview of the different passive dosing formats that have been applied to date is given, and the opportunities of passive dosing for the toxicity testing of single HOCs and complex mixtures discussed. In addition, the application of passive dosing in set-ups with metabolic transformation is reviewed. An overview of the cited publication is given in Table 1.

Because only a few studies have actually applied passive dosing in in vitro tests in well-plates, case studies that have used larger set-ups are also presented to illustrate what one can learn from these studies for in vitro testing and how the passive dosing formats might be modified for the smaller test systems that are typical for in vitro set-ups. Metabolic activation of some test compounds is required in certain toxicity bioassays. In those case studies covering metabolic transformation, not only toxicity bioassays requiring metabolic activation but also in vitro biotransformation and microbial degradation are covered. This is because the accurate determination of biotransformation rates is a key and emerging issue for the bioaccumulation assessment of chemicals and consequently for chemical regulation. Furthermore, a general understanding of the microbial degradation of contaminants can significantly contribute to our understanding of the fate of chemicals in the environment. Passive dosing is a useful tool for both of these aspects.

7.1 *Passive Dosing Formats Applied in Toxicity Testing*

In one of the first studies that applied passive dosing, C_{free} of halogenated aromatic compounds were controlled in an aquatic algal toxicity test by partitioning from C_{18} Empore disks [50]. Although stable C_{free} levels up to the chemicals' solubility limit were obtained, C_{18} Empore disks were found not to be an ideal passive dosing phase as the silica substrate which is the main component is known to dissolve at pH values above 7.5. Furthermore, the large size of the Empore disks limits their use in in vitro assays. Polystyrene polymer beads were applied by Bopp et al. [73, 74] for the passive dosing of PAHs in microtiter plates and measuring of EROD activity in a fish cell line. Here, the test compounds were first sorbed to the surface of the beads and the cells were then allowed to grow on the surfaces. This approach resulted in consistent concentration-response relationships but it is far from straightforward to translate the bead-sorbed amounts to a universally applicable parameter such as C_{free}. Stir-bars coated with Teflon or silicone have been used for dosing aquatic microalgae [45, 71]. As discussed above, silicone has ideal partitioning properties (absorptive partitioning, linear sorption isotherms, etc.) for passive dosing. In contrast, compounds adsorb to the Teflon surface, making it more challenging to convert the sorbed concentration into a corresponding value of C_{free}. In any case,

Table 1 Case studies covering the application of passive dosing in toxicity and metabolic transformation investigations. Indicated are the type of case study, the passive dosing format, the test compounds, the type of vessel in which the tests were performed, the type of bioassay, and the reference. Note that for case studies in which chemical mixture(s) were tested, single compounds were also investigated. n.s.: not stated in the respective study

Type of case study	Passive dosing format	Test compound(s)	Test vials	Bioassay	Reference (s)
Single compound(s)	C_{18} Empore disks	1,3,5-Tribromobenzene, HCB	n.s.	Algal (*Raphidocelis subcapitata*) growth inhibition	[55]
Single compounds, binary mixture	PDMS coating	PAHs	4-mL glass cuvettes, 96-well plates	Microtox assay (bioluminescence inhibition in *Vibrio fischeri*)	[47]
Single compound(s)	PDMS coating	Retene	20-mL test vials	Fish (*Oryzias latipes*) embryo toxicity	[48]
Single compound(s)	PDMS sheet	Benzo(*a*)pyrene, chlorobenzenes	24-well plates	EROD activity and cytotoxicity in fish (*Oncorhynchus mykiss*) cell lines	[18]
Single compound(s)	Silicone O-rings	PAHs	24-well plates	Cytotoxicity, formation of reactive oxygen species, interleukin-8 promoter induction, secretion of hemotactic protein-1 in human cells and cell lines (PBMC, A549, THP-1)	[21]
Single compound(s)	PDMS film/coating, liquid PDMS oil	PAHs	10-mL glass vials, 60-mL glass jars, 24-well plates	Fish (*Danio rerio*) embryo toxicity	[63]
Single compound(s)	Silicone O-rings	PAHs	24-well plates	Cytotoxicity, cytokine promoter induction and NF-αB binding sequence activation in different human cell lines (A459)	[60]
Single compound(s)	Silica gel	Polybrominated diphenylethers (BDE-47, BDE-99)	20-mL test vials	Larval development and mortality of copepods (*Nitocra spinipes*)	[64]
Single compound(s)	PDMS coating	PAHs	20-mL glass vials	Immobilisation of Daphniids (*Daphnia magna*)	[52]

(continued)

Table 1 (continued)

Type of case study	Passive dosing format	Test compound(s)	Test vials	Bioassay	Reference (s)
Single compound(s)	PDMS tubing	Benzyl butyl phthalate	24-well plates	Mortality of nematodes (*Caenorhabditis elegans*)	[65]
Single compound(s)	PDMS sheets	Chlorpyrifos	6- and 24-well plates	Gene expression in nematodes (*Caenorhabditis elegans*)	[66]
Single compound(s)	PDMS coating, silicone O-rings	Phenanthrene	Glass vials, Erlenmeyer flasks	Embryo and larval development of fish (*Danio rerio*)	[67]
Single compound(s)	PDMS coating	Benzo(*a*)pyrene	60-mL glass jars	Composition of intracellular metabolites in oligochaetes (*Lumbriculus variegatus*)	[68]
Single compound(s)	PDMS coating	PAHs	10-mL glass vials	Bioconcentration in and lethality of springtails (*Folsomia candida*)	[38]
Single compound(s)	PDMS coating	Phenanthrene	10-mL glass vials	Lethality of springtails (*Folsomia candida*)	[69]
Single compound(s)	PDMS coating	Pyrene	500-mL glass beakers	Immobilization of and enzymatic activity in Daphniids (*Daphnia magna*)	[70]
Single compounds	PDMS disks	Acenaphthene, phenanthrene, fluoranthene, benzo(*a*)pyrene	Glass cuvettes	Luminescence inhibition in the bacterium *Vibrio fisheri* (Microtox® bioassay)	[30]
Chemical mixture(s)	Teflon coated stirrer bars	Aroclors 1221, 1242, 1248, 1254	5-L glass vessel	Bioconcentration in algae (*Emiliana huxleyi*)	[71]
Chemical mixture(s)	Triolein inside of LDPE tubing	Phenanthrene, fluoranthene, chrysene, diazinon	600-mL glass vials	Lethality in insects (*Chironomus tentans*), crustaceans (H*yalella azteca, Daphnia magna*) and fish (*Pimephales promelas, Oryzias latipes*)	[72]
Chemical mixture(s)	Different biocompatible bead materials	PAHs, oil refinery effluent	48-well plates	EROD activity and cytotoxicity in fish (*Oncorhynchus mykiss*) cell lines (R1, RTL-W1)	[73]
Chemical mixture(s)	Polystyrene polymer beads	PAHs, groundwater sample	96-well plates	EROD activity and cytotoxicity in fish (*Oncorhynchus mykiss*) cell line (RTL-W1)	[74]

Chemical mixture(s)	Silicone stirrer bars	Pesticides, insecticides, pharmaceuticals, PCBs, PAHs, sediment extracts	n.s.	Algal (*Scenedesmus vacuolatus*) growth inhibition	[45]
Chemical mixture(s)	PDMS coating	Benzo(*k*)fluoranthene, sediment extract	24-(trans)well plates	DR-CALUX assay in rat cells (H4IIE Luc cells)	[29]
Chemical mixture(s)	PDMS coating	PAHs	100-mL glass jars	Lethality of amphipods (*Orchomonella pinguis*, *Corophium volutator*)	[75]
Chemical mixture(s)	Hanging drop	Benzene, toluene, ethylbenzene, xylenes	40-mL volatile organic analysis vials	Viability in human lung carcinoma cells (A549)	[76]
Chemical mixture(s)	PDMS coating	PAHs	10-mL glass vials	Lethality of Artemia nauplii (*Artemia franciscana*)	[77]
Chemical mixture(s)	PDMS coating	PAHs	10-mL glass vials	Immobilization of Daphniids (*Daphnia magna*)	[39]
Chemical mixture(s)	PDMS coating	PAHs	10-mL glass vials	Lethality of springtails (*Folsomia candida*)	[78]
Metabolic transformation	Silicone O-rings	PAHs	24-well plates	AMES fluctuation test (*Salmonella typhimurium* TA98)	[53, 79]
Metabolic transformation	Ethylene vinyl acetate coating	Chrysene, benzo(*a*)-pyrene, PCB 153	2-mL glass vials	Biotransformation in male Sprague–Dawley rat liver S9 homogenates	[80]
Metabolic transformation	Silicone O-rings	Phenanthrene, fluoranthene	4-mL glass vials	Biodegradation by bacterium *Sphigomonas paucimobilis* (EPA505)	[59]
Metabolic transformation	Polyoxymethylene strips	PCB 61, PCB 23	160-mL serum bottles	Biodegradation by bacterium *Dehalobium chlorocoercia*	[81]
Metabolic transformation	Ethylene vinyl acetate coating	Pyrene, chrysene, benzo(*a*) pyrene	2-mL glass vials	Biotransformation in rainbow trout (*Oncorhynchus mykiss*) liver S9	[82]
Metabolic transformation	PDMS coating	Phenanthrene, pyrene	20-mL glass vials	Biodegradation by bacterium *Mycobacterium gilvum*	[83]

for both formats the turbulence that accompanies stirring is unsuitable for many cell types and stir-bars can hardly be applied in microplates.

Given the advantages of silicone discussed in Sect. 6.1, most of the subsequent studies applied films, coatings, or other formats comprised of silicone polymer in small glass vessels or microplates for the passive dosing of various types of HOCs. Brown et al. [47] first introduced polydimethylsiloxane (PDMS) films as the partitioning phase for PAHs in a bacterial toxicity test. Rapid equilibrium partitioning and constant C_{free} of PAHs were obtained in glass cuvettes and polystyrene microplates coated with thin PDMS films. However, when low PDMS volumes were used in microplates, C_{free} was reduced because of irreversible adsorption of PAHs to the polystyrene walls outcompeting the buffering capability of the PDMS dosing phase. In a toxicity study with *Daphnia magna*, PAH exposure was controlled by casting a thick layer of PDMS into the base of small glass vessels [52]. Cast PDMS has also been used for the passive dosing of HOCs in studies with terrestrial springtails, a small collembola [38, 49, 69]. In this set-up, the springtails were exposed either via air or direct contact with the loaded PDMS polymer.

In general, PDMS coatings or films are prepared by mixing a prepolymer with a catalyst and casting it into the base or coating the walls of the vial or well. However, the preparation of such silicone coatings or films is time-consuming and practical training is required to produce homogenous films. Therefore, as a more practical alternative, PDMS sheets and tubes have been used for toxicity testing of HOCs in nematodes [65, 66] and disks of PDMS applied in the bacterial Microtox® bioassay [30].

One of the ways in which the test exposure concentrations can be assessed is by extraction of the PDMS followed by chemical analysis to obtain the polymer concentrations. For this, determination of the exact mass of PDMS is critical and here the use of commercially produced silicone O-rings as a passive dosing phase is advantageous because of their standardized format. They are also practical because, prior to starting the test, a large batch of O-rings can be loaded with test substance in one go and then stored. A further advantage of these silicone O-rings is that, for a range of HOCs, partition ratios from silicone to water are available, and can be used to calculate C_{free} for exposure confirmation if the chemical's concentration in the silicone is measured at the end of the test [60]. In in vitro toxicity tests with human cells and cell lines, these silicone O-rings demonstrated excellent passive dosing performance: concentrations of PAHs in silicone were highly reproducible, equilibrium partitioning was obtained within hours, and C_{free} was stable over more than 72 h [21]. The silicone O-rings further proved to be very versatile for tests with different cell types including primary cells, cell lines, and even adhering cells [21]. Butler et al. [67] used silicone O-rings as a passive dosing device in small glass vials but also in a flow-through test system to investigate the chronic toxicity of phenanthrene to embryo and larval life stages of zebrafish.

Irrespective of the nature of the passive dosing phase (e.g., silicone coatings, films, sheets, or O-rings as well as other materials), understanding the exact C_{free} exposure concentrations at the cell surface can be tricky when the cells can have direct contact with the dosing polymer or at least when they are not all equidistant from the dosing source [18]. To circumvent this, more sophisticated passive dosing systems for in vitro assays have been developed. Fish cells were grown on the membrane of well plate inserts and introduced into microplates to which loaded PDMS sheets had been added. In this way, direct contact of cells and the loaded polymer was prevented, and the toxicity of chlorobenzenes and benzo(*a*)pyrene could be measured at constant and well-defined C_{free} [18]. A similar approach was applied by Booij et al. [29], who coated the bottom of commercially available transwell plates with PDMS silicone. Adherent cells were grown on the membrane of transwell plate inserts, which were permeable to the test solutes whilst preventing direct contact with the loaded PDMS coating.

For volatile organic compounds, a hanging drop system has been developed for human lung carcinoma cells [76]. The cells were suspended in a drop of culture medium hanging on the inner side of the vial's lid during exposure via air. Nevertheless, testing volatile compounds in microplates remains challenging because cross-contamination of adjacent wells via the headspace might occur and volatile compounds might also get lost [21].

7.2 Case Studies I: Single Compounds

Toxic effects of concentrations up to several orders of magnitude above the chemical's aqueous solubility have been published, despite C_{free} being the driver for the toxic response. For example, in conventional fish embryo toxicity tests with static or semi-static renewal of the contaminated medium, excess concentrations of hydrophobic or easily degradable chemicals are often added to account for chemical losses and to allow the measurement of adverse effects in the test organisms. However, because C_{free} deviates from these above-saturation nominal concentrations, a chemical's toxicity is underestimated and consequently inaccurate [48]. Therefore, the major objectives of the first studies that applied passive dosing were to test the toxicity of single HOCs at their solubility limit by controlling and maintaining their maximum C_{free} throughout the test (e.g., [50]).

In these studies, the application of passive dosing often resulted in an increased test sensitivity compared with conventional dosing approaches. This higher sensitivity is obvious when comparing the effective concentrations measured in the passive dosing studies with literature data, and by the direct comparison of different dosing techniques. For example, when measuring the toxicity of an alkyl-substituted phenanthrene to fish embryos, lower effective concentrations were obtained with passive dosing compared with static and semi-static exposures [48]. Similarly, higher lethality and a more consistent concentration-related toxicity were observed in an aquatic toxicity test with benthic invertebrates when applying poorly water-soluble substances by passive dosing compared with a conventional semi-static test system [64]. In Bougeard et al. [79] the mutagenic activities of a number of nitro- and keto-substituted PAHs were determined in the Ames fluctuation test without metabolic activation using passive dosing from silicone O-rings and compared to solvent spiking. Although both approaches produced consistent concentration-response curves, the 50% effect concentrations based on the passive dosing derived C_{free} were 3 to 33 times lower than those calculated using the nominal concentrations from the solvent dosed experiments. This effect is even observed in relatively simple toxicity tests such as the Microtox® bioassay which takes place at 15°C, uses a medium without any sorbing components, and is of short duration. For example, the test sensitivity was increased by a factor of 3 for phenanthrene and 12 for fluoranthene when using passive dosing compared to solvent spiking [30]. In an aquatic toxicity test with nematodes, Roh et al. [66] demonstrated that the C_{free} of chlorpyrifos decreased by up to 20% of the initial value after dosing the test substance using a co-solvent. Furthermore, the resulting C_{free} depended both on the size of the microplates (6- vs 24-well plates) and the presence of the test organisms in the wells, resulting in greatly varying gene expression in the nematodes. In contrast, when passive dosing was applied, stable C_{free} values were obtained throughout the experiment and resulted in a quantitative concentration-response related gene expression.

These examples illustrate that passive dosing can improve the toxicity testing of chemicals but can also provide further possibilities for understanding the molecular processes involved in chemical stress because it allows for more sensitive markers to be investigated. When applying passive dosing in an in vitro test with human bronchial epithelial cells, Oostingh et al. [60] showed that PAHs can still lead to immunomodulatory effects despite no cytotoxicity being observed. Control of the lower exposure levels was particularly critical, because lower dissolved PAH concentrations often induced higher immunomodulatory responses when compared to the higher concentrations.

Another advantage of passive dosing in comparison with conventional spiking techniques is that effect data from in vitro studies can be extrapolated to the in vivo situation by improving the link between the in vitro exposure and internal exposure in cells or tissue. For example, the in vitro C_{free} can be directly related to C_{free} in tissues. Alternatively, equilibrium partitioning concentrations in the lipid fraction of the test cells can be obtained by measurement or modeling and then be compared to the lipid-normalized concentrations actually measured in tissue [84].

7.3 Case Studies II: Mixtures

In the environment, organisms are exposed to complex environmental mixtures of chemicals, whereas toxicity is still mainly tested for single compounds. The detrimental effects of a chemical mixture can be higher compared to single compounds, and a chemical risk assessment and management strategy focusing on single compounds might therefore underestimate the real environmental risk. However, the correct toxicity testing of HOC mixtures is more challenging than for single compounds because losses via sorption or volatilization are highly compound-specific. Consequently, not only the total concentrations but also the bioavailable mixture profile can change during the toxicity bioassay. Passive dosing can control the levels of the individual constituents and therefore also the mixture profile, but a prerequisite is that their partitioning behaviors are not mutually influenced when simultaneously dosing several compounds such that their solubilities remain additive. Fortunately, this seems to be the case. For example, the C_{free} of PAHs was found to be similar either when tested as single compounds or in a mixture [53].

Analogous to the situation observed for the toxicity testing of single compounds, passive dosing of mixtures results in an increased bioassay sensitivity compared to conventional solvent spiking. For example, the dioxin-like activity of sediment extracts was higher when administered by passive dosing compared to solvent spiking [29]. In addition, the concentration-dependent response of dilutions of sediment extracts was more reproducible after first loading silicone rods with varying amounts of the extracts and then using these in passive dosing mode in an algal test [46].

With regard to chemical mixtures, passive dosing has primarily been applied for the toxicity testing of PAHs (Table 1). Whilst the majority of studies have focused on artificial chemical mixtures (e.g., [39, 78]), only a few "real" environmental mixtures have been applied in toxicity tests using passive dosing. Rojo-Nieto et al. [77] recreated an analyzed seawater mixture comprised of seven PAHs in an aquatic toxicity test with *Artemia franciscana* nauplii. Direct recreation of the in situ mixture levels and profile did not result in toxicity, and the mixture had to be enriched by orders of magnitude before any toxicity was observed. In other studies, total sediment extracts have been loaded onto different passive dosing formats which were then introduced into the respective bioassays as partitioning phase to mimic partitioning better between the sediment and porewater as found in the field [29, 45, 72]. This partitioning step has been found to be rather important, because it can result in quite different compounds being identified as the key toxicity drivers compared to the situation when the extracts are directly spiked into the test using solvent. Heinis et al. [72] elucidated this theoretically for a binary mixture of diazinon and dichlordiphenyldichlorethene (DDE) in sediment interstitial waters. When sediment extracts were directly spiked into the bioassay, toxicity was mainly induced by DDE even though diazinon had a 100-fold greater toxicity in the native sediment compared to DDE. This resulted from the low bioavailability of DDE in the sediment pore water because of its strong partitioning to the sediment particles. Hence, when using a conventional solvent spike of the sediment extract the primary driver for toxicity would have been missed. Bandow et al. [46] fractionated sediment extracts as part of an effect-directed analysis procedure, and then loaded the extracts onto silicone coated stir-bars for passive dosing in an algal toxicity bioassay. Passive dosing resulted in polar compounds being identified as the key toxicants, whereas conventional solvent spiking of the extracts resulted in PAHs being identified as playing an important role. Therefore, in addition to its role in maintaining the mixture exposure regime the inclusion of a passive dosing step was important in recreating (at least to some extent) the in situ partitioning occurring in the sediment. Although not aimed at measuring toxicity, Teflon stir-bars loaded with Aroclor mixtures were used to dose PCBs passively in a 5-day bioconcentration experiment with aquatic algae, and resulted in constant concentrations being maintained, even in the face of uptake by the algal cells [71].

An interesting development is the combination of passive dosing with passive sampling. Here, the polymer is first used as sampling phase for accumulating the bioavailable portion of a chemical mixture, either directly in the environment or in an environmental sample, and subsequently introduced as the passive dosing phase into the bioassay for toxicity testing. This combination of passive sampling and passive dosing is discussed in detail in [85].

7.4 Case Studies III: Metabolic Transformation

Some toxicity bioassays require that a compound is metabolically activated before it can exert its toxicity [26]. Here, the inherently low aqueous solubility of HOCs leads to a number of challenges. Only low amounts can be added before the aqueous saturation level is reached, limiting the amount of compound that can be enzymatically metabolized. This in turn limits the amount of toxic metabolites produced and thus also the apparent sensitivity of the test. Simply adding excess compound can result in problems with cytotoxicity or low bioavailability because of slow dissolution kinetics. This can be solved by passive dosing because the high solubility of the HOC in the dosing polymer allows a large mass of test compound to be introduced to support high compound turnover in the aqueous phase, even when C_{free} is low – see (1). Thus, even at low non-cytotoxic concentrations passive dosing becomes a practical tool for introducing a sufficient mass of HOC into in vitro bioassays requiring metabolic activation to support increased turnover [59].

The concentration-dependent mutagenic activity of benzo(*a*)pyrene was measured in the Ames fluctuation test with metabolic activation for passive dosing and solvent spiking [53]. With solvent spiking, concentrations in the medium had to exceed the aqueous solubility of benzo(*a*)pyrene before any mutagenic activity was observed. In contrast, with passive dosing the concentration-response curves were more reproducible and shifted to lower concentrations by several orders of magnitude. This was because of the passive dosing supporting a higher production of the mutagenic metabolites.

A special case arises when passive dosing is used for studying the biotransformation processes of HOCs. In addition to taking the various abiotic losses into account, the HOC loss kinetics from the aqueous phases caused by the biotransformation process requires consideration. When applying passive dosing to study HOC biotransformation, two different approaches have been used. In the first, the time course in the dosing polymer and medium concentrations were measured in parallel set-ups with and without the biotransforming agent [80, 82]. Here, the passive dosing phase was dimensioned to ensure that polymer depletion could be measured over the duration of the experiment. A mass balance approach was then applied to derive the biotransformation kinetics, and passive dosing was found to give consistently higher biotransformation rates compared to those determined from the concentration decrease in the medium after solvent spiking. In the second approach, the passive dosing phase was dimensioned such that depletion was minimized [59]. Although the biotransformation loss process per se implies that the silicone dosing phase is progressively depleted, provided that this is kept within limits then it is valid to assume a constant C_{silicone} and thus constant C_{free}. C_{free} is now defined by steady-state rather than equilibrium partitioning considerations, and the steady-state value of C_{free} can be calculated provided the release kinetics from the silicone dosing phase and loss kinetics caused by the biotransformation are both known. The former was determined in an initial release experiment, whereas the biotransformation kinetics was inferred from the $^{14}CO_2$ production. Alternative endpoints for determining the biotransformation kinetics might include metabolite production or even measurement of the small depletion in the silicone concentrations.

8 Outlook for Passive Dosing

Passive dosing has made quite considerable progress in improving exposure control in toxicity tests but nevertheless a number of challenges still remain. So far, passive dosing has not yet or only rarely been applied in microplates with smaller well sizes, such as 96- and 384-well microplates, which are more commonly used for in vitro assays. Miniaturized passive dosing systems therefore urgently need to be developed and made available for the high-throughput toxicity testing of chemicals in such microplates. Smaller formats also lead to reduced consumption of test substance. In practice, this means the approach needs to be simplified, particularly with regard to the initial loading of the passive dosing phase. The exchange kinetics between the passive dosing and aqueous phases determine the speed of equilibration, and thus how effectively losses are compensated. The fastest passive dosing formats have equilibration times in the order of tens of minutes (e.g., [59]). Although this is sufficient for longer in vitro assays or when a pre-equilibration step can be incorporated, faster passive dosing formats would be advantageous for assays with very short response times. The application domain of silicone as a passive dosing polymer is mainly targeted towards HOCs because their exposure control in in vitro assays is particularly challenging. Nevertheless, extending passive dosing to include more polar and volatile organic compounds is useful because these face similar challenges to the HOCs when it comes to test losses, albeit via different mechanisms. This requires finding passive dosing polymers with higher affinities for these compounds, at the same time still exhibiting the desirable characteristics of inertness, linear partitioning, and low internal mass transfer resistance.

Passive dosing has recently been applied in studies investigating chemical exposure in combination with other abiotic stressors [68, 69]. For example, Schmidt et al. [69] developed a dual exposure system for assessing chemical and drought stress in springtails. For this purpose, they placed PDMS coated vials in glass jars containing saline solution for controlling the water vapor pressure. In this way, the combined effects of both stressors could be investigated. In the future, passive dosing might consequently form part of more complex set-ups looking at the interactions between multiple stressors to ensure a rigorous control of the HOC exposure. Furthermore, it might be applied to investigate the contribution of the particle-associated HOC fraction on toxicity as studied by Zhang et al. [70] with *Daphnia magna*.

Finally, introducing natural chemical mixtures into in vitro test systems by combining passive sampling and passive dosing is still in its infancy. However, it is envisaged that this approach can provide important knowledge with regard to the combined toxicity of mixtures in different environmental compartments as is discussed in ref. [85].

References

1. EU (2006) Regulation (EC) No. 1907/2006 of the European Parliament and of the Council of 18 December 2006 concerning the Registration, Evaluation, Authorisation and Restriction of Chemicals (REACH)
2. Hartung T, Rovida C (2009) Chemical regulators have overreached. Nature 460:1080–1081
3. Rovida C, Hartung T (2009) Re-evaluation of animal numbers and costs for in vivo tests to accomplish REACH legislation requirements for chemicals – a report by the transatlantic think tank for toxicology (t(4)). ALTEX 26:187–208
4. Dix DJ, Houck KA, Martin MT, Richard AM, Setzer RW, Kavlock RJ (2007) The ToxCast program for prioritizing toxicity testing of environmental chemicals. Toxicol Sci 95:5–12
5. Lilienblum W, Dekant W, Foth H, Gebel T, Hengstler JG, Kahl R, Kramer PJ, Schweinfurth H, Wollin KM (2008) Alternative methods to safety studies in experimental animals: role in the risk assessment of chemicals under the new European Chemicals Legislation (REACH). Arch Toxicol 82:211–236
6. Adeleye Y, Andersen M, Clewell R, Davies M, Dent M, Edwards S, Fowler P, Malcomber S, Nicol B, Scott A, Scott S, Sun B, Westmoreland C, White A, Zhang Q, Carmichael PL (2015) Implementing toxicity testing in the 21st century (TT21C): making safety decisions using toxicity pathways, and progress in a prototype risk assessment. Toxicology 332:102–111
7. National Research Council of the National Academy of Sciences (NRC) (2007) Toxicity testing in the 21st century: a vision and a strategy. The National Academies, Washington
8. EU (2009) Regulation (EC) No 1223/2009 of the European Parliament and of the Council of 30 November 2009 on cosmetic products
9. Escher BI, Allinson M, Altenburger R, Bain PA, Balaguer P, Busch W, Crago J, Denslow ND, Dopp E, Hilscherova K, Humpage AR, Kumar A, Grimaldi M, Jayasinghe BS, Jarosova B, Jia A, Makarov S, Maruya KA, Medvedev A, Mehinto AC, Mendez JE, Poulsen A, Prochazka E, Richard J, Schifferli A, Schlenk D, Scholz S, Shiraish F, Snyder S, Su G, Tang JYM, van der Burg B, van der Linden SC, Werner I, Westerheide SD, Wong CKC, Yang M, Yeung BHY, Zhang X, Leusch FDL (2014) Benchmarking organic micropollutants in wastewater, recycled water and drinking water with in vitro bioassays. Environ Sci Technol 48:1940–1956
10. van der Burg B, van der Linden S, Man H, Winter R, Jonker L, van Vugt-Lussenburg B, Brouwer A (2013) A panel of quantitative Calux® reporter gene assays for reliable high-throughput toxicity screening of chemicals and complex mixtures. In: Steinberg P (ed) High-throughput screening methods in toxicity testing. Wiley, Hoboken

11. Wilk-Zasadna I, Bernasconi C, Pelkonen O, Coecke S (2015) Biotransformation in vitro: an essential consideration in the quantitative in vitro-to-in vivo extrapolation (QIVIVE) of toxicity data. Toxicology 332:8–19
12. Escher BI, van Daele C, Dutt M, Tang JYM, Altenburger R (2013) Most oxidative stress response in water samples comes from unknown chemicals: the need for effect-based water quality trigger values. Environ Sci Technol 47:7002–7011
13. Dayeh VR, Schirmer K, Bols NC (2002) Applying whole-water samples directly to fish cell cultures in order to evaluate the toxicity of industrial effluent. Water Res 36:3727–3738
14. Johnson BT (2005) Microtox® acute toxicity test. In: Small-scale freshwater toxicity investigations, volume 1, toxicity test methods. Springer, Dordrecht, pp 69–105
15. BRD (2005) Abwasserabgabengesetz in der Fassung der Bekanntmachung vom 18. Januar 2005 (BGBl. I S. 114), das zuletzt durch Artikel 2 der Verordnung vom 2. September 2014 (BGBl. I S. 1474) geändert worden ist
16. Groothuis FA, Heringa MB, Nicol B, Hermens JLM, Blaauboer BJ, Kramer NI (2015) Dose metric considerations in in vitro assays to improve quantitative in vitro–in vivo dose extrapolations. Toxicology 332:30–40
17. Heringa MB, Schreurs RHMM, Busser F, van der Saag PT, van der Burg B, Hermens JLM (2004) Toward more useful in vitro toxicity data with measured free concentrations. Environ Sci Technol 38:6263–6270
18. Kramer NI, Busser FJM, Oosterwijk MTT, Schirmer K, Escher BI, Hermens JLM (2010) Development of a partition-controlled dosing system for cell assays. Chem Res Toxicol 23:1806–1814
19. Kramer NI, Krismartina M, Rico-Rico A, Blaauboer BJ, Hermens JLM (2012) Quantifying processes determining the free concentration of phenanthrene in basal cytotoxicity assays. Chem Res Toxicol 25:436–445
20. Schreiber R, Altenburger R, Paschke A, Kuster E (2008) How to deal with lipophilic and volatile organic substances in microtiter plate assays. Environ Toxicol Chem 27:1676–1682
21. Smith KEC, Oostingh GJ, Mayer P (2010) Passive dosing for producing defined and constant exposure of hydrophobic organic compounds during in vitro toxicity tests. Chem Res Toxicol 23:55–65
22. Gülden M, Schreiner J, Seibert H (2015) In vitro toxicity testing with microplate cell cultures: impact of cell binding. Toxicology 332:41–51
23. Stalter D, Dutt M, Escher BI (2013) Headspace-free setup of in vitro bioassays for the evaluation of volatile disinfection by-products. Chem Res Toxicol 26:1605–1614
24. Escher BI, Hermens JLM (2004) Internal exposure: linking bioavailability to effects. Environ Sci Technol 38:455A–462A
25. Endo SGK (2011) Serum albumin binding of structurally diverse neutral organic compounds: data and models. Chem Res Toxicol 24:2293–2301
26. Ku WW, Bigger A, Brambilla G, Glatt H, Gocke E, Guzzie PJ, Hakura A, Honma M, Martus HJ, Obach RS, Roberts S (2007) Strategy for genotoxicity testing – metabolic considerations. Mutat Res Genet Toxicol Environ Mutagen 627:59–77
27. Gülden M, Seibert H (2003) In vitro–in vivo extrapolation: estimation of human serum concentrations of chemicals equivalent to cytotoxic concentrations in vitro. Toxicology 189:211–222
28. Tanneberger K, Rico-Rico A, Kramer NI, Busser FJM, Hermens JLM, Schirmer K (2010) Effects of solvents and dosing procedure on chemical toxicity in cell-based in vitro assays. Environ Sci Technol 44:4775–4781
29. Booij P, Lamoree MH, Leonards PEG, Cenijn PH, Klamer HJC, van Vliet LA, Akerman J, Legler J (2011) Development of a polydimethylsiloxane film-based passive dosing method in the in vitro DR-Calux® assay. Environ Toxicol Chem 30:898–904
30. Smith KEC, Jeong Y, Kim J (2015) Passive dosing versus solvent spiking for controlling and maintaining hydrophobic organic compound exposure in the Microtox® assay. Chemosphere 139:174–180
31. Schirmer K (2006) Proposal to improve vertebrate cell cultures to establish them as substitutes for the regulatory testing of chemicals and effluents using fish. Toxicology 224:163–183

32. Poirier A, Lave T, Portmann R, Brun ME, Senner F, Kansy M, Grimm HP, Funk C (2008) Design, data analysis, and simulation of in vitro drug transport kinetic experiments using a mechanistic in vitro model. Drug Metab Dispos 36:2434–2444
33. Brinkmann M, Preuss TG, Hollert H (2015) Advancing in vitro-in vivo extrapolations of mechanism-specific toxicity data through toxicokinetic modeling. Adv Biochem Eng Biotechnol. doi:10.1007/10_2015_5015
34. Ferguson J (1939) The use of chemical potentials as indices of toxicity. Proc R Soc B Biol Sci 127:387–404
35. Mackay D, Arnot JA, Celsie A, Orazietti A, Parnis JM (2014) QSARs for aquatic toxicity: celebrating, extending and displaying the pioneering contributions of Ferguson, Konemann and Veith. SAR QSAR Environ Res 25:343–355
36. Mackay D, Arnot JA, Petkova EP, Wallace KB, Call DJ, Brooke LT, Veith GD (2009) The physicochemical basis of QSARs for baseline toxicity. SAR QSAR Environ Res 20:393–414
37. Mayer P, Reichenberg F (2006) Can highly hydrophobic organic substances cause aquatic baseline toxicity and can they contribute to mixture toxicity? Environ Toxicol Chem 25:2639–2644
38. Schmidt SN, Smith KEC, Holmstrup M, Mayer P (2013) Uptake and toxicity of polycyclic aromatic hydrocarbons in terrestrial springtails studying bioconcentration kinetics and linking toxicity to chemical activity. Environ Toxicol Chem 32:361–369
39. Smith KEC, Schmidt SN, Dom N, Blust R, Holmstrup M, Mayer P (2013) Baseline toxic mixtures of non-toxic chemicals: "Solubility Addition" increases exposure for solid hydrophobic chemicals. Environ Sci Technol 47:2026–2033
40. van Wezel AP, Opperhuizen A (1995) Narcosis due to environmental pollutants in aquatic organisms – residue-based toxicity, mechanisms, and membrane burdens. Crit Rev Toxicol 25:255–279
41. Flaten GE, Luthman K, Vaaskog T, Brandl M (2008) Drug permeability across a phospholipid vesicle-based barrier – 4. The effect of tensides, co-solvents and pH changes on barrier integrity and on drug permeability. Eur J Pharm Sci 34:173–180
42. Kais B, Schneider KE, Keiter S, Henn K, Ackermann C, Braunbeck T (2013) DMSO modifies the permeability of the zebrafish (*Danio rerio*) chorion – implications for the fish embryo test (FET). Aquat Toxicol 140:229–238
43. Lammer E, Kamp HG, Hisgen V, Koch M, Reinhard D, Salinas ER, Wendler K, Zok S, Braunbeck T (2009) Development of a flow-through system for the fish embryo toxicity test (FET) with the zebrafish (*Danio rerio*). Toxicol In Vitro 23:1436–1442
44. van der Valk J, Brunner D, De Smet K, Svenningsen AF, Honegger P, Knudsen LE, Lindl T, Noraberg J, Price A, Scarino ML, Gstraunthaler G (2010) Optimization of chemically defined cell culture media – replacing fetal bovine serum in mammalian in vitro methods. Toxicol in Vitro 24:1053–1063
45. Bandow N, Altenburger R, Lübcke-von Varel U, Paschke A, Streck G, Brack W (2009) Partitioning-based dosing: an approach to include bioavailability in the effect-directed analysis of contaminated sediment samples. Environ Sci Technol 43:3891–3896
46. Bandow N, Altenburger R, Streck G, Brack W (2009) Effect-directed analysis of contaminated sediments with partition-based dosing using green algae cell multiplication inhibition. Environ Sci Technol 43:7343–7349
47. Brown RS, Akhtar P, Akerman J, Hampel L, Kozin IS, Villerius LA, Klamer HJC (2001) Partition controlled delivery of hydrophobic substances in toxicity tests using poly(dimethylsiloxane) (PDMS) films. Environ Sci Technol 35:4097–4102
48. Kiparissis Y, Akhtar P, Hodson PV, Brown RS (2003) Partition-controlled delivery of toxicants: a novel in vivo approach for embryo toxicity testing. Environ Sci Technol 37:2262–2266
49. Mayer P, Holmstrup M (2008) Passive dosing of soil invertebrates with polycyclic aromatic hydrocarbons: limited chemical activity explains toxicity cutoff. Environ Sci Technol 42:7516–7521

50. Mayer P, Wernsing J, Tolls J, de Maagd PGJ, Sijm D (1999) Establishing and controlling dissolved concentrations of hydrophobic organics by partitioning from a solid phase. Environ Sci Technol 33:2284–2290
51. Turcotte D, Akhtar P, Bowerman M, Kiparissis Y, Brown RS, Hodson PV (2011) Measuring the toxicity of alkyl-phenanthrenes to early life stages of medaka (*Oryzia latipes*) using partition-controlled delivery. Environ Toxicol Chem 30:487–495
52. Smith KEC, Dom N, Blust R, Mayer P (2010) Controlling and maintaining exposure of hydrophobic organic compounds in aquatic toxicity tests by passive dosing. Aquat Toxicol 98:15–24
53. Smith KEC, Heringa MB, Uytewaal M, Mayer P (2013) The dosing determines mutagenicity of hydrophobic compounds in the Ames II assay with metabolic transformation: passive dosing versus solvent spiking. Mutat Res Genet Toxicol Environ Mutagen 750:12–18
54. Rusina TP, Smedes F, Klanova J, Booij K, Holoubek I (2007) Polymer selection for passive sampling: a comparison of critical properties. Chemosphere 68:1344–1351
55. Mayer P, Vaes WHJ, Hermens JLM (2000) Absorption of hydrophobic compounds into the poly(dimethylsiloxane) coating of solid-phase microextraction fibers: high partition coefficients and fluorescence microscopy images. Anal Chem 72:459–464
56. Kwon JH, Wuethrich T, Mayer P, Escher BI (2007) Dynamic permeation method to determine partition coefficients of highly hydrophobic chemicals between poly(dimethylsiloxane) and water. Anal Chem 79:6816–6822
57. Difilippo ELERP (2010) Assessment of PDMS-water partition coefficients: implications for passive environmental sampling of hydrophobic organic compounds. Environ Sci Technol 44:6917–6925
58. Endo S, Droge STJ, Goss KU (2011) Polyparameter linear free energy models for polyacrylate fiber-water partition coefficients to evaluate the efficiency of solid-phase microextraction. Anal Chem 83:1394–1400
59. Smith KEC, Rein A, Trapp S, Mayer P, Karlson UG (2012) Dynamic passive dosing for studying the biotransformation of hydrophobic organic chemicals: microbial degradation as an example. Environ Sci Technol 46:4852–4860
60. Oostingh GJ, Smith KEC, Tischler U, Radauer-Preiml I, Mayer P (2015) Differential immunomodulatory responses to nine polycyclic aromatic hydrocarbons applied by passive dosing. Toxicol In Vitro 29:345–351
61. Reichenberg F, Smedes F, Jonsson JA, Mayer P (2008) Determining the chemical activity of hydrophobic organic compounds in soil using polymer coated vials. Chem Cent J 2:10
62. Booij K, Smedes F, van Weerlee EM (2002) Spiking of performance reference compounds in low density polyethylene and silicone passive water samplers. Chemosphere 46:1157–1161
63. Seiler TB, Best N, Fernqvist MM, Hercht H, Smith KEC, Braunbeck T, Mayer P, Hollert H (2014) PAH toxicity at aqueous solubility in the fish embryo test with *Danio rerio* using passive dosing. Chemosphere 112:77–84
64. Breitholtz M, Ricklund N, Bengtsson BE, Persson NJ (2007) Silica gel as a particulate carrier of poorly water-soluble substances in aquatic toxicity testing. Aquat Toxicol 82:251–264
65. Kwon HC, Roh JY, Lim D, Choi J, Kwon JH (2011) Maintaining the constant exposure condition for an acute *Caenorhabditis elegans* mortality test using passive dosing. Environ Health Toxicol 26:e2011015
66. Roh JY, Lee H, Kwon JH (2014) Changes in the expression of cyp35a family genes in the soil nematode *Caenorhabditis elegans* under controlled exposure to chlorpyrifos using passive dosing. Environ Sci Technol 48:10475–10481
67. Butler JD, Parkerton TF, Letinski DJ, Bragin GE, Lampi MA, Cooper KR (2013) A novel passive dosing system for determining the toxicity of phenanthrene to early life stages of zebrafish. Sci Total Environ 463:952–958
68. Agbo SO, Keinanen M, Keski-Saari S, Lemmetyinen J, Akkanen J, Leppänen MT, Mayer P, Kukkonen JVK (2013) Changes in *Lumbriculus variegatus* metabolites under hypoxic

exposure to benzo(*a*)pyrene, chlorpyrifos and pentachlorophenol: Consequences on biotransformation. Chemosphere 93:302–310

69. Schmidt SN, Holmstrup M, Damgaard C, Mayer P (2014) Simultaneous control of phenanthrene and drought by dual exposure system: the degree of synergistic interactions in springtails was exposure dependent. Environ Sci Technol 48:9737–9744
70. Zhang X, Xia X, Li H, Zhu B, Dong J (2015) Bioavailability of pyrene associated with suspended sediment of different grain sizes to *Daphnia magn*a as investigated by passive dosing devices. Environ Sci Technol 49:10127–10135
71. Gerofke A, Komp P, McLachlan MS (2004) Stir bar contamination: a method to establish and maintain constant water concentrations of poorly water-soluble chemicals in bioconcentration experiments. Water Res 38:3411–3419
72. Heinis LJ, Highland TL, Mount DR (2004) Method for testing the aquatic toxicity of sediment extracts for use in identifying organic toxicants in sediments. Environ Sci Technol 38:6256–6262
73. Bopp SK, Bols NC, Schirmer K (2006) Development of a solvent-free, solid-phase in vitro bioassay using vertebrate cells. Environ Toxicol Chem 25:1390–1398
74. Bopp SK, McLachlan MS, Schirmer K (2007) Passive sampler for combined chemical and toxicological long-term monitoring of groundwater: the ceramic toximeter. Environ Sci Technol 41:6868–6876
75. Engraff M, Solere C, Smith KEC, Mayer P, Dahllof I (2011) Aquatic toxicity of PAHs and PAH mixtures at saturation to benthic amphipods: linking toxic effects to chemical activity. Aquat Toxicol 102:142–149
76. Liu FF, Escher BI, Were S, Duffy L, Ng JC (2014) Mixture effects of benzene, toluene, ethylbenzene, and xylenes (BTEX) on lung carcinoma cells via a hanging drop air exposure system. Chem Res Toxicol 27:952–959
77. Rojo-Nieto E, Smith KEC, Perales JA, Mayer P (2012) Recreating the seawater mixture composition of HOCs in toxicity tests with *Artemia franciscana* by passive dosing. Aquat Toxicol 120:27–34
78. Schmidt SN, Holmstrup M, Smith KEC, Mayer P (2013) Passive dosing of polycyclic aromatic hydrocarbon (PAH) mixtures to terrestrial springtails: linking mixture toxicity to chemical activities, equilibrium lipid concentrations, and toxic units. Environ Sci Technol 47:7020–7027
79. Bougeard C, Gallampois C, Brack W (2011) Passive dosing: an approach to control mutagen exposure in the Ames fluctuation test. Chemosphere 83:409–414
80. Lee YS, Otton SV, Campbell DA, Moore MM, Kennedy CJ, Gobas FAPC (2012) Measuring in vitro biotransformation rates of super hydrophobic chemicals in rat liver S9 fractions using thin-film sorbent-phase dosing. Environ Sci Technol 46:410–418
81. Lombard NJ, Ghosh U, Kjellerup BV, Sowers KR (2014) Kinetics and threshold level of 2,3,4,5-tetrachlorobiphenyl dechlorination by an organohalide respiring bacterium. Environ Sci Technol 48:4353–4360
82. Lee YS, Lee DHY, Delafoulhouze M, Otton SV, Moore MM, Kennedy CJ, Gobas FAPC (2014) In vitro biotransformation rates in fish liver S9: effect of dosing techniques. Environ Toxicol Chem 33:1885–1893
83. Tejeda-Agredano MC, Mayer P, Ortega-Calvo JJ (2014) The effect of humic acids on biodegradation of polycyclic aromatic hydrocarbons depends on the exposure regime. Environ Pollut 184:435–442
84. Jahnke A, McLachlan MS, Mayer P (2008) Equilibrium sampling: partitioning of organochlorine compounds from lipids into polydimethylsiloxane. Chemosphere 73:1575–1581
85. Jahnke A, Witt G, Schäfer S, Haase N, Escher BI (2015) Combining passive sampling with toxicological characterization of complex mixtures of pollutants from the aquatic environment. Adv Biochem Eng Biotechnol. doi:10.1007/10_2015_5014

Adv Biochem Eng Biotechnol (2017) 157: 293–318
DOI: 10.1007/10_2015_5015

Published online: 13 September 2016

Advancing In Vitro–In Vivo Extrapolations of Mechanism-Specific Toxicity Data Through Toxicokinetic Modeling

Markus Brinkmann, Thomas G. Preuss, and Henner Hollert

Abstract International legislation, such as the European REACH regulation (registration, evaluation, authorization, and restriction of chemicals), mandates the assessment of potential risks of an ever-growing number of chemicals to the environment and human health. Although this legislation is considered one of the most important investments in consumer safety ever, the downside is that the current testing strategies within REACH rely on extensive animal testing. To address the ethical conflicts arising from these increased testing requirements,

---

M. Brinkmann (✉)
Department of Ecosystem Analysis, Institute for Environmental Research, Aachen Biology and Biotechnology – ABBt, RWTH Aachen University, Worringerweg 1, 52074 Aachen, Germany
e-mail: markus.brinkmann@bio5.rwth-aachen.de

T.G. Preuss
Chair of Environmental Biology and Chemodynamics, Institute for Environmental Research, Aachen Biology and Biotechnology – ABBt, RWTH Aachen University, Worringerweg 1, 52074 Aachen, Germany

Bayer CropScience AG, Monheim am Rhein, Germany

H. Hollert (✉)
Department of Ecosystem Analysis, Institute for Environmental Research, Aachen Biology and Biotechnology – ABBt, RWTH Aachen University, Worringerweg 1, 52074 Aachen, Germany

College of Resources and Environmental Science, Chongqing University, 1 Tiansheng Road Beibei, Chongqing 400715, China

College of Environmental Science and Engineering and State Key Laboratory of Pollution Control and Resource Reuse, Tongji University, 1239 Siping Road, Shanghai, China

State Key Laboratory of Pollution Control and Resource Reuse, School of the Environment, Nanjing University, Nanjing, China
e-mail: henner.hollert@bio5.rwth-aachen.de

decision-makers, such as the European Chemicals Agency (ECHA), are committed to Russel and Burch's 3R principle (i.e., reduction, replacement, refinement) by demanding that animal experiments should be substituted with appropriate alternatives whenever possible. A potential solution of this dilemma might be the application of in vitro bioassays to estimate toxic effects using cells or cellular components instead of whole organisms. Although such assays are particularly useful to assess potential mechanisms of toxic action, scientists require appropriate methods to extrapolate results from the in vitro level to the situation in vivo. Toxicokinetic models are a straightforward means of bridging this gap. The present chapter describes different available options for in vitro-in vivo extrapolation (IVIVE) of mechanism-specific effects focused on fish species and also reviews the implications of confounding factors during the conduction of in vitro bioassays and their influence on the optimal choice of different dose metrics.

Keywords Bioassay, IVIVE, PBPK, PBTK, Predictive toxicology, Toxicokinetics

Contents

1 Introduction

The emission of anthropogenic chemicals into the environment is a key determinant for water quality and an issue of increasing public and scientific interest. To protect prospectively the environment and ultimately also us humans from the negative consequences of exposure to environmental chemicals, legislation of varying rigor such as the European REACH regulation (which concerns the registration, evaluation, authorisation, and restriction of chemicals) or the United States Toxic Substances Control Act (TSCA) has been established around the world [1–3].

Unlike in previous national legislation, the responsibility to guarantee that chemicals produced in or imported to the European Union are safe in use is assigned solely to industry under REACH following the guiding principle "No data, no market" [2]. To meet this mandate, producers and importers are obliged to register chemicals in a central database of the European Chemicals Agency (ECHA), along with information on their physicochemical properties and the risk of possible human or environmental health effects. Apart from information on the potential exposure to a chemical, estimating such risks requires information on its toxicity. In the aquatic risk assessment process, toxicity data for representative species of all trophic levels, i.e., destruents (bacteria), producers (algae), and invertebrate and vertebrate consumers (daphnids and fish, respectively), must be provided. With increasing production volumes, these trophic levels need to be covered with different testing requirements [4, 5]. Regardless of the production volume, carcinogenic, mutagenic, and reprotoxic (CMR) substances, persistent, bioaccumulative, and toxic (PBT), and very persistent and very bioaccumulative (vPvB) substances need to be identified and authorized by ECHA [6, 7].

REACH has been estimated to concern approximately 30,000 compounds out of 100,000 chemicals already in use in Europe [8]. Up to May 2015, 13,149 unique substances have been registered [9]. The downside of REACH is that it potentially requires an enormous number of animal experiments [10]. To address this ethical conflict of interest, ECHA is committed to Russell and Burch's 3R principle (i.e., reduction, replacement, refinement) by requiring animal experiments to be substituted with appropriate alternatives whenever possible [11–14]. Non-experimental methods, such as quantitative structure-activity relationships (QSARs), read-across, grouping, or weight-of-evidence approaches are mostly based on previous knowledge about a chemical, and attempt to predict its toxicological effects based on physicochemical characteristics or by assuming that similar chemical structures result in similar effects [12–14]. Experimental animal alternatives, mostly in vitro bioassays, use cells or preparations of biological materials outside their biological context to study the effects of chemicals on biological processes without performing experiments on live animals [15]. The results generated using in vitro bioassays generally cannot easily be transposed to the reaction of whole organisms in vivo [16], which is one reason why they are currently not as widely accepted in regulatory ecotoxicology as would be desirable from an ethical perspective.

To overcome these current limitations, reliable and robust methods for quantitative in vitro-in vivo extrapolation (IVIVE) are urgently needed to face the challenge of increased testing requirements. IVIVE can be roughly subdivided into two distinct areas: (1) IVIVE of pharmaco-/toxicokinetics (PK/TK), i.e., the fate of a chemical within an animal's body and (2) IVIVE of pharmaco-/toxicodynamics (PD/TD), i.e., the effects of a chemical at the site of action [16].

Extrapolations of PK/TK processes generally utilize in vitro bioassays to generate experimental data on individual aspects regarding the processes of absorption, distribution, metabolism, and excretion (ADME). On the one hand these might include the study of active transport phenomena, e.g., at intestinal epithelia using the heterogeneous human epithelial colorectal adenocarcinoma cell line Caco-2 [17] or at the hepatobiliary interface using sandwich-cultured hepatocytes (SCH) assays [18, 19]. On the other hand, they might comprise in vitro assays with hepatocytes or liver subcellular fractions (microsomes or S9 fractions) to study the metabolic clearance of a chemical from the system through biotransformation enzymes.

By definition, IVIVE of PK/TK requires the application of quantitative PK/TK models that describe the "*baseline disposition*" of a chemical, i.e., its disposition under conditions not affected by active transport or biotransformation phenomena, to be able subsequently to extrapolate the relevance of the process studied in vitro to the in vivo level [20]. The IVIVE of toxicokinetics is currently a very active field of dedicated scientific research, particularly for the assessment of a chemical's bioaccumulation potential [21, 22]. Currently, laborious and expensive exposure studies with fish under flow-through conditions are required to determine the bioconcentration factor (BCF) of a compound as the metric of central regulatory importance with regard to bioaccumulation [23, 24]. Toxicokinetic models for fish typically work sufficiently well for neutral organic substances with low to intermediate *n*-octanol–water partitioning coefficients ($\log K_{\mathrm{ow}}$) ranging from 1.5 to 4.5 [25]. If a chemical is readily biotransformed in fish, the actual measured accumulation of that chemical would be lower than predicted by the model. IVIVE of biotransformation can add this extra information to the model and thus has the potential to obviate the need for animal experiments in the context of bioaccumulation assessments. As proof of the importance of such protocols, the Organization for Economic Co-operation and Development (OECD) is currently conducting a project (project 3.13) to establish a new test guideline for in vitro determination of hepatic biotransformation in fish. Nonetheless, this chapter does not go into further detail concerning methods to extrapolate PK/TK processes and parameters from in vitro to in vivo, but capitalizes on methods to extrapolate mechanism-specific effects, i.e., PD/TD processes, from in vitro to in vivo by use of toxicokinetic models.

Following the introduction of the already mentioned Russell and Burch's 3R principle [11], the development of in vitro alternatives to animal experiments has been an active and rapidly progressing field in toxicological research. It is obvious that in vivo outcomes cannot necessarily be directly predicted from effects in vitro. However, the results of in vitro bioassays for mechanism-specific endpoints in particular have often been demonstrated to be highly correlated with the results of in vivo injection studies in rats and mice [26–29]. Unlike in toxicology, chemicals in ecotoxicological research with fish are most often administered through the aqueous phase. Because of differences in physicochemical properties of different chemicals, they can be absorbed at different rates and accumulated to various extents in different tissues and organs [30]. For many studies of this type, no correlation was observed between in vitro and in vivo data [31–33].

It had already been acknowledged in the early 1990s by the critical body residue (CBR) concept that the internal chemical concentration in the organism is a central factor for acute toxicity [34, 35]. Later, this methodology was extended to be able to relate the effects of a chemical to its corresponding concentration in the target tissue; this concept is commonly referred to as the "tissue residue approach for toxicity assessment" (TRA); [36]. Both CBR and TRA are important improvements of our mechanistic understanding of differences in toxicity of chemicals and the sensitivity of different species. Nonetheless, a major disadvantage is that both concepts are based on either whole-body or tissue-specific BCFs, respectively, which is why they can only be applied under equilibrium conditions [30]. In contrast, toxicokinetic models can be used to predict kinetically the bioconcentration of chemicals and have been demonstrated to be particularly useful as tools for "retrospective" or "reverse" toxicokinetics, i.e., the prediction of toxicokinetics if the analytical information provided with the originally published toxicity data was insufficient [37, 38].

The present chapter presents confounding factors and dose metric considerations which need to be acknowledged when conducting or interpreting in vitro bioassays and IVIVE, summarizes recent approaches to apply toxicokinetic models to problems of IVIVE, and provides examples on how IVIVE can be of practical use in chemical risk assessments of the twenty-first century.

2 Confounding Factors and Dose Metrics Used for In Vitro Testing

When conducting in vitro to in vivo extrapolations, it is evident that not only are the concentrations of chemicals in whole organisms time-variable and variable between different organs and tissues but also the concentration of a chemical test item in in vitro bioassays may, depending on its physicochemical properties, also follow complex temporal variations and differ significantly from the nominal concentration [39, 40]. Theoretically, the most relevant fraction of a chemical for toxicity assessments is the target dose/concentration, often referred to as the biologically effective dose (BED), i.e., the dose or concentration of a chemical reaching the biological site of action [41]. Practically, however, this concentration is difficult to determine experimentally, which is why surrogate dose metrics are used in in vitro research.

Figure 1 illustrates a number of processes which affect the effectively (freely) available chemical concentrations in exposure media, which is acknowledged by the majority of scientists as the only fraction of a chemical readily available for uptake into organisms and cells [40, 43–45], and the freely dissolved internal concentration in cells, which probably shows the greatest correlation with the target dose.

These confounding factors are reflected to a different extent by the most frequently applied dose metrics, i.e., measures of the chemical dose or concentration relative to different reference values, among others the amount of chemical added per volume of exposure medium (nominal concentration), the amount of chemical determined analytically in the exposure medium (total concentration), or the unbound concentration in the medium (freely available concentration). Figure 2 (top) illustrates three different approaches forming the theoretical foundation of the different dose metrics. When nominal concentrations are used as the dose metric,

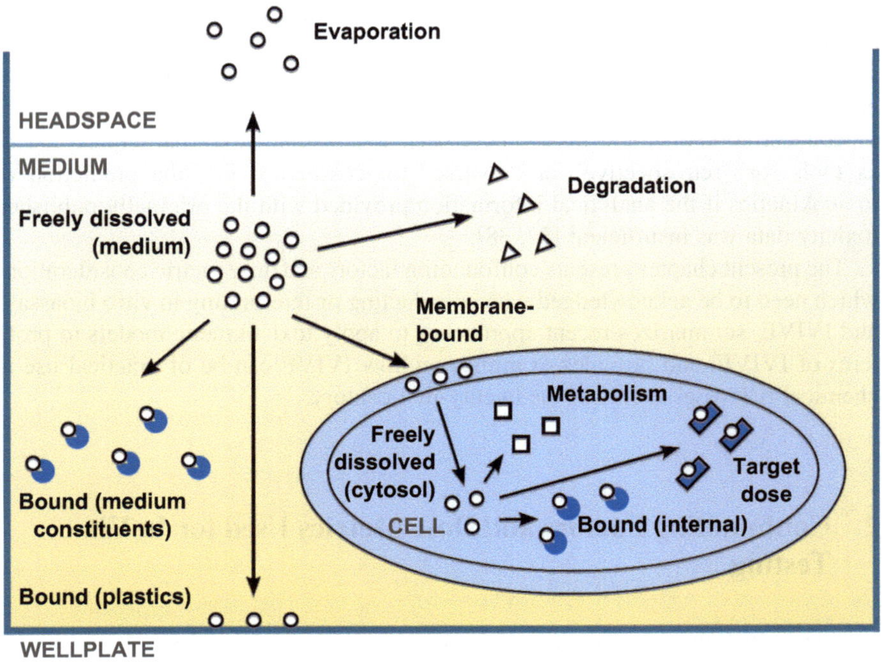

Fig. 1 Illustration of the processes which determine the amount of chemical freely available for uptake into cells and reaching the biological target site, i.e., the target dose. Chemicals within the medium may be subject to evaporation, degradation, as well as binding to plastics or constituents of cell culture media. Within the cell, the chemical might partition into the membrane, be metabolized or bind to cellular constituents. Adapted from Groothuis et al. [39] and Heringa et al. [42]

fractions of the compound that dissipated through binding to plastic materials such as pipette tips and multiwell plates [46–48], evaporation [49–51], degradation, and binding to constituents of the cell culture medium [48, 52] are not accounted for. Choosing total concentration as the dose metric accounts for losses through volatilization, degradation, and binding to plastics, but not for the fraction bound to proteins and other constituents of the exposure media. The latter fraction is only accounted for by the freely available concentration. These discrepancies between nominal, total, and freely available compound concentrations also result in differences of the fraction available for uptake into the cell, and consequently to different measured effect concentrations (Fig. 2, bottom).

To be able to account for these differences appropriately, it appears advisable always to measure or control the freely available concentration of a chemical test item when conducting in vitro bioassays. Analytical methods to measure freely available concentrations comprise equilibrium dialysis, ultrafiltration, centrifugation, and solid-phase micro extraction (SPME) [39, 53, 54]. Furthermore, passive dosing techniques –described in Chap. 5017 in more detail – have been developed to maintain relatively stable free concentrations of the chemicals of interest in

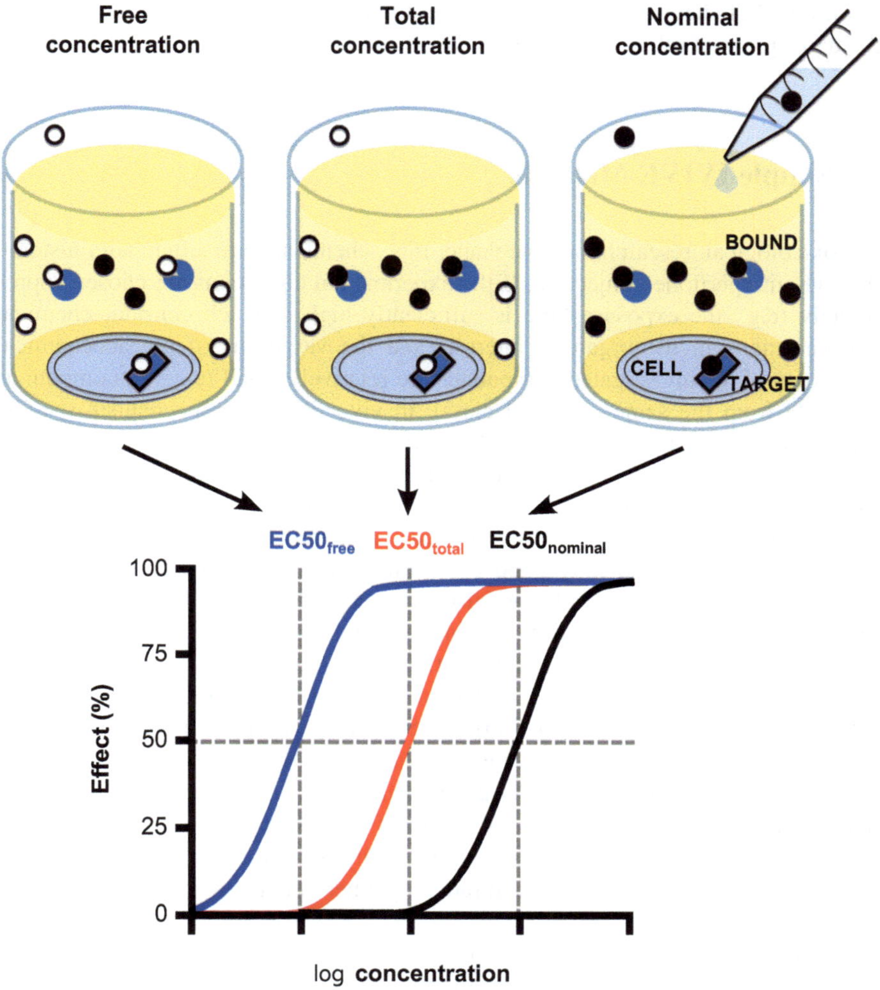

Fig. 2 Illustration of different dose metrics often used in toxicological and ecotoxicological in vitro bioassays, i.e., nominal concentrations, total concentrations, and free concentrations, and the hypothetical influence of applying these different dose metrics on the resulting median effect concentrations (EC_{50}s). *Filled circles*: fraction of molecules included in the dose metric. *Open circles*: fraction of molecules not included in the dose metric. Redrawn from Groothuis et al. [39] and Escher and Hermens [43]

exposure media [55, 56]. Although the analytical determination of the freely available concentration is feasible from an experimental point of view, to maintain the high throughput capability of the different in vitro bioassays it would be desirable to use computational models to predict the free concentration instead if measuring it. Several mathematical approaches for estimating the freely available compound concentration have been proposed, most of which are based on the partitioning of chemicals to the protein and lipid fraction of exposure media and/or the description of other routes of dissipation [46, 48, 57].

3 Simple IVIVE Methods

In toxicological research on mice and rats, chemicals are often administered through intraperitoneal injection. If the experimental conditions are chosen appropriately (e.g., the exposure time is sufficiently high to reach constant chemical concentrations in the organ of interest) and the variations of physicochemical properties of the investigated compounds are relatively small (e.g., all compounds originate from the same chemical class), in vivo EC_{50}s for mechanism-specific effects may be linearly correlated with in vitro EC_{50}s [26–29]. This assumption was also confirmed to be valid for such effects following intraperitoneal injections in fish by a collection of literature data from our own group (cf. Fig. 6a; [58]).

Castano et al. [59] reviewed cytotoxicity data from fish and mammalian cell lines and found a reasonably good correlation with acute toxicity in fish. They speculate that this good correlation, also between different cell lines, results from the unspecific mode of action responsible for baseline cytotoxicity/narcosis [60, 61]. Following this line of argument, narcotic chemicals cause acute toxicity by unspecifically interfering with biological macromolecules and lipid membranes which are common to all cells and organs. Schirmer et al. [62] thus concluded that cytotoxicity assays with fish cell lines could be a reasonably predictive alternative for the fish acute toxicity test. As detailed in the previous section, Gülden and Seibert [40] found that the predictive power of such correlations is even enhanced when the effective concentrations in cytotoxicity assays are calculated based on freely available chemical concentrations rather than on nominal or total concentrations.

A multi-national research project organized by the Scandinavian Society of Cell Toxicology in the early 1990s under the title "Multicenter Evaluation of In Vitro Cytotoxicity" (MEIC) found a similar correlation between cytotoxicity in mammalian/human cell lines and acutely lethal concentrations in blood [63, 64].

It should be emphasized, however, that such correlative methods are based on a mathematical rather than on a mechanistic foundation, which is why the applicability domain of these methods for IVIVE needs to be evaluated carefully on a substance-by-substance basis. Furthermore, these methods only account for the toxicokinetics in both, cells and animals, to a very limited extent (mainly by choosing specific exposure conditions and durations), which is why they cannot be used for IVIVE of the effects of time-variable exposures or to extrapolate beyond the calibrated range of compounds and/or organisms. In the subsequent sections, we describe how toxicokinetic modeling can be applied to overcome these shortcomings.

4 IVIVE Using Toxicokinetic Modeling

As with the free concentration in cells, one frequently overlooked factor that determines the difference between the reactions of in vitro systems compared to in vivo systems is the by far more complex toxicokinetics in whole organisms, i.e., the processes of absorption, distribution, metabolism, and excretion (ADME). These processes result in complex temporal variations of a compound's concentration at the target site, and in differences of the internal concentrations between the various organs and tissues [43, 65]. When comparing the toxicokinetics of a compound among different species and genera, differences arise from variations in body size, total lipid content, biotransformation capacity, and/or respiratory strategy [66–69]. For example, Nyman et al. [70] experimentally demonstrated the importance of toxicokinetics for interspecies variations in sensitivity of the aquatic invertebrates *Gammarus pulex*, *Gammarus fossarum*, and *Lymnaea stagnalis* exposed to the pesticide diazinon. *L. stagnalis* accumulated a higher whole-body concentration of diazinon than the two gammarids on the basis of whole-body concentrations, but less in target tissues (i.e., the nervous system), thereby explaining the greater tolerance of *L. stagnalis* to diazinon. The same underlying principle has been previously demonstrated by Meador [71], who found that inter-species variation in the acute toxicity of tributyltin to four marine invertebrate and one marine fish species were related to differences in the concentrations in the target organ.

Unlike in the two mentioned examples, it is not always possible to measure the tissue concentrations in organisms, or even directly at the target site [72]. Toxicokinetic models, which are quantitative mathematical descriptions of the ADME processes in biota, are thus increasingly used and valued as powerful tools in ecotoxicology [73, 74].

4.1 Compartmental Toxicokinetic Models

Toxicokinetic models often describe organisms based on one of two strategies: in one-compartment models, the chemical concentration is assumed to be equal throughout the organism, whereas multi-compartment models assume that organisms are composed of different compartments (usually corresponding to organs or tissues) which may differ in their characteristics and the resulting chemical concentrations [75]. Furthermore, they can be differentiated between equilibrium and kinetic models [76], as well as empirical and mechanistic models [30] – all of which have certain advantages and disadvantages. The most widely used toxicokinetic models in aquatic ecotoxicology, probably also because they are recommended by the international guideline OECD 305 [23], are empirical kinetic one-compartment models. Figure 3 depicts a conceptual representation of such a model, which considers the major routes of uptake and elimination. Similar models are frequently

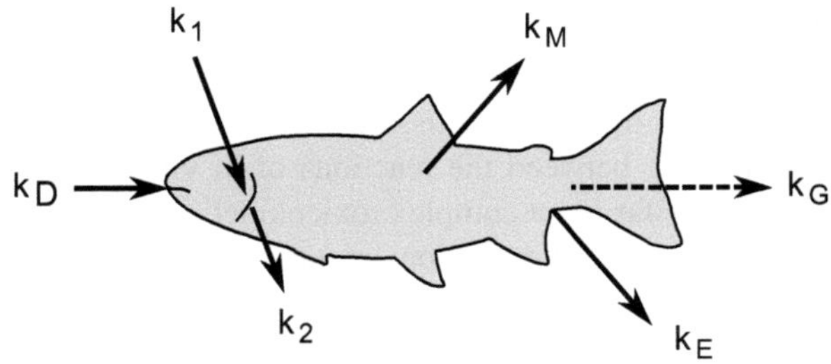

Fig. 3 Conceptual representation of a one-compartment model for fish considering the major routes of chemical uptake and elimination. k_D: dietary uptake rate constant; k_1: gill uptake rate constant; k_2: gill elimination rate constant; k_M: metabolic transformation rate constant; k_E: fecal egestion rate constant; k_G: growth dilution rate constant. Redrawn from Arnot and Gobas [77]

applied in toxicological research on mammals and in pharmacological research on mammals and humans [78], where they have been used with good success for IVIVE [79].

Such models are developed by fitting mathematical equations, e.g., (1), to experimental data of the time-dependence of the chemical concentration in fish exposed to a certain compound [75, 80]. The presented example only takes into consideration uptake and elimination through aqueous routes of exposure, i.e., pure bioconcentration.

$$\frac{d}{dt}C_{\mathrm{int}}(t) = k_1 \cdot C_{\mathrm{w}}(t) - k_2 \cdot C_{\mathrm{int}}(t), \tag{1}$$

where $C_{\mathrm{int}}(t)$ is the internal concentration in the fish per unit body mass, $C_{\mathrm{w}}(t)$ is the chemical concentration in the water per volume, k_1 is the uptake rate constant (volume per unit body mass and time), and k_2 is the elimination rate constant per unit time.

These models can be used with great confidence to interpolate internal chemical concentrations, but they are suitable neither for extrapolation beyond the range of measured values with regard to exposure conditions, species or routes of exposure, nor for predicting a chemical's concentration in specific target organs or tissues [72, 75, 81].

4.2 Physiologically-Based Toxicokinetic Models

Many of the shortcomings of empirical kinetic one-compartment models can be addressed by physiologically-based toxicokinetic (PBTK) models, which are often referred to as physiologically-based pharmacokinetic (PBPK) models in pharmacological research [72]. This model type is based on the physiology of animals or humans rather than on descriptive mathematics, thus providing higher confidence for extrapolations beyond the range of measured concentrations in a toxicokinetic experiment [75, 82]. Organs and tissues are explicitly represented as individual compartments or as tissue groups within PBTK models, each of which is characterized by its volume (fraction of total body weight), its total lipid and water contents (fraction of tissue wet weight), and the blood flow to the compartment (Fig. 4). Uptake and disposition, i.e., changes of chemical concentrations in each of

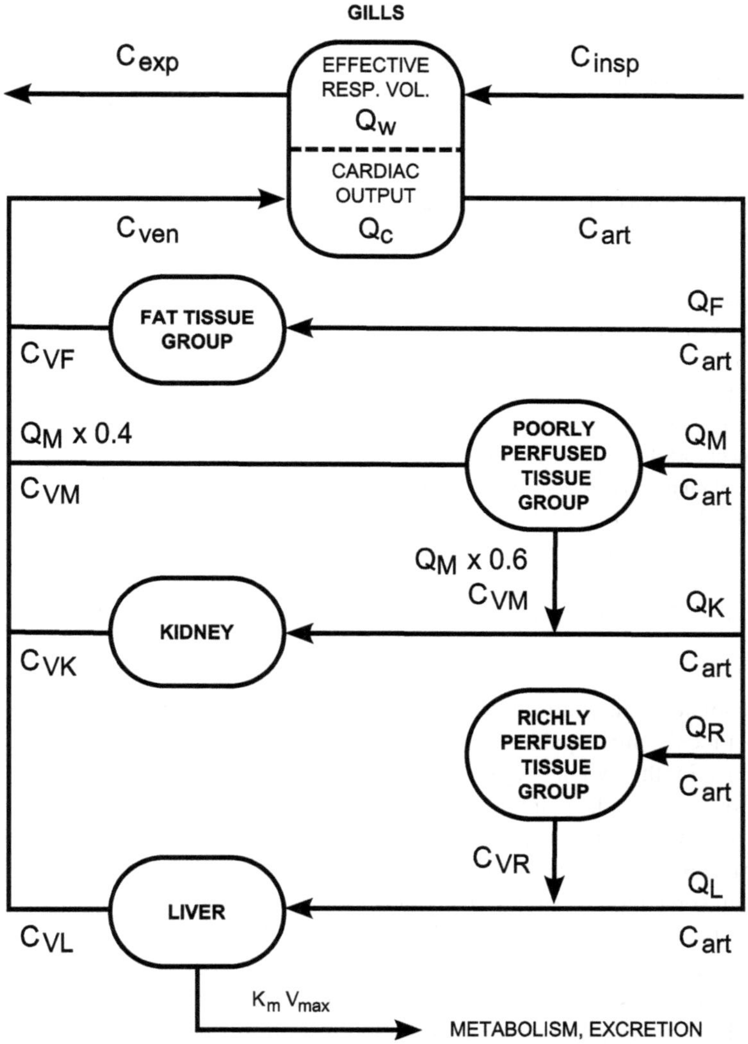

Fig. 4 Conceptional representation of the PBTK model for rainbow trout developed by Nichols et al. [81]. C_{insp}: inspired chemical concentration; C_{exp}: expired chemical concentration; C_{art}: chemical concentration in arterial blood; C_{ven}: chemical concentration in venous blood; Q_F, Q_M, Q_K, Q_R, and Q_L: arterial blood flow to fat tissue group, poorly perfused tissue group, kidney, richly perfused tissue group and liver, respectively (fraction of Q_c); C_{VF}, C_{VM}, C_{VK}, C_{VR}, and C_{VL}: chemical concentration in venous blood leaving fat tissue group, poorly perfused tissue group, kidney, richly perfused tissue group and liver, respectively; K_m: Michaelis–Menten constant of saturable metabolism, V_{max}: maximum velocity of saturable metabolism

Table 1 Compilation of examples of PBPK/TK models for different genera and species

Genus/species	References
Humans	[84–87]
Mammals	
Cattle (*Bos taurus*)	[88]
Sheep (*Ovis aries*)	[89]
Domestic pig (*Sus scrofa domesticus*)	[90]
Rat (*Rattus norvegicus*)	[91]
Mice (*Mus musculus*)	[92]
Syrian hamster (*Mesocricetus auratus*)	[93]
Harbor porpoise (*Phocoena phocoena*)	[94]
Birds	
Chicken/laying hen (*Gallus gallus domesticus*)	[95]
American kestrel (*Falco sparverius*)	[96]
Fish	
Dogfish shark (*Squalus acanthias*)	[97]
Rainbow trout (*Oncorhynchus mykiss*)	[58, 75, 81, 98, 99]
Brook trout (*Salvelinus fontinalis*)	[100]
Lake trout (*Salvelinus namaycush*)	[101]
Atlantic salmon (*Salmo salar*)	[102]
Channel catfish (*Ictalurus punctatus*)	[103]
Fathead minnows (*Pimephales promelas*)	[75]
Tilapia (*Oreochromis mossambicus*)	[104]
Zebrafish (*Danio rerio*)	[105]
Japanese medaka (*Oryzias latipes*)	[106]
Striped mullet (*Mugil cephalus*)	[107]

these compartments, are described by a number of differential equations. Thus, PBTK models are capable of predicting the concentrations of neutral organic pollutants in the whole organism and in different tissues at any time during exposure [16, 83]. Depending on the complexity of the underlying ADME processes and the available experimental data for parameterization and calibration, the level of complexity and sophistication of different PBTK models varies greatly. Although some models are relatively generic in nature and can be applied to a large variety of chemicals, the applicability domain of other models is relatively narrow, e.g., limited to only one specific chemical [72]. The explicit representation of organs and tissues and the high level of mechanistic complexity of PBTK models render them suitable tools for numerous applications in the context of chemical risk assessment and particularly for IVIVE. PBTK models have been developed for a range of different organisms and species, which are exemplarily summarized in Table 1.

Only a limited number of studies have so far used PBTK/PBPK models for IVIVE [16, 108]. De Jongh et al. [109] used in vitro data on biotransformation and tissue-blood partitioning to calibrate a PBTK model for eight neurotoxic compounds (benzene, toluene, lindane, acrylamide, parathion/oxon, caffeine, diazepam, and phenytoin). Subsequently, in vivo neurotoxicity was estimated from in vitro neurotoxicity studies by use of this calibrated model and compared to in vivo data from the literature. This study demonstrated the possibilities and limitations of this approach for the eight reference compounds: although predictions were generally accurate for compounds with low neurotoxicity (approximately twofold deviation from measured values), the accuracy was lower for compounds with higher neurotoxic potency (with deviations up to tenfold). Nonetheless, the study laid the foundation for and defined the direction of research using PBTK models for IVIVE.

Verwei et al. [110] investigated seven compounds with well-described in vivo effects on development. These chemicals were tested in the embryonic stem cell test (EST), which qualitatively classified 5-fluorouracil, methotrexate, retinoic acid, 2-ethoxyacetic acid, and 2-methoxyacetic acid correctly with regard to their in vivo embryotoxic potential. The embryotoxicity of 2-methoxyethanol and 2-ethoxyethanol was underestimated because these compounds require metabolic activation, which is not accounted for in the EST. Next, the authors used a PBTK model to extrapolate the in vitro effect concentrations to the in vivo level. A comparison of the resulting predicted effect values with effect levels measured in rodents resulted in correct predictions for 2-methoxyethanol, 2-ethoxyethanol, methotrexate, and retinoic acid by use of the IVIVE method, although the embryotoxicity of 5-fluorouracil was overestimated. A very similar approach was used by Louisse et al. [111] with good success to predict the developmental toxicity of four different glycol ethers.

The following sections describe how PBTK models for fishes can be used in ecotoxicological research for IVIVE and cross-species extrapolation of bioaccumulation and toxicity, and potentially even in combination with the adverse outcome pathway (AOP) concept.

5 Example: IVIVE of Receptor-Mediated Effects in Rainbow Trout

In a recent study published by our own group, we approached the question of whether the results of in vitro bioassays using primary fish hepatocytes for two receptor-mediated effects can be predictive of effects in rainbow trout in vivo [58]. Endpoints comprised the induction of 7-ethoxyresorufin-*O*-deethylase (EROD) activity which is mediated via the cytosolic aryl hydrocarbon receptor (AhR), and the estrogen receptor (ER)-mediated induction of Vitellogenin (Vtg) expression. EROD activity is a common biomarker of exposure to dioxin-like chemicals (DLCs), whereas Vtg is a biomarker for estrogenic effects in fish,

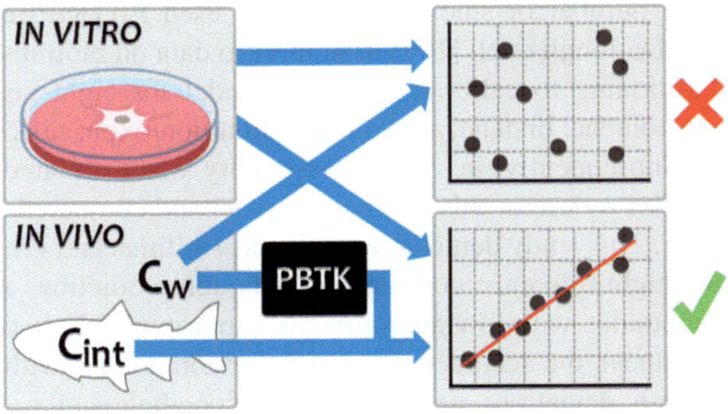

Fig. 5 Outline of the IVIVE study of Brinkmann et al. [58], in which a PBTK model for rainbow trout was used as a tool for reversed toxicokinetics. In vivo EC_{50}s for EROD and Vtg induction based on aqueous concentrations were recalculated to internal concentrations in the liver of exposed fish and then correlated with in vitro data generated by use of fish hepatocytes. Reprinted with permission from Brinkmann et al. [58]. Copyright 2014 American Chemical Society

belonging to the much wider group of endocrine disrupting effects that are a central mode of action (MOA) under REACH [1, 112]. To answer the question raised above, a quantitative framework for IVIVE applying a PBTK model for rainbow trout originally developed by Nichols et al. [82], with modifications by Stadnicka et al. [75], was used (cf. Fig. 4). Five compartments (richly perfused tissues, poorly perfused tissues, liver, kidney, and fat) were explicitly represented in the model. The accuracy of the predictions of the reimplemented model was verified by use of a dataset published by Stadnicka et al. [75].

The original model was extended for the option to simulate injections and an algorithm for saturable metabolism [98]. A comprehensive dataset for the two above-mentioned receptor-mediated MOAs in rainbow trout (EROD and Vtg), was collected, which comprised both in vitro and in vivo data. Using in vivo EC50 values from the literature, the corresponding internal concentrations in the whole body and the liver were calculated using the PBTK model. Both measured and modeled in vivo EC50s were then correlated with the respective in vitro EC50 values (Fig. 5).

Following this approach, it was possible to demonstrate that predicted concentrations of different DLCs in the liver of fish at the corresponding aqueous in vivo EC_{50} showed an excellent correlation with in vitro EC_{50} values. This observation was established on a robust data basis for hepatic activities of EROD (Fig. 6), and confirmed with a smaller and thus weaker dataset for Vtg induction (Fig. 7). Shortly after publication of the research presented in this section, a publication by Stadnicka-Michalak et al. [113] demonstrated that the same methodology was also applicable to predict the acute toxicity of chemicals in fish from cytotoxicity experiments with fish cells with good success.

Together with the results of the present study, this mechanistic link between in vitro alternatives with the corresponding in vivo experiments with fish can be considered an important step towards a broader acceptance of acute and

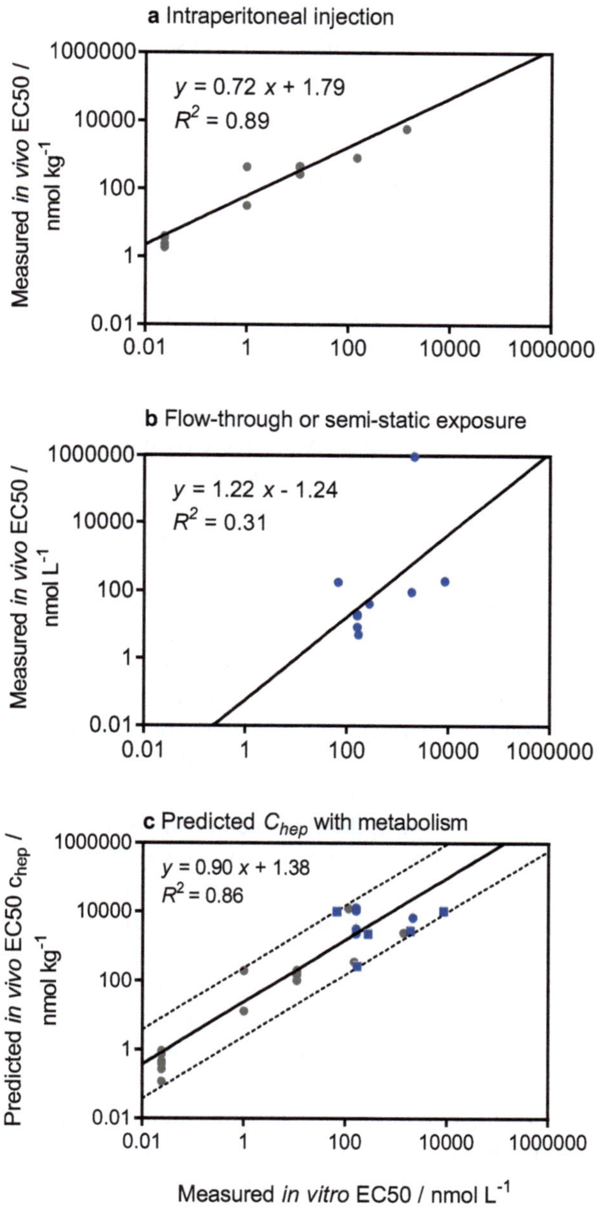

Fig. 6 Correlation between in vitro and in vivo data for EROD induction in rainbow trout on the basis of experimental in vivo EC_{50}s following intraperitoneal injection (**a**) or aqueous exposure (**b**), as well in vivo EC_{50}s from both datasets (**a, b**) recalculated to EC_{50}s-based internal hepatic concentrations (IEC_{50}s) by use of the PBTK model (**c**). *Solid lines* represent linear regression line, and *dashed lines* indicate a tenfold difference from the regression line. The coefficient of determination (R^2) and equations for the regression lines using log-transformed data are provided in the graphs. Modified with permission from Brinkmann et al. [58]. Copyright 2014 American Chemical Society

mechanism-specific cell-based bioassays in aquatic risk assessment, and has the potential to result in a major reduction of animals used for toxicity tests with fish.

6 Cross-Species Extrapolation

The next logical step when developing methods for IVIVE is applying the same methods to extrapolate between different organisms, e.g., species of fishes. In the context of most regulatory frameworks, bioaccumulation is considered an inherent substance property that is independent of the actual chemical concentration in the environment [114]. Nonetheless, bioaccumulation in some cases should be viewed with special emphasis on environmental exposure of biota, particularly because bioaccumulation represents the link between the environmental concentration of a chemical and its internal concentration in exposed wildlife [115, 116]. The internal concentration in the target tissue is a key aspect of inter-species differences in sensitivity because it represents the compound fraction which ultimately provokes the biological effects [34, 43].

To be able to account for differences in bioconcentration and toxicokinetics of chemicals between different species of fishes, several approaches have been proposed. Probably the most frequently used method to predict concentrations in biota is the equilibrium partitioning model [117]. In this model it is assumed that the internal concentration of a chemical in an organism depends solely on its concentration in the water phase and the whole-body total lipid content of the organism [118]. There are several factors not taken into account by this simple practitioner's model, including active transport, the influence of the diffusion behavior through cell membranes, different rates of metabolism in various organisms, accumulation behavior of the metabolites, accumulation in specific organs and tissues, special chemical properties such as amphiphilic or ionogenic substances leading to multiple equilibrium processes, uptake and depuration kinetics, and the remaining level of parent compounds or metabolites after depuration [117, 119]. Many different models have been developed to overcome these limitations, including models based on bioenergetics and food web accumulation, and the life-cycle of different organisms [120–123].

The PBTK modeling approaches presented within this chapter attempt to overcome the limitations of the equilibrium partitioning model by specifically considering a number of physiological processes which are the mechanistic foundation for inter-species differences in toxicokinetics. In this way, such models, although based on the partitioning of chemicals into the lipid fraction of an organism, provide fairly

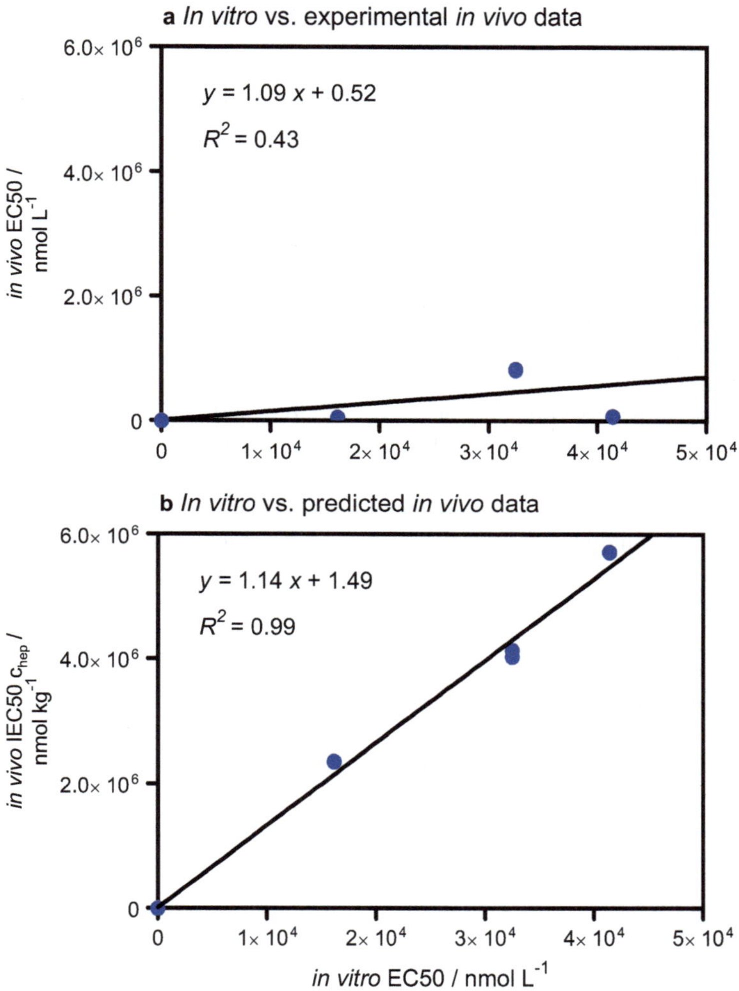

Fig. 7 Correlation between in vitro and in vivo data for Vtg induction in rainbow trout. The correlations were either based on experimental in vivo EC_{50}s (**a**) or derived through EC_{50}s based on modeled internal hepatic in vivo concentrations (IEC_{50}s) (**b**). *Solid line* represents the linear regression line. The coefficient of determination (R^2) and when applicable, the equation for the regression line using log-transformed data are provided in the graphs. Reprinted with permission from Brinkmann et al. [58]. Copyright 2014 American Chemical Society

exact estimates of accumulation and elimination rates, not only bioconcentration factors (BCFs). Furthermore, because organs and tissues are explicitly represented within their structure, PBTK models are powerful tools for predicting a chemical's distribution in exposed organisms. Developing and combining a variety of different PBTK models for different species of fishes would, apart from increased capabilities for IVIVE, also result in powerful options for cross-species extrapolation [124].

7 Example: Integration with the AOP Concept

Both regulators and industry are faced with the challenge to assess the environmental and human health risks associated with an ever-increasing number of chemicals and simultaneously reducing costs, animal use, and time required for chemical testing. To face this challenge, there has been an increasing effort to use mechanistic data (in vivo and in vitro) in support of chemical risk assessments, such as molecular biology methods and omics techniques [125]. This type of data can be generated more rapidly and cost-effectively [126]. One recent approach proposed to integrate such information in the risk assessment process of chemicals is that of the adverse outcome pathway (AOP). AOPs are conceptual frameworks that establish biologically plausible links between molecular-level perturbation of a biological system and an adverse outcome at a level of biological organization of regulatory relevance [125]. AOPs are applicable across species and are not chemical specific, but rather describe the progression from a molecular initiating event (MIE, first interaction of a chemical with a molecular target) that groups of chemicals have in common (e.g., binding to hormone receptors) to an apical outcome (e.g., disruption of reproduction or development). Thus, AOPs allow assessing toxicity across groups of chemicals and species without the need to test each chemical in each species [127]. It has recently been emphasized by Groh et al. [128] that PBTK models are highly useful tools to link toxicokinetic information to the mechanistic knowledge represented by AOPs. Specifically, PBTK models could be used to establish the cause-effect chain between external exposure, internal exposure, and MIEs. This combination surely results in quantitative models for predictive toxicology with a broad applicability domain in chemical risk assessment. It is useful to achieve an overall reduction of animal experiments, at the same time reducing the uncertainties associated with the current risk assessment strategies.

8 Conclusions

We conclude that toxicokinetic models, particularly those based on the physiology of an animal rather than on descriptive mathematics, are one piece of the puzzle which results in the development of scientifically sound integrated testing and risk assessment strategies. Toxicokinetic modeling today already plays an important role as a tool to deepen our understanding of processes that result in differences in uptake and disposition of chemicals in different species, life stages, and under varying environmental conditions. Numerous studies have demonstrated that such models can be conveniently used for extrapolating the results of mechanistic in vitro bioassays to the in vivo level, concerning both effects and biotransformation rates of a chemical. The next logical step is to synergize toxicokinetic models with the enormous amount of toxicological data generated using molecular and omics techniques, and with adverse outcome pathways (AOPs). The resulting advanced approaches are of enormous value to regulators and industry, and significantly reduce the uncertainties of the risk assessment process, at the same time being more economic and reducing the need for animal testing.

Acknowledgements The authors acknowledge funding through the project "DioRAMA – Assessment of the dioxin-like activity in sediments and fish for sediment evaluation "that was financed by the German Federal Ministry of Transport, Building and Urban Development, as well as a personal scholarship of the German National Academic Foundation ("Studienstiftung des deutschen Volkes") which was granted to the first author.

References

1. Schwarzman MR, Wilson MP (2009) New science for chemicals policy. Science 326:1065–1066
2. EC (2006) Regulation no. 1907/2006 of the European Parliament and of the Council concerning the registration, evaluation, authorization and restriction of chemicals
3. Schwarzenbach RP, Escher BI, Fenner K, Hofstetter TB, Johnson CA, von Gunten U, Wehrli B (2006) The challenge of micropollutants in aquatic systems. Science 313:1072–1077
4. ECHA (2008) Guidance on information requirements and chemical safety assessment. European Chemicals Agency, Helsinki, Finland
5. Schulte C, Tietjen L, Bambauer A, Fleischer A (2012) Five years REACH – lessons learned and first experiences. I. an authorities' view. Environ Sci Eur 24:31
6. Hansson SO, Rudén C (2006) Priority setting in the REACH system. Toxicol Sci 90:304–308
7. Zarfl C, Matthies M (2013) PBT borderline chemicals under REACH. Environ Sci Eur 25:11
8. Wolf O, Delgado L (2003) The impact of REACH on innovation in the chemical industry. European Commission, Joint Research Centre, report EUR 20999
9. ECHA (2015) Registered substances. European Chemicals Agency (ECHA); retrieved 28 May 2015, from http://echa.europa.eu/information-on-chemicals/registered-substances; last updated: 24 April 2015, Helsinki
10. Hartung T, Rovida C (2009) Chemical regulators have overreached. Nature 460:1080–1081
11. Russell WMS, Burch RL (1959) The principles of humane experimental technique; special edition, reprinted 1992. UFAW, London
12. ECHA (2011) The use of alternatives to testing on animals for the REACH regulation. European Chemicals Agency, Helsinki, Finland
13. Spielmann H, Sauer UG, Mekenyan O (2011) A critical evaluation of the 2011 ECHA reports on compliance with the REACH and CLP regulations and on the use of alternatives to testing on animals for compliance with the REACH regulation. Altern Lab Anim 39:481–493
14. Gilbert N (2011) Data gaps threaten chemical safety law. Nature 475:150–151
15. Lilienblum W, Dekant W, Foth H, Gebel T, Hengstler JG, Kahl R, Kramer PJ, Schweinfurth H, Wollin KM (2008) Alternative methods to safety studies in experimental animals: role in the risk assessment of chemicals under the new European chemicals legislation (REACH). Arch Toxicol 82:211–236
16. Yoon M, Campbell J, Andersen M, Clewell H (2012) Quantitative in vitro to in vivo extrapolation of cell-based toxicity assay results. Crit Rev Toxicol 42:633–652
17. Pinto M, Robine-Leon S, Appay MD, Kedinger M, Triadou N, Dussaulx E, Lacroix B, Simon-Assmann P, Haffen K, Fogh J, Zweibaum A (1983) Enterocyte-like differentiation and polarization of the human colon carcinoma cell line Caco-2 in culture. Biol Cell 47:323–330

18. Liu X, Chism JP, LeCluyse EL, Brouwer KR, Brouwer KL (1999) Correlation of biliary excretion in sandwich-cultured rat hepatocytes and in vivo in rats. Drug Metab Dispos 27:637–644
19. Swift B, Pfeifer ND, Brouwer KL (2010) Sandwich-cultured hepatocytes: an in vitro model to evaluate hepatobiliary transporter-based drug interactions and hepatotoxicity. Drug Metab Rev 42:446–471
20. Balaz S (2009) Modeling kinetics of subcellular disposition of chemicals. Chem Rev 109:1793–1899
21. Nichols JW, Fitzsimmons PN, Burkhard LP (2007) In vitro–in vivo extrapolation of quantitative hepatic biotransformation data for fish. II. Modeled effects on chemical bioaccumulation. Environ Toxicol Chem 26:1304–1319
22. Nichols JW, Schultz IR, Fitzsimmons PN (2006) In vitro–in vivo extrapolation of quantitative hepatic biotransformation data for fish: I. A review of methods, and strategies for incorporating intrinsic clearance estimates into chemical kinetic models. Aquat Toxicol 78:74–90
23. OECD 305 (2012) Test no. 305: bioaccumulation in fish: aqueous and dietary exposure. OECD Publishing, Paris
24. Schlechtriem C, Fliedner A, Schafers C (2012) Determination of lipid content in fish samples from bioaccumulation studies: contributions to the revision of guideline OECD 305. Environ Sci Eur 24:13
25. Nichols J (2002) Modeling the uptake and disposition of hydrophobic organic chemicals in fish using a physiologically based approach. In: Verhaar HM, de Raat WK, Krüse J (eds) The practical applicability of toxicokinetic models in the risk assessment of chemicals. Springer, Netherlands, pp 109–133
26. Mason G, Zacharewski T, Denomme MA, Safe L, Safe S (1987) Polybrominated dibenzo-*p*-dioxins and related compounds: quantitative in vivo and in vitro structure-activity relationships. Toxicology 44:245–255
27. Safe S, Bandiera S, Sawyer T, Zmudzka B, Mason G, Romkes M, Denomme MA, Sparling J, Okey AB, Fujita T (1985) Effects of structure on binding to the 2,3,7,8-TCDD receptor protein and AHH induction - halogenated biphenyls. Environ Health Perspect 61:21–33
28. Fricker G, Drewe J, Huwyler J, Gutmann H, Beglinger C (1996) Relevance of p-glycoprotein for the enteral absorption of cyclosporin A: in vitro-in vivo correlation. Br J Pharmacol 118:1841–1847
29. Dahan A, Hoffman A (2006) Use of a dynamic in vitro lipolysis model to rationalize oral formulation development for poor water soluble drugs: correlation with in vivo data and the relationship to intra-enterocyte processes in rats. Pharm Res 23:2165–2174
30. Mackay D, Fraser A (2000) Bioaccumulation of persistent organic chemicals: mechanisms and models. Environ Pollut 110:375–391
31. Segner H, Navas JM, Schäfers C, Wenzel A (2003) Potencies of estrogenic compounds in in vitro screening assays and in life cycle tests with zebrafish in vivo. Ecotox Environ Saf 54:315–322
32. Van den Belt K, Berckmans P, Vangenechten C, Verheyen R, Witters H (2004) Comparative study on the in vitro/in vivo estrogenic potencies of 17β-estradiol, estrone, 17-α-ethynylestradiol and nonylphenol. Aquat Toxicol 66:183–195
33. Legler J, Zeinstra LM, Schuitemaker F, Lanser PH, Bogerd J, Brouwer A, Vethaak AD, de Voogt P, Murk AJ, van der Burg B (2002) Comparison of in vivo and in vitro reporter gene assays for short-term screening of estrogenic activity. Environ Sci Technol 36:4410–4415
34. McCarty LS, Mackay D (1993) Enhancing ecotoxicological modeling and assessment body residues and modes of toxic action. Environ Sci Technol 27:1718–1728
35. Barron MG, Anderson MJ, Lipton J, Dixon DG (1997) Evaluation of critical body residue QSARS for predicting organic chemical toxicity to aquatic organisms. SAR QSAR Environ Res 6:47–62

36. Meador JP, Adams WJ, Escher BI, McCarty LS, McElroy AE, Sappington KG (2011) The tissue residue approach for toxicity assessment: findings and critical reviews from a society of environmental toxicology and chemistry Pellston workshop. Integr Environ Assess Manage 7:2–6
37. Takacs AR (1995) Ancillary approaches to toxicokinetic evaluations. Toxicol Pathol 23:179–186
38. Wetmore BA, Wambaugh JF, Ferguson SS, Sochaski MA, Rotroff DM, Freeman K, Clewell HJ, Dix DJ, Andersen ME, Houck KA, Allen B, Judson RS, Singh R, Kavlock RJ, Richard AM, Thomas RS (2012) Integration of dosimetry, exposure, and high-throughput screening data in chemical toxicity assessment. Toxicol Sci 125:157–174
39. Groothuis FA, Heringa MB, Nicol B, Hermens JL, Blaauboer BJ, Kramer NI (2013) Dose metric considerations in in vitro assays to improve quantitative in vitro–in vivo dose extrapolations. Toxicology 332:30–40
40. Gülden M, Seibert H (2005) Impact of bioavailability on the correlation between in vitro cytotoxic and in vivo acute fish toxic concentrations of chemicals. Aquat Toxicol 72:327–337
41. Paustenbach DJ (2000) The practice of exposure assessment: a state-of-the-art review. J Toxicol Environ Health B 3:179–291
42. Heringa MB, Schreurs RH, Busser F, Van Der Saag PT, Van Der Burg B, Hermens JL (2004) Toward more useful in vitro toxicity data with measured free concentrations. Environ Sci Technol 38:6263–6270
43. Escher BI, Hermens JLM (2004) Internal exposure: linking bioavailability to effects. Environ Sci Technol 38:455A–462A
44. Hervé F, Urien S, Albengres E, Duché J-C, Tillement J-P (1994) Drug binding in plasma. Clin Pharmacokinet 26:44–58
45. Gülden M, Mörchel S, Seibert H (2001) Factors influencing nominal effective concentrations of chemical compounds in vitro: cell concentration. Toxicol In Vitro 15:233–243
46. Riedl J, Altenburger R (2007) Physicochemical substance properties as indicators for unreliable exposure in microplate-based bioassays. Chemosphere 67:2210–2220
47. Schirmer K, Chan AGJ, Greenberg BM, Dixon DG, Bols NC (1997) Methodology for demonstrating and measuring the photocytotoxicity of fluoranthene to fish cells in culture. Toxicol In Vitro 11:107–119
48. Kramer NI, Krismartina M, Rico-Rico Á, Blaauboer BJ, Hermens JLM (2012) Quantifying processes determining the free concentration of phenanthrene in basal cytotoxicity assays. Chem Res Toxicol 25:436–445
49. Knöbel M, Busser FJM, Rico-Rico Á, Kramer NI, Hermens JLM, Hafner C, Tanneberger K, Schirmer K, Scholz S (2012) Predicting adult fish acute lethality with the zebrafish embryo: relevance of test duration, endpoints, compound properties, and exposure concentration analysis. Environ Sci Technol 46:9690–9700
50. Hinger G, Brinkmann M, Bluhm K, Sagner A, Takner H, Eisenträger A, Braunbeck T, Engwall M, Tiehm A, Hollert H (2011) Some heterocyclic aromatic compounds are Ah receptor agonists in the DR-CALUX assay and the EROD assay with RTL-W1 cells. Environ Sci Pollut Res 18:1297–1304
51. Brinkmann M, Maletz S, Krauss M, Bluhm K, Schiwy S, Kuckelkorn J, Tiehm A, Brack W, Hollert H (2014) Heterocyclic aromatic hydrocarbons show estrogenic activity upon metabolization in a recombinant transactivation assay. Environ Sci Technol 48:5892–5901
52. Kramer NI, van Eijkeren JCH, Hermens JLM (2007) Influence of albumin on sorption kinetics in solid-phase microextraction: consequences for chemical analyses and uptake processes. Anal Chem 79:6941–6948
53. Oravcova´ J, Böhs B, Lindner W (1996) Drug-protein binding studies new trends in analytical and experimental methodology. J Chromatogr B Biomed Sci Appl 677:1–28
54. Heringa MB, Hermens JLM (2003) Measurement of free concentrations using negligible depletion-solid phase microextraction (nd-SPME). Trends Analyt Chem 22:575–587

55. Kramer NI, Busser FJM, Oosterwijk MTT, Schirmer K, Escher BI, Hermens JLM (2010) Development of a partition-controlled dosing system for cell assays. Chem Res Toxicol 23:1806–1814
56. Smith KEC, Oostingh GJ, Mayer P (2010) Passive dosing for producing defined and constant exposure of hydrophobic organic compounds during in vitro toxicity tests. Chem Res Toxicol 23:55–65
57. Gülden M, Seibert H (2003) In vitro–in vivo extrapolation: estimation of human serum concentrations of chemicals equivalent to cytotoxic concentrations in vitro. Toxicology 189:211–222
58. Brinkmann M, Eichbaum K, Buchinger S, Reifferscheid G, Bui T, Schäffer A, Hollert H, Preuss TG (2014) Understanding receptor-mediated effects in rainbow trout: in vitro–in vivo extrapolation using physiologically based toxicokinetic models. Environ Sci Technol. doi:10.1021/es4053208
59. Castaño A, Bols N, Braunbeck T, Dierickx P, Halder M, Isomaa B, Kawahara K, Lee L, Mothersill C, Pärt P (2003) The use of fish cells in ecotoxicology. The report and recommendations of ECVAM workshop 47. Alternatives to laboratory animals: ATLA 31, 317
60. Ekwall B (1980) Preliminary studies on the validity of in vitro measurement of drug toxicity using HeLa cells II. Drug toxicity in the MIT-24 system compared with mouse and human lethal dosage of 52 drugs. Toxicol Lett 5:309–317
61. Schultz TW (1989) Nonpolar narcosis: a review of the mechanism of action for baseline aquatic toxicity. Aquat toxicol hazard assess 12:104–109
62. Schirmer K (2006) Proposal to improve vertebrate cell cultures to establish them as substitutes for the regulatory testing of chemicals and effluents using fish. Toxicology 224:163–183
63. Clemedson C, Ekwall B (1999) Overview of the final MEIC results: I. The in vitro--in vitro evaluation. Toxicol In Vitro 13:657–663
64. Ekwall B (1999) Overview of the final MEIC results: II. The in vitro–in vivo evaluation, including the selection of a practical battery of cell tests for prediction of acute lethal blood concentrations in Humans 1. Toxicol In Vitro 13:665–673
65. Escher BI, Hermens JLM (2002) Modes of action in ecotoxicology: their role in body burdens, species sensitivity, QSARs, and mixture effects. Environ Sci Technol 36:4201–4217
66. Baird DJ, Van den Brink PJ (2007) Using biological traits to predict species sensitivity to toxic substances. Ecotoxicol Environ Saf 67:296–301
67. Rubach MN, Ashauer R, Maund SJ, Baird DJ, Van den Brink PJ (2010) Toxicokinetic variation in 15 freshwater arthropod species exposed to the insecticide chlorpyrifos. Environ Toxicol Chem 29:2225–2234
68. Buchwalter D, Jenkins J, Curtis L (2002) Respiratory strategy is a major determinant of [3H] water and [14C] chlorpyrifos uptake in aquatic insects. Can J Fish Aquat Sci 59:1315–1322
69. Livingstone D (1998) The fate of organic xenobiotics in aquatic ecosystems: quantitative and qualitative differences in biotransformation by invertebrates and fish. Comp Biochem Physiol A Mol Integr Physiol 120:43–49
70. Nyman A-M, Schirmer K, Ashauer R (2014) Importance of toxicokinetics for interspecies variation in sensitivity to chemicals. Environ Sci Technol 48:5946–5954
71. Meador JP (1997) Comparative toxicokinetics of tributyltin in five marine species and its utility in predicting bioaccumulation and acute toxicity. Aquat Toxicol 37:307–326
72. Krishnan K, Peyret T (2009) Physiologically based toxicokinetic (PBTK) modeling in ecotoxicology. Ecotoxicology modeling. Springer, Newyork, pp 145–175
73. US-EPA (2006) Approaches for the application of physiologically based pharmacokinetic (PBPK) models and supporting data in risk assessment (Final Report). US Environmental Protection Agency, Washington

74. Chiu WA, Barton HA, DeWoskin RS, Schlosser P, Thompson CM, Sonawane B, Lipscomb JC, Krishnan K (2007) Evaluation of physiologically based pharmacokinetic models for use in risk assessment. J Appl Toxicol 27:218–237
75. Stadnicka J, Schirmer K, Ashauer R (2012) Predicting concentrations of organic chemicals in fish by using toxicokinetic models. Environ Sci Technol 46:3273–3280
76. Landrum PF, Lydy MJ, Lee H (1992) Toxicokinetics in aquatic systems: model comparisons and use in hazard assessment. Environ Toxicol Chem 11:1709–1725
77. Arnot JA, Gobas FAPC (2004) A food web bioaccumulation model for organic chemicals in aquatic ecosystems. Environ Toxicol Chem 23:2343–2355
78. Mager DE, Wyska E, Jusko WJ (2003) Diversity of mechanism-based pharmacodynamic models. Drug Metab Dispos 31:510–518
79. Punt A, Brand W, Murk AJ, van Wezel AP, Schriks M, Heringa MB (2013) Effect of combining in vitro estrogenicity data with kinetic characteristics of estrogenic compounds on the in vivo predictive value. Toxicol In Vitro 27:44–51
80. Hendriks AJ, van der Linde A, Cornelissen G, Sijm DT (2001) The power of size. 1. Rate constants and equilibrium ratios for accumulation of organic substances related to octanol-water partition ratio and species weight. Environ Toxicol Chem 20:1399–1420
81. Nichols JW, McKim JM, Andersen ME, Gargas ML, Clewell Iii HJ, Erickson RJ (1990) A physiologically based toxicokinetic model for the uptake and disposition of waterborne organic chemicals in fish. Toxicol Appl Pharmacol 106:433–447
82. Nichols JW, McKim JM, Lien GJ, Hoffman AD, Bertelsen SL (1991) Physiologically based toxicokinetic modeling of three waterborne chloroethanes in rainbow trout (*Oncorhynchus mykiss*). Toxicol Appl Pharmacol 110:374–389
83. Louisse J, Verwei M, Woutersen RA, Blaauboer BJ, Rietjens IM (2012) Toward in vitro biomarkers for developmental toxicity and their extrapolation to the in vivo situation. Expert Opin Drug Metab Toxicol 8:11–27
84. Timchalk C, Nolan R, Mendrala A, Dittenber D, Brzak K, Mattsson J (2002) A physiologically based pharmacokinetic and pharmacodynamic (PBPK/PD) model for the organophosphate insecticide chlorpyrifos in rats and humans. Toxicol Sci 66:34–53
85. Barrett J, Della Casa Alberighi O, Läer S, Meibohm B (2012) Physiologically based pharmacokinetic (PBPK) modeling in children. Clin Pharmacol Ther 92:40–49
86. Bois FY, Jamei M, Clewell HJ (2010) PBPK modelling of inter-individual variability in the pharmacokinetics of environmental chemicals. Toxicology 278:256–267
87. Jongeneelen FJ, Berge WFT (2011) A generic, cross-chemical predictive PBTK model with multiple entry routes running as application in MS excel; design of the model and comparison of predictions with experimental results. Ann Occup Hyg 55:841–864
88. Freijer JI, van Eijkeren JCH, Sips AJAM (1999) Model for estimating initial burden and daily absorption of lipophilic contaminants in cattle. RIVM Report 643810005. Dutch National Institute for Public Health and the Environment (RIVM). pp. 59
89. Craigmill A (2003) A physiologically based pharmacokinetic model for oxytetracycline residues in sheep. J Vet Pharmacol Ther 26:55–63
90. Buur JL, Baynes RE, Craigmill AL, Riviere JE (2005) Development of a physiologic-based pharmacokinetic model for estimating sulfamethazine concentrations in swine and application to prediction of violative residues in edible tissues. Am J Vet Res 66:1686–1693
91. Tardif R, Lapare S, Krishnan K, Brodeur J (1993) Physiologically based modeling of the toxicokinetic interaction between toluene and m-xylene in the rat. Toxicol Appl Pharmacol 120:266–273
92. Garg A, Balthasar JP (2007) Physiologically-based pharmacokinetic (PBPK) model to predict IgG tissue kinetics in wild-type and FcRn-knockout mice. J Pharmacokinet Pharmacodyn 34:687–709
93. Thrall KD, Vucelick ME, Gies RA, Benson J (2000) Comparative metabolism of carbon tetrachloride in rats, mice, and hamsters using gas uptake and PBPK modeling. J Toxicol Environ Health A 60:531–548

94. Weijs L, Yang RS, Das K, Covaci A, Blust R (2013) Application of Bayesian population physiologically based pharmacokinetic (PBPK) modeling and Markov chain Monte Carlo simulations to pesticide kinetics studies in protected marine mammals: DDT, DDE, and DDD in harbor porpoises. Environ Sci Technol 47:4365–4374
95. Van Eijkeren JC, Zeilmaker MJ, Kan C, Traag WA, Hoogenboom L (2006) A toxicokinetic model for the carry-over of dioxins and PCBs from feed and soil to eggs. Food Addit Contam 23:509–517
96. Nichols JW, Bennett RS, Rossmann R, French JB, Sappington KG (2010) A physiologically based toxicokinetic model for methylmercury in female American kestrels. Environ Toxicol Chem 29:1854–1867
97. Bungay P, Dedrick R, Guarino A (1976) Pharmacokinetic modeling of the dogfish shark (*Squalus acanthias*): distribution and urinary and biliary excretion of phenol red and its glucuronide. J Pharmacokinet Biopharm 4:377–388
98. Law FCP, Abedini S, Kennedy CJ (1991) A biologically based toxicokinetic model for pyrene in rainbow trout. Toxicol Appl Pharmacol 110:390–402
99. Brinkmann M, Eichbaum K, Kammann U, Hudjetz S, Cofalla C, Buchinger S, Reifferscheid G, Schüttrumpf H, Preuss T, Hollert H (2014) Physiologically-based toxicokinetic models help identifying the key factors affecting contaminant uptake during flood events. Aquat Toxicol 152:38–46
100. Nichols JW, Jensen KM, Tietge JE, Johnson RD (1998) Physiologically based toxicokinetic model for maternal transfer of 2,3,7,8-tetrachlorodibenzo-*p*-dioxin in brook trout (*Salvelinus fontinalis*). Environ Toxicol Chem 17:2422–2434
101. Lien GJ, McKim JM, Hoffman AD, Jenson CT (2001) A physiologically based toxicokinetic model for Lake trout (*Salvelinus namaycush*). Aquat Toxicol 51:335–350
102. Brocklebank JR, Namdari R, Law F (1997) An oxytetracycline residue depletion study to assess the physiologically based pharmokinetic (PBPK) model in farmed Atlantic salmon. Can Vet J 38:645
103. Nichols JW, McKim JM, Lien GJ, Hoffman AD, Bertelsen SL, Gallinat CA (1993) Physiologically-based toxicokinetic modeling of three waterborne chloroethanes in channel catfish, *Ictalurus punctatus*. Aquat Toxicol 27:83–111
104. Liao C-M, Liang H-M, Chen B-C, Singh S, Tsai J-W, Chou Y-H, Lin W-T (2005) Dynamical coupling of PBPK/PD and AUC-based toxicity models for arsenic in tilapia *Oreochromis mossambicus* from Blackfoot disease area in Taiwan. Environ Pollut 135:221–233
105. Péry ARR, Devillers J, Brochot C, Mombelli E, Palluel O, Piccini B, Brion F, Beaudouin R (2013) A physiologically based toxicokinetic model for the zebrafish Danio rerio. Environ Sci Technol 48:781–790
106. Lien GJ, McKim JM (1993) Predicting branchial and cutaneous uptake of 2,2′,5,5-′-tetrachlorobiphenyl in fathead minnows (*Pimephales promelas*) and Japanese medaka (*Oryzias latipes*): rate limiting factors. Aquat Toxicol 27:15–31
107. Woofter RT, Brendtro K, Ramsdell JS (2005) Uptake and elimination of brevetoxin in blood of striped mullet (Mugil cephalus) after aqueous exposure to Karenia brevis. Environ Health Perspect 113:11–16
108. Caldwell JC, Evans MV, Krishnan K (2012) Cutting edge PBPK models and analyses: providing the basis for future modeling efforts and bridges to emerging toxicology paradigms. J Toxicol 2012
109. DeJongh J, Forsby A, Houston JB, Beckman M, Combes R, Blaauboer BJ (1999) An integrated approach to the prediction of systemic toxicity using computer-based biokinetic models and biological in vitro test methods: overview of a prevalidation study based on the ECITTS1 project. Toxicol In Vitro 13:549–554
110. Verwei M, van Burgsteden JA, Krul CAM, van de Sandt JJM, Freidig AP (2006) Prediction of in vivo embryotoxic effect levels with a combination of in vitro studies and PBPK modelling. Toxicol Lett 165:79–87

111. Louisse J, de Jong E, van de Sandt JJM, Blaauboer BJ, Woutersen RA, Piersma AH, Rietjens IMCM, Verwei M (2010) The use of in vitro toxicity data and physiologically based kinetic modeling to predict dose–response curves for in vivo developmental toxicity of glycol ethers in rat and man. Toxicol Sci 118:470–484
112. Hecker M, Hollert H (2011) Endocrine disruptor screening: regulatory perspectives and needs. Environ Sci Eur 23:15
113. Stadnicka-Michalak J, Tanneberger K, Schirmer K, Ashauer R (2014) Measured and modeled toxicokinetics in cultured fish cells and application to in vitro-in vivo-toxicity extrapolation. PLoS One 9, e92303
114. Gobas FAPC, de Wolf W, Burkhard LP, Verbruggen E, Plotzke K (2009) Revisiting bioaccumulation criteria for POPs and PBT assessments. Integr Environ Assess Manag 5:624–637
115. Swackhamer DL, Needham LL, Powell DE, Muir DCG (2009) Use of measurement data in evaluating exposure of humans and wildlife to POPs/PBTs. Integr Environ Assess Manag 5:638–661
116. Schafer S, Buchmeier G, Claus E, Duester L, Heininger P, Korner A, Mayer P, Paschke A, Rauert C, Reifferscheid G, Rudel H, Schlechtriem C, Schroter-Kermani C, Schudoma D, Smedes F, Steffen D, Vietoris F (2015) Bioaccumulation in aquatic systems: methodological approaches, monitoring and assessment. Environ Sci Eur 27:5
117. Van der Oost R, Beyer J, Vermeulen NP (2003) Fish bioaccumulation and biomarkers in environmental risk assessment: a review. Environ Toxicol Pharmacol 13:57–149
118. Van der Kooij L, Van De Meent D, Van Leeuwen C, Bruggeman W (1991) Deriving quality criteria for water and sediment from the results of aquatic toxicity tests and product standards: application of the equilibrium partitioning method. Water Res 25:697–705
119. Franke C (1996) How meaningful is the bioconcentration factor for risk assessment? Chemosphere 32:1897–1905
120. Gobas FA (1993) A model for predicting the bioaccumulation of hydrophobic organic chemicals in aquatic food-webs: application to Lake Ontario. Ecol Model 69:1–17
121. Morrison HA, Gobas FA, Lazar R, Whittle DM, Haffner GD (1997) Development and verification of a benthic/pelagic food web bioaccumulation model for PCB congeners in western Lake Erie. Environ Sci Technol 31:3267–3273
122. Sijm DT, Seinen W, Opperhuizen A (1992) Life-cycle biomagnification study in fish. Environ Sci Technol 26:2162–2174
123. Thomann RV, Connolly JP (1984) Model of PCB in the Lake Michigan lake trout food chain. Environ Sci Technol 18:65–71
124. Brinkmann M, Schlechtriem C, Reininghaus M, Eichbaum K, Buchinger S, Reifferscheid G, Hollert H, Preuss TG (2016) Cross-species extrapolation of uptake and disposition of neutral organic chemicals in fish using a multispecies physiologically-based toxicokinetic model framework. Environ Sci Tech 50:1914–1923
125. Ankley G, Bennett R, Erickson R, Hoff D, Hornung M, Johnson R, Mount D, Nichols J, Russom C, Schmieder P, Serrano J, Tietge J, Villeneuve D (2010) Adverse outcome pathways: a conceptual framework to support ecotoxicology research and risk assessment. Environ Toxicol Chem 29:730–741
126. Villeneuve DL, Garcia-Reyero N (2011) Vision & strategy: predictive ecotoxicology in the 21st century. Environ Toxicol Chem 30:1–8
127. Kramer VJ, Etterson MA, Hecker M, Murphy CA, Roesijadi G, Spade DJ, Spromberg JA, Wang M, Ankley GT (2011) Adverse outcome pathways and ecological risk assessment: bridging to population-level effects. Environ Toxicol Chem 30:64–76
128. Groh KJ, Carvalho RN, Chipman JK, Denslow ND, Halder M, Murphy CA, Roelofs D, Rolaki A, Schirmer K, Watanabe KH (2015) Development and application of the adverse outcome pathway framework for understanding and predicting chronic toxicity: I challenges and research needs in ecotoxicology. Chemosphere 120:764–777

Index

Y

Z

MIX
Papier aus verantwortungsvollen Quellen
Paper from responsible sources
FSC® C105338
FSC
www.fsc.org

If you have any concerns about our products,
you can contact us on
ProductSafety@springernature.com

In case Publisher is established outside the EU,
the EU authorized representative is:
Springer Nature Customer Service Center GmbH
Europaplatz 3, 69115 Heidelberg, Germany

Printed by Libri Plureos GmbH
in Hamburg, Germany